Natural Computing Series

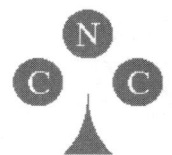

Series Editors: G. Rozenberg
Th. Bäck A.E. Eiben J.N. Kok H.P. Spaink
Leiden Center for Natural Computing

Springer
Berlin
Heidelberg
New York
Hong Kong
London
Milan
Paris
Tokyo

Gabriel Ciobanu
Grzegorz Rozenberg (Eds.)

Modelling in Molecular Biology

With 106 Figures and 14 Tables

 Springer

Editors

Gabriel Ciobanu

Romanian Academy of Sciences, Institute of Computer Science
700506 Iasi, Romania
gabriel@info.uaic.ro

Grzegorz Rozenberg

Leiden University, Institute of Advanced Computer Science
Niels Bohrweg 1, 2333 CA Leiden, The Netherlands
rozenber@liacs.nl

Series Editors

G. Rozenberg (Managing Editor)
rozenber@liacs.nl

Th. Bäck, J. N. Kok, H. P. Spaink

Leiden Center for Natural Computing, Leiden University
Niels Bohrweg 1, 2333 CA Leiden, The Netherlands

A. E. Eiben
Vrije Universiteit Amsterdam, The Netherlands

Library of Congress Control Number: 2004106401

ACM Computing Classification (1998): G.2, I.6, F.1.2, I.2, J.3

ISSN 1619-7127
ISBN 3-540-40799-5 Springer-Verlag Berlin Heidelberg New York

Springer-Verlag is a part of Springer Science+Business Media
springeronline.com

© Springer-Verlag Berlin Heidelberg 2004
Printed in Germany

Cover design: KünkelLopka, Heidelberg
Typesetting: Digital data supplied by author
Printed on acid-free paper   45/3142GF – 5 4 3 2 1 0

Preface

This volume is based on the workshop "Modelling in Molecular Biology" that took place in 2002 in Singapore. The main goal of the workshop was to present models/methods used in solving some fundamental problems in biosciences. The volume consists of a selection of papers presented at the workshop as well as of some other papers that are included so that the presentation of the theme of the workshop is broader and more balanced. As a matter of fact we feel that the collection of papers comprising this volume represents a wide spectrum of quite diverse ideas and trends.

The paper by D.A. Beard et al. explores the common thesis that understanding the behaviour of large interacting systems of many enzymes and reactants underlies the modelling and simulation of whole-cell systems. Moreover, the models need to represent the basic stoichiometry, with balanced chemical reactions and the conservation of mass, energy and charge. The authors discuss the stoichiometric and then kinetic details of approaches to modelling and simulation of biochemical systems. P.R.A. Campos et al. are concerned with models of evolution and adaptation (which is essential for precise understanding of molecular phylogeny). In particular, their paper is concerned with the rate of adaptation of asexual organisms (which is important because it influences the speed of the assumed molecular clock). It is known that for such organisms the rate of adaptation does not steadily increase with the increasing rate of advantageous mutations, and this paper studies the mutual interference of two advantageous mutants that are each initially present in only a single organism. The authors derive a phenomenological description of the mutants' fixation probabilities in the interference regime. G. Ciobanu exploits the potential of π-calculus (a process algebra invented for the purpose of studying concurrent systems) for modelling and simulation of biological systems. More specifically, he is using π-calculus to describe the kinetics of the sodium-potassium exchange pump, emphasizing the software verification of some properties of the pump. The paper by O. Demin et al. describes a new approach for constructing large-scale kinetic network models. Then, in the framework of this approach, it suggests a novel way to collect and mine

large-scale experimental data, and a way of using these data to build and
verify kinetic models. The paper shows how to apply these models for various
practical problems of biotechnology, bioengineering and biomedicine.

The paper by A. Ehrenfeucht et al. is concerned with modelling of the gene
assembly process that takes place in ciliates (which are single-cell organisms).
Gene assembly is the most involved DNA processing in living organisms (that
we know of), and authors discuss three models of this process, all three on
different abstraction levels. In particular, the paper demonstrates that all three
abstraction levels are equivalent as far as the operational modelling of gene
assembly is concerned. M. Hagiya reviews major research trends presented
at the DNA Based Computers Meeting (DNA8) in Sapporo, Japan in 2002
to illustrate the main areas of molecular programming – the term used by
the author for the research concerned with establishing systematic design
principles for molecules and molecular systems with information processing
capabilities. S. Ji argues that in order to really understand the functioning
of the living cell one needs a computer model of it, and that such a model
should be based *not* on a traditional physics approach utilizing differential
equations but rather on a fuzzy logic-based approach employing fuzzy if-then
rules. The paper by N. Kam et al. presents a novel approach to modelling
biological phenomena. It is using the methodology consisting of the language
of live sequence charts with the play-in/play-out process (that was developed
by computer scientists for the purpose of designing and analysing reactive
systems) to model the well-characterized process of cell fat acquisition during
C. elegans vulval development.

C. Martín-Vide and G. Păun are concerned with membrane systems (also
called P systems) – a model of computation inspired by the role that biological membranes play in the functioning of the living cell. In particular, they
discuss the class of membrane systems with symport/antiport rules which is
motivated by the symport and antiport mechanisms of transport of ions and
molecules through membranes. S. Motta and V. Brusic discuss the immune
system considered here as a network of cells, molecules, and organs whose
primary tasks are to defend the organism from pathogens and maintain its
integrity. They give a brief introduction to the biology of the immune system and describes several approaches used in the mathematical modelling of
the immune system. It is hoped that huge experimental data sets produced
by genomics, proteomics, and molecular biology efforts will ultimately be integrated with mathematical models of the immune system at the organism
level to produce models of the whole organism. The paper by A. Regev and
E. Shapiro begins with the general problem of abstraction for biomolecular
systems, which must be relevant, computable, understandable and extensible.
The paper argues for the "molecule-as-computation" paradigm that satisfies
these properties, and it presents an extensive and in-depth study of π-calculus
as an abstraction for biomolecular systems. B.M.R. Stadler and P.F. Stadler
explore the idea that central notions in evolutionary biology are intrinsically
topological. They discuss a mathematical framework that derives the concept

of phenotypic characters and homology from the topological structure of the phenotype space. This structure in turn is determined by the genetic operators and their interplay with the properties of the genotype-phenotype map. Finally, the paper by Y. Zhou and B. Mishra centers around the proposition that the theory of "evolution by duplication" is likely to be a central ("and most elegant") among the fundamental dogmas of biological sciences. Based on this theory the paper explores and surveys various connections between biology, mathematics and computer science in order to reveal simple, and yet deep models of life itself.

The research area of modelling in molecular biology is genuinely interdisciplinary, and (as seen from the above) this volume reflects this feature very well – the main trends and ideas are coming from mathematics, computer science, statistics, chemistry, and biology.

The editors are grateful to the contributors and to the referees. Special thanks are due to Limsoon Wong and to the Institute for Mathematical Sciences of the National University of Singapore, and to Springer-Verlag, in particular Mrs. I. Mayer, for the pleasant and constructive cooperation.

The Editors
April 2004

Contents

Stoichiometric Foundation of Large-Scale Biochemical System Analysis

Daniel A. Beard[1], Hong Qian[2], and James B. Bassingthwaighte[3]

[1] Bioengineering, University of Washington, Seattle, WA
   dbeard@bioeng.washington.edu
[2] Applied Mathematics and Bioengineering, University of Washington, Seattle
   qian@amath.washington.edu
[3] Bioengineering, University of Washington, Seattle, WA
   jbb@bioeng.washington.edu

1 Introduction

The traditional approach to unraveling functions of a biochemical system is to study isolated enzymes and/or complexes, and to determine their kinetic mechanisms for catalyzing given biochemical reactions along with estimates of the associated parameter values [41, 47]. While this reductionist approach has been fruitful, the buzzwords of the present are *integration* and *systems*. One of the important tasks in current computational biology is to assimilate and integrate the behaviour of interacting systems of many enzymes and reactants. Understanding of such systems lays the foundation for modelling and simulation of whole-cell systems, a defining goal of the current era of biomedical science.

In this chapter we discuss approaches to modelling biochemical systems, with an emphasis on the basic concepts and techniques used in building large-scale integrated models of biochemical reaction networks. We consider the vices and virtues of the available methods; we speculate on what approaches are most reasonable for large-scale cellular modelling.

How far current technology is from a reasonable quantification of whole-cell biochemistry depends on what level of detail one considers. At the simplest level (considering only reaction stoichiometry), whole-genome metabolic models of several single-celled organisms have been developed [2, 22, 45, 46, 48]. At the more detailed level of kinetic modelling, models of the relatively simple metabolism of the red blood cell represent some of the most ambitious attempts to date at modelling whole-cell metabolism [23, 27, 28, 53].

While there is no one single approach to biochemical reaction network modelling deemed superior, all models have to satisfy a set of basic criteria. Recently, one of us has proposed the concept of "sustainable conservative cell" [5]. It is argued that all biochemical systems models need to properly represent the basic stoichiometry, with balanced chemical reactions, and the conservation of mass, energy, and charge. It is along this line that we carry on our discussion.

2 Stoichiometric Organization of Biochemical Systems

We group approaches to modelling and simulation of biochemical systems into three hierarchical levels of detail: (1) stoichiometric, in which only the stoichiometry of the reaction network is known; (2) kinetic, in which detailed kinetic mechanisms and associated parameters are known for a reaction system; and (3) distributed, in which, along with detailed kinetics, information on the heterogeneous spatial organization of a biological system is considered. Stoichiometric rules are outlined in this section; kinetic approaches are described in following sections. Spatially distributed systems, based on reaction-diffusion modelling, are reviewed elsewhere [3,4].

The stoichiometry of a reaction network constrains the allowable metabolic fluxes according to mass balance and thermodynamics. In stoichiometric systems, such as a system of chemical reactions, the reactant concentrations change according to:

$$dc/dt = SJ, \tag{1}$$

where $S \in \Re^{M \times N}$ is the stoichiometric matrix [10, 11], c is the vector of concentrations of M reactants in the systems, and J is the vector of N fluxes. Equation (1) constrains the dc/dt vector to a subspace of $\Re^M$ as follows:

$$L dc/dt = 0, \tag{2}$$

where L is the left null space of S, e.g., $LS = 0$. Equation (2) defines linear combinations of concentrations that are constant: $Lc = k$, where k is a constant vector of length equal to the dimension of the left null space. The matrix L is a so-called conservation matrix; the consequences of Eq. (2) have been extensively examined by Alberty [1].

In the steady state, $dc/dt = 0$, and the fluxes obey flux balance:

$$SJ = 0. \tag{3}$$

Equation (3) is the mass balance constraint that serves as the centerpiece of flux-balance analysis (FBA). In FBA, which has been applied to large-scale metabolic systems with promising success [6, 22, 46], Eq. (3) is used in concert with some biological objective function (such as biomass production, or growth) that is assumed to be effectively optimized. Flux balance is discussed in greater detail in Sect. 3.4.

In addition to mass balance, the stoichiometry of a system constrains the fluxes according to the laws of thermodynamics. It has been shown that by introducing the right null space, R, of the stoichiometric matrix S one can derive an energy-balance law [6, 32]:

$$\sum_j R_{jk} \sum_i S_{ij} \mu_i = \sum_j R_{jk} \Delta \mu_j = 0, \tag{4}$$

where μ_i is the chemical potential of the ith chemical reactant, and $\Delta \mu_j = \sum_i S_{ij} \mu_i$ is the chemical potential difference of the jth reaction. The right null space satisfies $\mathbf{SR} = 0$. The Second Law of Thermodynamics can be interpreted as

$$J_j \Delta \mu_j \leq 0 , \tag{5}$$

where the equality holds when and only when the reaction is in equilibrium. The thermodynamic constraint on the flux is expressed as: for a flux vector $\mathbf{J}$ to be feasible, there must exist a vector $\boldsymbol{\mu}$ for which Eqs. (4) and (5) are satisfied. In practice, this constraint is difficult to implement. As an alternative, we have introduced an algorithm that is based on the sign structure of the null space, which is the subject of a forthcoming publication [7].

The stoichiometric conditions on thermodynamically allowable and mass-balanced fluxes are a set of mathematical rules that should be followed by any reaction system of a given stoichiometry. These rules alone, however, are not sufficient information to understand and predict the behaviour of living metabolic systems, because the stoichiometric conditions fail to constrain systems to behave in a unique way. In fact, it is this inherent unconstrained flexibility of metabolic systems that contributes to their robust ability to maintain Claude Bernard's *milieu interior*.

To model, with specificity and accuracy, the dynamic variables—reactant concentrations and their rates of change—in a living system, requires adopting mechanistic models that relate these variables to the fluxes among them. In doing so, it is paramount to remember that the foundational principles, the basic stoichiometric rules, must apply, regardless of the formulation of the detailed reaction mechanisms. As *La vie n'est autre chose qu'un phénomène physique* (Life is nothing else than a physical phenomenon) [25], our models of living things should not violate fundamental principles of physical chemistry.

3 Theory and Modelling of Biochemical Systems

3.1 Enzyme Mechanisms

Standard approaches to modelling biochemical kinetics begin with mass action relationships [41]. For example, for the simple unimolecular reaction

$$S \underset{k_{-1}}{\overset{k_{+1}}{\rightleftharpoons}} P, \tag{6}$$

we have rate equations

$$\frac{d[S]}{dt} = -\frac{d[P]}{dt} = -k_{+1}[S] + k_{-1}[P] \tag{7}$$

where $[S]$ and $[P]$ are the concentrations[1] of species S and P. It is important to remember that Eq. (7) is really based on two physiochemical principles. First, consider the stoichiometry

$$\frac{d[S]}{dt} = -\frac{d[P]}{dt} = -J_1 + J_2 \,, \tag{8}$$

and, second, the rate law according to mass action: $J_1 = k_1[S]$, $J_2 = k_2[P]$. As we shall show below, while the form of the rate law can vary, the stoichiometry is more fundamental.

The above case does not consider the existence of enzyme-catalyzed intermediates that can enhance the turnover between S and P. The well-known Michaelis–Menten model incorporates a substrate–enzyme intermediate step:

$$\xrightarrow{J} S + E \underset{k_{-1}}{\overset{k_{+1}}{\rightleftharpoons}} SE \underset{k_{-2}}{\overset{k_{+2}}{\rightleftharpoons}} E + P \xrightarrow{J} \,. \tag{9}$$

Assuming the system to be in steady state (the net turnover from S to P is constant and balanced by a flux J of S into and P out of the system) we arrive at the Michaelis–Menten law for the flux:

$$J = J_1 - J_2 = \frac{E_o\left(k_{+2}[S]/K_{M,S} - k_{-1}[P]/K_{M,P}\right)}{1 + [S]/K_{M,S} + [P]/K_{M,P}}, \tag{10}$$

where E_o is the total concentration (free plus bound) of enzyme present, $K_{M,S} = (k_{+2} + k_{-1})/k_{+1}$, and $K_{M,P} = (k_{+2} + k_{-1})/k_{-2}$. This model predicts a chemical equilibrium ($J = 0$) when $[P]/[S] = J_1/J_2 = K_{eq} = (k_{+1}k_{+2})/(k_{-1}k_{-2})$. This is known as the Haldane relation [41].

In Eq. (9), as in Eq. (6), all steps are considered reversible, allowing for a physically realistic finite equilibrium. The action of a catalyst as in Eq. (9) does not change the overall thermodynamic equilibrium associated with the system represented in Eq. (6). When $k_{-2} = 0$, we have the familiar irreversible Michaelis–Menten scheme, for which the $K_{M,P}$ terms disappear from Eq. (10). In this case, the resulting equilibrium constant is infinite, and it is no longer possible to characterize the thermodynamics of the overall reaction $S \rightleftharpoons P$.

In practice, the kinetics of many enzyme-catalyzed biochemical reactions are more complex than the single-step Michaelis–Menten model described above. In 1956, King and Altman [24] introduced a systematic method for obtaining the form of steady-state flux laws from diagrams of a given enzyme mechanism (i.e., the collection of intermediate states and the various routes of transition between them). In a concise and accessible chapter, Cornish-Bowden [12] outlines the King–Altman method along with several approaches to simplifying the approach for complex mechanisms. A more comprehensive

---

[1] Strictly speaking, if the activity of a species changes with concentration, then the effective rate constants in Eq. (7) are no longer independent of concentration, resulting in a nonlinear system.

treatment is found in [41]. With modern computers and symbolic algebra packages, the exercise of deriving steady-state flux expressions from complex mechanisms may be done automatically.

When nonlinear expressions such as Eq. (10) can be obtained for all of the reactions of a given system, a complete model in the form of Eq. (1) can be constructed. Large-scale examples of such systems can be found in [23, 27, 28, 53].

3.2 Biochemical Systems Theory: the S-System Approach

While the systems approach has entered the general consciousness of biochemistry only in recent years [9], the original idea can be traced to the early 1970s or even earlier. M.A. Savageau and his coworkers, through Biochemical Systems Theory (BST), have continuously championed the cause [34–37, 50]. Throughout its development, BST has been applied to many different biological research areas such as immunology, molecular genetics, even epidemiology and population dynamics. The approach, however, is not without controversy.

The most important contributions of the BST are inherent recognitions of: (1) the importance of a network of biochemical reactions to cellular functions, (2) the nonlinearity in the governing dynamics of biochemical systems, and (3) the complexities as emerging properties of a reaction network requiring an approach based on systems science.

BST grows out of the realization that the mechanism for an enzyme reaction is often hard to obtain. Even when the mechanism is clearly worked out, it is usually quite complex with many intermediate steps between the substrate binding and the product release. As an approximation, Michaelis–Menten theory assumes that all the intermediate forms of enzyme–substrate complexes remain in steady state. (This assumption can be mathematically justified if the total enzyme concentration is sufficiently less than that of the substrate [29].) This assumption greatly simplifies the multiple steps of an enzyme reaction, and yields a compact rate law in terms of a ratio of polynomials in the substrate and modifier concentrations (rational functions, e.g., Eq. 10). Initial applications of BST revealed some interesting properties of the kinetic systems in terms of the roots of the polynomial functions [34]. A closely related approach called kinetic polynomial [26] also appears in the literature on reaction systems with heterogeneous catalysis [51] in which the reaction order for catalysts can be greater than unity (i.e., nonlinear).

Rational functions, however, are not necessarily easy mathematical objects to work with. In contrast, the so-called power-law representation can be easily manipulated mathematically; it is also a mathematical form consistent with the law of mass action. Hence, it has gradually become a core constituent of BST [49]. However, writing an enzymatic reaction in an aggregated form in a single rate equation, and using a power-law to approximate the turnover rate, requires some justification. This justification is extensively discussed in [49], a very readable paper. Using Michaelis–Menten as the reference, it shows how the power-law approximation is more accurate than a linear approximation near a steady state.

However, some difficulties are associated with the power-law representation when dealing with reversible reactions catalyzed by an enzyme. In Michaelis–Menten theory, the product is considered to be a competitor for the substrate (the $[P]$ terms in the denominator of Eq. 10). Hence, the flux should be proportional to $(c_Y)^\gamma$ in the power-law formulation, with $\gamma < 0$, where γ is a parameter and c_Y is the concentration of the product. This means one cannot represent initial or transient states where $c_Y \to 0$. In addition, it is not clear how to impose the Haldane relationship, which is necessary to develop a thermodynamically valid kinetic theory [41].

The promise of BST is as a theory for large-scale systems. Yet, further difficulties arise when dealing with networks of interacting biochemical reactions. In BST, the dynamics of a biochemical network can be represented by an elegant system of differential equations known as S-systems [38] where "S" stands for *synergetic*.

For a simple reaction system,

$$X_2 \rightleftharpoons X_1 \rightleftharpoons X_3, \tag{11}$$

the S-system representation has the form

$$
\begin{aligned}
\frac{dc_1}{dt} &= -J_{12} - J_{13} + J_{21} + J_{31} \\
&= -\alpha_2 c_1^{g_2} - \alpha_3 c_1^{g_3} + \beta_2 c_2^{h_2} + \beta_3 c_3^{h_3} \tag{12} \\
&\approx -\alpha c_1^g + \beta c_2^{h_{12}} c_3^{h_{13}}, \tag{13}
\end{aligned}
$$

where the α's, g's, and h's in the above expressions are scalar parameters. In Eq. (13), the sum of many power-law terms is further approximated by two single power-law terms, one positive and one negative [35].

While the S-system form is both elegant and convenient, there are at least two major disadvantages for the rate law as in Eq. (13). First, the identity of different reactions participating in J has disappeared. Therefore the stoichiometric structure of a reaction network is gone. Knowing everything about Eq. (13) still does not provide us with any information on the individual fluxes in $X_2 \rightleftharpoons X_1$ and $X_1 \rightleftharpoons X_3$. Second, because the stoichiometry has not been preserved, this system of differential equations in general will not satisfy the basic stoichiometric conservation laws, as described above. For the example in Eq. (11), the stoichiometric constraint $d(c_1 + c_2 + c_3)/dt = 0$ should be imposed. Although this constraint remains at the level of Eq. (12), it is lost in the final reduced power-law representation (Eq. 13). This means that computational simplicity is achieved only by loss of physiochemical accuracy.

3.3 Metabolic Control Analysis

Metabolic control analysis (MCA) is concerned with the perturbation of a steady state of a metabolic network in response to changing enzyme concentrations (i.e., gene expression level) and substrate concentrations. MCA is not a tool for dynamic modelling of biochemical systems. More precisely, in MCA we are interested in characterizing the relationships between arbitrary enzyme i and the arbitrary flux J_k, in steady state. Even when the enzyme i is not catalyzing reaction k, its activity may influence the flux J_k through network interactions in a biochemical system.

To probe the quantitative relationship between enzyme i and flux J_k near a steady state, the natural choice is sensitivity analysis from standard statistics:

$$C_i^k = \frac{e_i}{J_k} \frac{\partial J_k}{\partial e_i} \, , \tag{14}$$

where e_i is the activity (or activity coefficient times concentration) of enzyme i. Because of the network connectivity, there are mathematical relations that these flux control coefficients C_i^k have to satisfy. One of such relation, called the summation rule, is the central piece of MCA.

For a given metabolic network, one can compute the flux control coefficients if all the enzyme mechanisms and rate constants are known or, alternatively, one can experimentally determine the flux control coefficients from perturbation and measurement on a biochemical reaction system. The former is a simple mathematical problem so we shall focus on the latter. More discussions of the details can be found in [17–19, 52].

While the traditional MCA emphasizes steady states and mathematical derivations, recent developments by Reder, Delgado and Liao [14, 33] focus on obtaining flux control coefficients from measurements on transient, linear relaxation of concentrations to steady state. The key result needed for the analysis is the summation rule:

$$\sum_{i=1}^{N} C_i^k \left(\frac{J_i(t)}{J_i^*} \right) = 1 \, , \tag{15}$$

where $J_i(t)$ is the flux in any linear transient and $J_i^* = J_i(\infty)$ is the corresponding flux in steady state.

The key step in the dynamic MCA (dMCA) is to obtain the transient fluxes $J_i(t)$ from measurements on the rate of concentration change dc_k/dt. This problem is an under-determined linear algebra problem without unique solution since (Eq. 1)

$$\frac{dc_k}{dt} = \sum_{i=1}^{N} S_{ki} J_i \, , \tag{16}$$

and the right null space of the stoichiometric matrix $\mathbf{S}$ has non-trivial solution. This is precisely the mathematical problem confronted by the FBA.

In dMCA, the steady state $\mathbf{J}^*$ is assumed to be known. Based on this information, one can computationally obtain $\mathbf{J}$ from a given set of observed concentration changes $d\mathbf{c}/dt \neq 0$. An optimality condition assumes that the solution to Eq. (16) occurs at the shortest possible distance from the steady state $\mathbf{J}^*$. The resulting fluxes can be obtained by starting with arbitrary solution to Eq. (16), denoted $\widetilde{\mathbf{J}}$, and then constructing

$$\mathbf{J} = \widetilde{\mathbf{J}} - \mathbf{J}^\perp + \mathbf{J}^* \,, \tag{17}$$

where $\mathbf{J}^\perp$ is the projection of $\widetilde{\mathbf{J}}$ in the null space[2] of $\mathbf{S}$.

It has been shown [14, 33] that

$$\left\{ \frac{J_i(t) - J_i^*}{J_i^*} \Big| i = 1, \ldots, N; t \geq 0 \right\} \,, \tag{18}$$

as a set of vectors in linear space, expands exactly the same subspace of the elasticity coefficients,

$$\epsilon_k^i = \frac{c_k}{J_i^*} \frac{\partial J_i^*}{\partial c_k},$$

the second key quantity in MCA. Hence, the solution to Eq. (15) gives the complete set of flux control coefficients. This result in fact serves as an alternative proof for Eq. (15), the central piece of dMCA.

3.4 Flux Balance Analysis and Stoichiometric Network Theory

The recent surge of the flux-balance-based computational analyses of metabolic networks is due to the unifying ability in integrating null-space analysis of stoichiometric matrices, biological hypothesis-driven constraint-based optimization, and available bioinformatic data on cell metabolism.

The theoretical tools for mathematical analysis of the null space of stoichiometric matrices trace back to the work done by chemical kineticists in 1970s and 1980s [10, 11]. A quite complete mathematical approach based on the flux cone and its generating vectors[3] (with certain modifications called internal representation in [10], elementary modes in [40], and extreme pathways in [39]) was developed. Clarke's stoichiometric network analysis (SNA) [10, 11] combines the null-space analysis with the analysis of dynamic stability, representing a viable approach for extending FBA beyond steady-state applications.

A powerful algorithm for computing the generating vectors is presented by Schuster et al. [39, 40] and is based on a German reference [31]. The approach can be traced back to Fourier [16] and is also known as the Fourier–Motzkin method [13]. We have summarized the basic idea in the Appendix to help understand the algorithm. We point out that R.J. Duffin [16], who had trained some of the greatest pure and applied mathematicians of the

---

[2] The null space of $\mathbf{S}$ is defined as the subspace of $\Re^N$ for which $\mathbf{SJ} = 0$ is satisfied.

[3] The null space of $\mathbf{S}$ lies within an abstract mathematical entity called a feasible "cone" in linear analysis. The "generating vectors" are the edges of the feasible cone in $\Re^N$.

20th century, [8, 30, 43], had devoted his lifework to the theories of nonlinear electrical networks in terms of abstract algebraic topology and differential geometry, as well as practical results in mathematical programing and optimization. Duffin's work is a treasure which proves highly relevant to current constraint-based modelling of biochemical networks.

A key concept in stoichiometric analysis is setting a convention for what is a flux. If one takes all reaction fluxes to be unidirectional, $J_i \geq 0$, then reversible reactions must be represented using separated forward and backward fluxes. This approach, combined with the equality $\mathbf{SJ} = 0$, cogently defines a polyhedral flux cone in the positive quadrant. While this convention for fluxes is mathematically satisfactory, it obscures the nature of chemical reactions. In Fig. 1A the linear combination of the two cycles on the left is <u>not</u> identified as equivalent to the single cycle on the right. For a reversible reaction with forward and backward fluxes J_ℓ^+ and J_ℓ^-, the net flux, which is what one needs in the FBA, is $J_\ell = J_\ell^+ - J_\ell^-$. So, this convention introduces unnecessary degeneracy and indeterminacy.

Different conventions for J's can leads to different conclusions. If one takes all the J's to be irreversible and unidirectional, then the cycles on the left-hand side of Fig. 1A in fact represent a compound mode with two independent cycles (this is not an elementary mode, as defined by Schuster and Hilgetag [40], since one can eliminate the enzyme associated with the reversible step and still preserve a flux-balanced steady state; neither is it an extreme pathway since it is a linear combination of two extreme pathways) and it is not equivalent to the single cycle on the right-hand side. However, if one takes all J's to be reversible as convention, then the sum of cycle I and II gives cycle III, which is itself elementary (Fig. 1B). Thus, in terms of the reversible flux convention, one will also be interested in finding a minimal cycle base for a null space [20, 21].

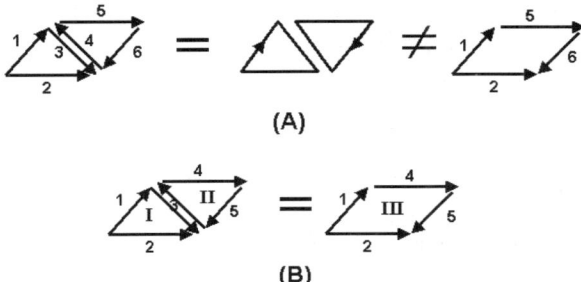

(A)

(B)

Fig. 1. In (A), the flux convention is set for all unidirectional fluxes. There are in total six dimensions. Therefore we have $(1, 1, 1, 1, 1, 1) = (1, 1, 1, 0, 0, 0) + (0, 0, 0, 1, 1, 1)$ $\neq (1, 1, 0, 0, 1, 1)$. The mathematics is quite different for (B) in which there are only five fluxes. Hence $(1, 1, 1, 0, 0) + (0, 0, -1, 1, 1) = (1, 1, 0, 1, 1)$.

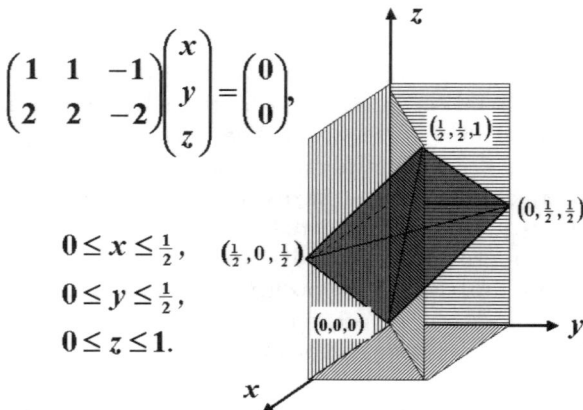

$$\begin{pmatrix} 1 & 1 & -1 \\ 2 & 2 & -2 \end{pmatrix} \begin{pmatrix} x \\ y \\ z \end{pmatrix} = \begin{pmatrix} 0 \\ 0 \end{pmatrix},$$

$$0 \le x \le \tfrac{1}{2},$$
$$0 \le y \le \tfrac{1}{2},$$
$$0 \le z \le 1.$$

Fig. 2. The LPP shown on the left defines a polytope, a plane, with vertices $(0,0,0)$, $(1/2, 0, 1/2)$, $(0, 1/2, 1/2)$, and $(1/2, 1/2, 1)$. Note that the last vertex is not on the generating vectors, i.e., the edge of the cone defined by the equality in the positive octant.

Finally, when combining the null-space analysis (i.e., characterization of the polyhedral flux cone) and a linear objective function, plus a corresponding second set of inequalities $J_\ell \le J_\ell^*$, searching for an optimal solution becomes a linear programming problem (LPP). (The second set of inequalities is necessary for the subspace on which the objective function is not a constant.) It is important to point out, however, that the extreme points of the polytope associated with the LPP, $\mathbf{SJ} = 0$, $0 \le J_\ell \le J_\ell^*$, $\max{(\mathbf{b} \cdot \mathbf{J})}$, do not all coincide with the extreme rays of the cone (see Fig. 2 for a simple counter example). Furthermore, if a nonlinear objective function is introduced (e.g., the minimal heat dissipation of the network), then the optimal solution in general will lie inside the polytope. Therefore, having a minimal cycle base will be invaluable for nonlinear optimization problems.

4 Large-Scale Model Building

4.1 Modular Principles

The need to develop computational models encompassing major portions of biochemical and genetic regulatory networks and their control is evident. Integrative models are the key to providing context for the individual reactions and for eliciting an understanding of the influences of the various components upon whole-system behaviour. The methods for which the theory was described in Sect. 3 are reduced to practice, for example, in formulating a virtual cell model, or in constructing models of more limited expanse, such as cellular energetics.

Small portions of a metabolic system can be described in detailed form. For a system such as glycolysis, it is useful to develop several representations providing different levels of complexity, or accuracy, or computational speed. Having the highest level of precise detail is not compatible with the fastest computing of the solutions, yet speed is required to allow widespread exploration to gain insight, develop predictions, or optimize the fits of model solutions to experimental data. Making compromises for specific purposes is therefore essential.

Larger, more all-encompassing models are best composed of smaller modules, each of which has been previously validated by comparisons with data, and verified for computational accuracy. The individual modules must adhere to a pre-chosen standard and provide a scientifically accurate representation of the system, using semantics compatible with those of the larger system. Individual modules are best developed and maintained by individual investigators or groups who are expert in the particular science. Models are merely working hypotheses that must be kept at the forefront of the field if they are to be useful as tools for experiment design and for data analysis. Leaving them in model repositories tended by technical staff relegates them to obsolescence in a short time. Certain principles and practices should be upheld in order that modules be actively sustainable. Here, we enumerate the essential principles that we attempt to adhere to in constructing the "eternal" or "sustainable cell" model [5]. The list begins with the science, and extends to matters of style, convenience, and dissemination to the scientific community.

1. Write model code to conserve mass, charge, volume, energy, and redox state.
2. Define all variables and parameters, along with name and units.
3. For linking purposes, identify all inputs and outputs.
4. Identify all assumptions and approximations.
5. Identify all information sources.
6. Write the code for maximal computational speed.
7. Provide operations manuals and tutorials for developed models.
8. Publish models on the web, so that they can be run or downloaded.
9. Establish open forum discussion of models and modules.

Speed, only number 6 on our list, is vital: one needs to compute at the speed of thought in order to facilitate exploration and the gaining of insight into biological processes. Speed is also critical in the use of models as tools to analyse experimental data through automated optimization procedures.

4.2 Linking of Modules to Compose Larger Models

In the list above, the first five are essential to using a module as a component in a more comprehensive system. The first, conservation, provides some assurance of self-consistency; furthermore if each building block fulfills conservation requirements, then determination of conservation for a multi-component model is simplified. The second, definitions and units, defines content. The third, the list of all the inputs and outputs, automatically provides the minimal list of links to other modules; the list of links must be extended to include variables which are common to two modules and variables which influence the behaviour of a component of another module, e.g., the effect of cAMP on contractility. The fourth, identifying assumptions and approximations, introduces new possibilities into the composite model, such as the ability to ask "does the combining of modules allow for the elimination of certain approximations or assumptions?" and "what additional assumptions might become implicit in the combining of modules?" Such assumptions must be made explicit. The remaining three points are essential for documentation and dissemination.

The process of linking modules is suggested by the composite model in Fig. 3. The upper panel shows a set of modules for intermediary metabolism that are linked with known stoichiometry. Consider each to be a separate model, each the evolutionary result of years of research. Linking them through the stoichiometric relationships of input substrates and output products appears straightforward, and results in the integrated model shown in the lower panel. It appears as if the sanctity of the individual modules remains intact, and in fact it is almost this simple. Problems arise, not so much in accounting for NADH and other slowly changing variables, but in accounting for the influences of rapidly changing calcium and hydrogen ion concentrations in some circumstances.

The argument that models are best developed and maintained by a group working in the particular field leads to contradictions when models become modules of larger system models. The expertise required to develop and maintain a given model may not be available in the group that chooses it as the best version of the desired component of a higher level integrative model. From the technical point of view, putting together two models from different sources is not too difficult when both submodels or modules can be described by ODEs (ordinary differential equations). Using a simulation system like JSim (from nsr.washington.edu), or Madonna (from www.berkeleymadonna.com/index.html), or SAAM (from www.saam.com), or Gepasi (from www.gepasi.org), one simply combines the two modules into a common piece of source code, combining all those equations which have common variables. The process sounds simple, but the problem is that one of the originating groups is not necessarily expert in the field of the other, so that unless all the common variables have been defined with the same names in both modules their identity will not be easily recognized. Clear semantics requires lists of synonyms.

The second issue is how to build a composite model out of modules while maintaining the identity of the code of the module so that it can be replaced automatically when the originating group improves the module. Ideally the composite model should be reconfigured whenever it is judged that improvements have been achieved in any particular module. Automating this is possible when common variables are named identically; it is also possible, but

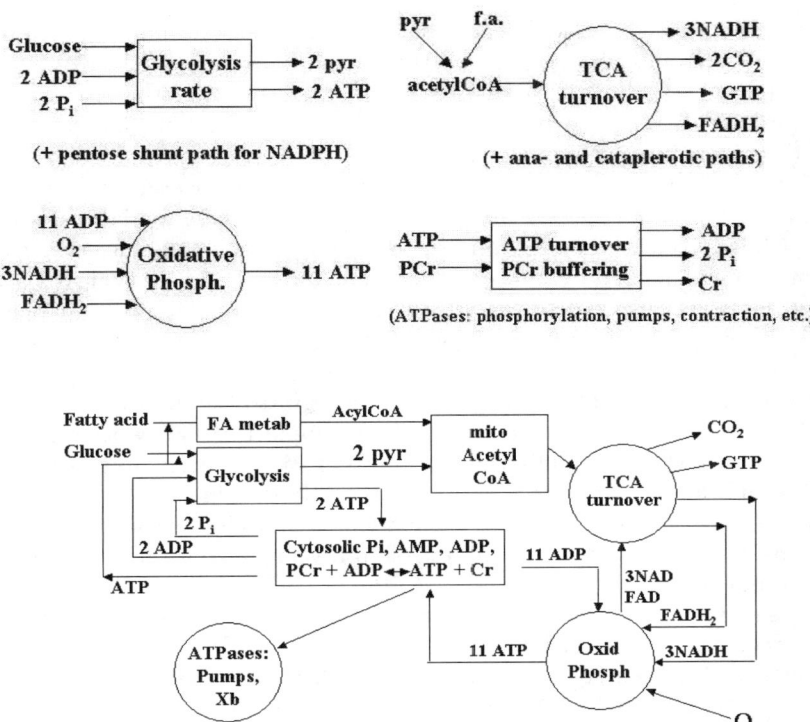

Fig. 3. Building composite modules from pre-built modules in intermediary metabolism. *Upper panel*: Individual models for glycolysis, Krebs (TCA) cycle, oxidative phosphorylation and nucleotide energetics showing their main inputs and outputs. *Lower panel*: Combining the modules from the upper panel, and adding fatty acid metabolism, gives a composite model of intermediary metabolism. Stoichiometric balance is maintained, and the modules remain distinct from each other.

requires human intervention, to define equivalences, when the variable names in each module are not identical. There is a trade-off here: when the variable names are identical, the combining of the modules can be automated because the equations for the common variable can be automatically combined, as has been achieved by Gary Raymond in our laboratory. The cost is that the two source codes are now intermixed. For computational reasons this is good since it minimizes the numbers of different variables and facilitates solving the whole system simultaneously. But what it costs is that a composite model composed of a large set of modules must be entirely formed anew when a module is to be replaced.

Separating modules from one another can help to obtain computational speed. Solving a large set of simultaneous equations as a whole gives high accuracy when the system is non-stiff and linear, but is computationally costly when it is composed of nonlinear equations and/or has dramatically different time constants in different parts of the model. Then one would like to solve separately those submodels that have rate or time constants that are relatively slow and to solve at higher frequencies those submodels that have time constants orders of magnitude faster. Allowing difference in time steps from one module to another greatly reduces the "stiffness" of the overall system and increases computation speed. This argument favors keeping modules separated even while linked in the composite model, and enhances the incentive to use automated methods for composite model building. Last but not least, attentions should be directed to the numerical accuracy of composite models for which fractional steps and exponential splitting could be introduced [42, 44].

A situation in which modular separation can be maintained occurs whenever the common variables change slowly relative to the internal rates of the modules to which they are relevant. An example is ATP, which is in such large concentration normally that its concentration changes only very slowly even with dramatic changes in circumstances. By treating ATP as an external variable from the point of view of the individual modules its concentration can be considered constant during a time step: by preserving the fluxes of ATP into or out of the relevant modules, ATP concentration can still be correctly represented to a high degree of accuracy without solving for it at the high rates required for the fast modules. Then modules can be computed separately, bringing their solutions together at relatively long time intervals. This same approach lends itself to parallel computation of the different modules on different CPUs. Since computation time is a major factor for metabolic and electrophysiologic cell models and a huge issue for integrated organ models, such approaches need much further development to achieve maximal efficiency in computation and to enhance progress in investigation.

5 Summary and Conclusions

To summarize, we view biochemical systems as represented at the basic level as networks of given stoichiometry. Whether the steady-state or the kinetic behaviour of a given system is explored, the stoichiometry constrains the feasible behaviour, according to mass balance and the laws of thermodynamics. Study of the feasible behaviour of a system, given a set of stoichiometric constraints, forms the basis of flux-balance analysis, and, more generally, stoichiometric network theory.

To obtain precise information beyond what stoichiometry can tell, one must develop a kinetic representation of the transformations occurring in a given system. The classical enzyme-kinetic approach is often used to build kinetic models that satisfy the stoichiometric rules. The drawback of this approach lies in the inherent uncertainly involved in assigning kinetic mechanisms to biochemical fluxes and in identifying the associated parameter values.

As an alternative to classical enzyme kinetics, the biochemical systems theory alleviates some of these drawbacks. At the heart of BST, the S-system is a versatile and powerful phenomenological nonlinear dynamic equation system with many applications in science and engineering. However, as a modelling paradigm for metabolic reaction networks, it lacks the structure for introducing several fundamental, physiochemical elements of biochemical reactions: stoichiometry, mass and energy balance, and reversibility. These difficulties prevent a true integration of mechanistic studies of individual enzymes and simple metabolic reactions into a complex network system. Therefore, as it stands, BST has not fulfilled the need of current systems biochemistry.

Though many of the useful tools in biochemical analysis (enzyme kinetics, metabolic control analysis, stoichiometric network theory) have been well established in the literature for decades, new developments in theory continue to enhance the usefulness of these tools. In particular, recent additions of thermodynamic considerations into stoichiometric network theory [6,32], and the development of a dynamic metabolic control analysis [14,33], hold promise.

Whether or not these new tools provide the key to improving our understanding of the operation of whole-cell systems remains to be seen. In any case, we can be confident that whatever future technologies prove useful for biochemical systems analysis, those approaches will not conflict with the basic stoichiometric principles. As we move up the hierarchical ladder of system complexity, toward biophysically realistic representations of large-scale systems, the foundation must remain intact.

6 Appendix

Let $\mathbf{b}_1$, $\mathbf{b}_2, \ldots, \mathbf{b}_n$ be an n-dimensional, orthonormal base set. In terms of this base set, n-dimensional vectors $\mathbf{a}_1$, $\mathbf{a}_2, \ldots, \mathbf{a}_m$ can be written as column vectors. Matrix $\{a_{ij}|0 \leq i \leq n, 0 \leq j \leq m\}$ appended with the vectors $\mathbf{b}$'s is called a tableau [16]:

$$\begin{pmatrix} \mathbf{b}_1 & a_{11} & a_{21} & \cdots & a_{m1} \\ \mathbf{b}_2 & a_{12} & a_{22} & \cdots & a_{m2} \\ \mathbf{b}_3 & a_{13} & a_{23} & \cdots & a_{m3} \\ \vdots & \vdots & \vdots & & \vdots \\ \mathbf{b}_n & a_{1n} & a_{2n} & \cdots & a_{mn} \end{pmatrix}. \tag{19}$$

If one carries out row operation for the above tableau, say

$$\begin{pmatrix} \mathbf{b}_1 & a_{11} & a_{21} & \cdots & a_{m1} \\ \mathbf{b}_2 + c\mathbf{b}_1 & a_{12} + ca_{11} & a_{22} + ca_{21} & \cdots & a_{m2} + ca_{m1} \\ \mathbf{b}_3 & a_{13} & a_{23} & \cdots & a_{m3} \\ \vdots & \vdots & \vdots & & \vdots \\ \mathbf{b}_n & a_{1n} & a_{2n} & \cdots & a_{mn} \end{pmatrix}, \tag{20}$$

then the first column on the left is no longer a set of orthonormal vectors. However, the entries (i, j) of the matrix on the right still gives the inner product $(\mathbf{b}'_i, \mathbf{a}_j)$ where $\mathbf{b}'_i$ is the current ith vector on the left, after the operation. If one carries out Gauss–Jordan elimination by row operation and reaches

$$
\begin{pmatrix}
\mathbf{b}'_1 & 1 \ 0 \ \dots \ 0 \\
\mathbf{b}'_2 & 0 \ 1 \ \dots \ 0 \\
\mathbf{b}'_3 & * \ * \ \dots \ * \\
\vdots & \vdots \ \vdots \qquad \vdots \\
\mathbf{b}'_n & 0 \ 0 \ \dots \ *
\end{pmatrix},
\tag{21}
$$

then one has $(\mathbf{b}'_1, \mathbf{a}_1) = 1$ and $(\mathbf{b}'_1, \mathbf{a}_2) = (\mathbf{b}'_1, \mathbf{a}_3) = \dots = (\mathbf{b}'_1, \mathbf{a}_m) = 0$. In other words, vector $\mathbf{b}'_1$ lies in the linear subspace defined by the intersection of hyperplanes $(\mathbf{x}, \mathbf{a}_2) = (\mathbf{x}, \mathbf{a}_3) = \dots = (\mathbf{x}, \mathbf{a}_m) = 0$; and $(\mathbf{b}'_1, \mathbf{a}_1) = 1$.

One notices that the above algorithm is closely related to that for inverting a matrix and computing its null space. For a system of linear equations, the null space plays a fundamental role in the solution to and characterization of the problem. The null space of a matrix $\mathbf{S}$ can be expressed in terms of the loop matrix $\mathbf{R}$: $\mathbf{SR} = 0$. Analogously, for a system of linear inequalities [15], the *solvent matrix* $\mathbf{Y} = \{Y_{ij}|Y_{ij} \geq 0, \sum_j S_{ij}Y_{jk} = 0\}$ plays the fundamental role in the solution to and characterization of the convex system.

The above method can be used to compute the generating vectors of the polyhedral cone defined by a set of linear inequalities $(\mathbf{x}, \mathbf{a}_\ell) \geq 0$, $(\ell = 1, 2, \dots, m)$. In this case, proposed by Fourier, pairwise eliminations with positive multipliers and only additions are carried out [13, 16] to preserve the inequalities, and one stops when all the nonzero entries are positive. For a column with P, Q, and R number of positive, negative, and zero terms respectively, the number of pairwise eliminations is determined by the *expansion number* $PQ + R$ where $P + Q + R = n$ [15]. An efficient pairwise elimination algorithm is designed to have a rule for selection of an optimal order for the elimination steps [16, 31].

References

1. Alberty, R.A. (1991) Equilibrium compositions of solutions of biochemical species and heats of biochemical reactions. *Proc. Natl. Acad. Sci. USA*, **88**, 3268-3271.
2. Bailey, J.E. (1991) Toward a science of metabolic engineering. *Science*, **252**, 1668-1675.
3. Bassingthwaighte, J.B. & Goresky, C.A. (1984) Modeling in the analysis of solute and water exchange in the microvasculature, in *Handbook of Physiology. Sect. 2, The Cardiovascular System Vol IV, The Microcirculation*. E.M. Renkin and C.C. Michel eds., Am. Physiol. Soc., Bethesda, MD. pp. 549-626.
4. Bassingthwaighte, J.B., Goresky, C.A. & Linehan, J.H., eds. (1998) *Whole organ approaches to cellular metabolism: permeation, cellular uptake, and product formation*. Springer, New York.

5. Bassingthwaighte, J.B. (2001) The modeling of a primitive 'sustainable' conservative cell. *Phil. Trans. R. Soc. Lond. A.*, **359**, 1055-1072.

6. Beard, D.A., Liang, S.-D. & Qian, H. (2002) Energy balance for analysis of complex metabolic networks. *Biophys. J.*, **83**, 79-86.

7. Beard, D.A. & Qian, H. (2003) Thermodynamic constraints for cellular metabolic analysis. Manuscript in preparation.

8. Bott, R. & Duffin, R.J. (1949) Impedance synthesis without use of transformers. *J. Appl. Phys.*, **20**, 816.

9. Chong, L. & Ray, L.B. (2002) Whole-istic biology. *Science*, **295**, 1661-1661.

10. Clarke, B.L. (1980) Stability of complex reaction networks. *Adv. Chem. Phys.*, **43**, 1-215.

11. Clarke, B.L. (1988) Stoichiometric network analysis. *Cell Biophys.*, **12**, 237-253.

12. Cornish-Bowden, A. (1976) *Fundamentals of Enzyme Kinetics* 3rd Ed., Butterworths, London. Chapter 4.

13. Dantzig, G.B. & Eaves, B.C. (1973) Fourier-Motzkin elimination and its dual. *J. Combin. Theory A*, **14**, 288-297.

14. Delgado, J.P. & Liao, J.C. (1991) Identifying rate-controlling enzymes in metabolic pathways without kinetic parameters. *Biotechnol. Prog.*, **7**, 15-20.

15. Dines, L.L. (1919) Systems of linear inequalities. *Ann. Math.*, **20**, 191-199.

16. Duffin, R.J. (1974) On Fourier's analysis of linear inequality systems. *Math. Progm. Study*, **1**, 71-95.

17. Fell, D.A. & Sauro, H.M. (1985) Metabolic control and its applications. *Eur. J. Biochem.*, **148**, 555-561.

18. Giersch, G. (1988) Control analysis of metabolic networks. I. Homogeneous functions and the summation theorems for control coefficients. *Eur. J. Biochem.*, **174**, 509-513.

19. Giersch, G. (1988) Control analysis of metabolic networks. II. Total differentials and general formulation of the connectivity relations. *Eur. J. Biochem.*, **174**, 515-519.

20. Golynski, A. & Horton, J.D. (2001) A polynomial time algorithm to find the minimal cycle basis of a regular matroid. Preprint, http://www.stfx.ca/academic/mathcs/apics2001/discrete/Joseph.pdf.

21. Horton, J.D. (1987) A polynomial-time algorithm to find the shortest cycle basis of a graph. *SIAM J. Comput.*, **16**, 358-366.

22. Ibarra, R.U., Edwards, J.S. & Palsson, B.O. (2002) Escherichia coli K-12 undergoes adaptive evolution to achieve in silico predicted optimal growth. *Nature*, **420**, 186-189.

23. Kauffman, K.J., Pajerowski, J.D., Jamshidi, N., Palsson, B.O. & Edwards, J.S. (2002) Description and analysis of metabolic connectivity and dynamics in the human red blood cell. *Biophys. J.*, **83**, 646-662.

24. King, E.L. & Altman, C. (1956) A schematic method of deriving the rate laws for enzyme-catalyzed reactions. *J. Phys. Chem.*, **60**, 1375-1381.

25. Lamarck, J.B. (1830) *Systeme Analytique des Connaissances Positives de l'Homme*, J.B. Baillire, Paris.

26. Lazman, M.Z. & Yablonskii, G.S. (1991) Kinetic polynomial: a new concept of chemical kinetics. In *Patterns and Dynamics in Reactive Media*. R. Aris, D.G. Aronson, and H.L. Swinney, Eds., Springer, New York, pp. 117-149.

27. Mulquiney, P.J., Bubb, W.A. & Kuchel, P.W. (1999) Model of 2,3-bisphosphoglycerate metabolism in the human erythrocyte based on detailed enzyme kinetic equations: in vivo kinetic characterization of 2,3-bisphosphoglycerate synthase/phosphatase using 13C and 31P NMR. *Biochem. J.*, **342**, 576-580.

28. Mulquiney, P.J. & Kuchel, P.W. (1999) Model of 2,3-bisphosphoglycerate metabolism in the human erythrocyte based on detailed enzyme kinetic equations: equations and parameter refinement. *Biochem. J.*, **342**, 581-596.

29. Murray, J.D. (2002) *Mathematical Biology I: An Introduction.* 3rd Ed., Springer, New York.

30. Nasar, S. (1998) *A Beautiful Mind: A Biography of John Forbes Nash, Jr*, Simon & Schuster, New York.

31. Nozicka, F., Guddat, J., Hollatz, H. & Bank, B. (1974) *Theorie der linearen parametrischen Optimierung*, Akademie-Verlag, Berlin.

32. Qian, H., Beard, D.A. & Liang, S.-D. (2002) Stoichiometric network theory for nonequilibrium biochemical systems. *Eur. J. Biochem.*, **270**, 415-421.

33. Reder, C. (1988) Metabolic control theory: a structural approach. *J. Theor. Biol.*, **135**, 175-201.

34. Savageau, M.A. (1969) Biochemical systems analysis. I. *J. Theor. Biol.*, **25**, 365-369.

35. Savageau, M.A. (1969) Biochemical systems analysis. II. *J. Theor. Biol.*, **25**, 370-379.

36. Savageau, M.A. (1970) Biochemical systems analysis. III. *J. Theor. Biol.*, **26**, 215-226.

37. Savageau, M.A. (1976) *Biochemical Systems Analysis: A Study of Function and Design in Molecular Biology.* Addison-Wesley, Reading, MA.

38. Savageau, M.A. & Voit, E.O. (1987) Recasting nonlinear differential equations as S-systems: a canonical nonlinear form. *Math. Biosci.*, **87**, 83-115.

39. Schilling, C.H., Letscher, D. & Palsson, B.O. (2000) Theory for the systemic definition of metabolic pathways and their use in interpreting metabolic function from a pathway-oriented perspective. *J. Theor. Biol.*, **203**, 229-248.

40. Schuster, S. & Hilgetag, C. (1994) On elementary flux modes in biochemical reaction systems at steady-state. *J. Biol. Syst.*, **2**, 165-182.

41. Segel, I.H. (1975) *Enzyme Kinetics.* Wiley Interscience, New York.

42. Sheng, Q. (1994) Global error estimates for exponential splitting. *IMA J. Numer. Anal.*, **14**, 27-56.

43. Smale, S. (1972) On the mathematical foundations of electrical circuit theory. *J. Diff. Geom.*, **7**, 193-210.

44. Strang, G. (1968) On the construction and comparison of differential schemes. *SIAM J. Numer. Anal.*, **5**, 506-517.

45. Stephanopoulos, G. (1994) Metabolic engineering. *Curr. Opin. Biotechnol.*, **5**, 196-200.

46. Stelling, J., Klamt, S., Bettenbrock, K., Schuster, S. & Gilles, E.D. (2002) Metabolic network structure determines key aspects of functionality and regulation. *Nature*, **420**, 190-193.

47. Stryer, L. (1981) *Biochemistry.* 2nd Ed., W.H. Freeman, San Francisco.

48. Van Dien, S.J. & Lidstrom, M.E. (2002) Stoichiometric model for evaluating the metabolic capabilities of the facultative methylotroph Methylobacterium extorquens AM1, with application to reconstruction of C(3) and C(4) metabolism. *Biotechnol. Bioeng.*, **78**, 296-312.

49. Voit, E.O. & Savageau, M.A. (1987) Accuracy of alternative representations for integrated biochemical systems. *Biochemistry*, **26**, 6869-6880.

50. Voit, E.O. (2000) *Computational Analysis of Biochemical Systems: A Practical Guide for Biochemists and Molecular Biologists*. Cambridge Univ. Press, New York.

51. Wei, J. & Prater, C.D. (1962) The structure and analysis of complex reaction systems. *Adv. Catal.*, **13**, 203-392.

52. Westhoff, H.V. & Chen, Y.D. (1984) How does enzyme activity control metabolic concentrations? *Eur. J. Biochem.*, **142**, 425-430.

53. Wiback, S.J. & Palsson, B.O. (2002) Extreme pathway analysis of human red blood cell metabolism. *Biophys J.*, **83**, 808-818.

Modelling Stochastic Clonal Interference

Paulo R.A. Campos[1], Christoph Adami[1,2], and Claus O. Wilke[1]

[1] Digital Life Laboratory 136-93, California Institute of Technology, Pasadena, CA
   prac@caltech.edu
[2] Jet Propulsion Laboratory 126-347, California Institute of Technology, USA
   adami@caltech.edu

Summary. We study the competition between several advantageous mutants in an asexual population (clonal interference) as a function of the time between the appearance of the mutants Δt, their selective advantages, and the rate of deleterious mutations. We find that the overall probability of fixation (the probability that at least one of the mutants becomes the ancestor of the entire population) does not depend on the time interval between the appearance of these mutants, and equals the probability that a genotype bearing all of these mutations reaches fixation. This result holds also in the presence of deleterious mutations, and for an arbitrary number of competing mutants. We also show that if mutations interfere, an increase in the mean number of fixation events is associated with a decrease in the expected fitness gain of the population.

1 Introduction

Evolution, according to [6], is the unifying concept that pulls together all the different strands of biology. Indeed, while evolution provides the framework to understand the otherwise bewildering panoply of adapted forms, it also allows us to understand the patterns that mutation and selection leave in the molecules of life, namely DNA and proteins. One of the central problems of the branch of biology that is devoted to the systematics of the tree of all living things, molecular evolution, concerns the rate of adaptation of individual organisms and species. A precise understanding of molecular phylogeny requires accurate models of evolution and adaptation. In this contribution, we model the adaptation of chromosomes that do not undergo recombination, or more generally, the rate of adaptation of asexual organisms.

The main observation that we study here is that the rate of adaptation of asexual organisms, or of regions of low recombination in the genomes of sexual organisms, does not steadily increase with increasing rate of advantageous mutations. When two or more advantageous mutations appear in different organisms at approximately the same time, then only one of them can go to fixation, i.e., become shared by all members of the population, while the others will be lost. This loss of potentially beneficial mutations limits the rate of adaptation to the rate at which individual mutants can go to fixation. By contrast, in sexual organisms, several mutations can recombine and thus go to fixation together. This interference of advantageous mutations has long been recognized as a potential disadvantage of asexual populations [13,19,29]. Recently, several groups have worked on an exact quantification of the interference effect, both in theoretical studies [1,15,16,27,30] and in experimental studies with bacteria [5,33,34] and viruses [4,28]. A good quantitative understanding of the interference effect is necessary in order to assess the influence that interference has on the patterns of molecular evolution and variation in large populations.

There are two separate dynamics that both contribute to the overall interference effect. First, if two advantageous mutants are both present in sufficiently high concentrations, such that loss to drift can be neglected, then they compete deterministically, and the mutant with the higher selective advantage will replace the other one. Second, if at least one advantageous mutant is still very rare, then we have to consider the influence of other mutants' presence on the chance that this mutant is lost to drift. To date, there is no single theory that takes into account both dynamics to their full extent. [16] calculated the speed of adaptation as a function of population size and beneficial mutation rate, under the assumption that clonal interference can be neglected during the initial phase of drift. They assumed that the probability that a mutant is not lost to drift corresponds to one minus the standard probability of fixation (as calculated for example by [12,17,21]) of that mutant. [30] modified the calculations of Gerrish and Lenski to include beneficial mutations that arise in genomes bearing one or more deleterious mutations, but also did not address the effect of interference of other beneficial mutations during the initial phase of drift. In general, the influence of deleterious mutations on the probability of fixation has been studied extensively [3,20,26,32], but there are very few studies that consider the effect of interfering beneficial mutations. The problem of interfering advantageous mutations is that there exists no simple theoretical framework with which their influence on drift can be described accurately. [1] derived an approximation that allowed him to calculate the probability of fixation of a mutant in a population that is undergoing a selective sweep. However, his approximation is valid only for very small selective advantages. [27] studied a similar situation with numerical simulations, and applied their results to codon bias and levels of polymorphism in molecular evolution.

Here, we study the mutual interference of two advantageous mutants that are each initially present in only a single organism. We derive a phenomenological description of the mutants' fixation probabilities in the interference regime. This phenomenological description, which is in essence an interpolation between limiting cases that can be described with standard theory, agrees very well with numerical simulations. We also find that the expected fitness increase of the population is maximized if the mutant with higher selective advantage arrives earlier, even though this order of appearance leads, on average, to fewer mutants that go to fixation.

2 Model

We consider three distinct sequence types: wild type, advantageous mutant 1, and advantageous mutant 2, with fitness values 1, $1 + s_1$, and $1 + s_2$, respectively. Thus, s_1 denotes the selective advantage of mutants of type 1, and s_2 is the selective advantage of mutants of type 2. For simplicity, we assume that $s_2 > s_1$. Initially, the population is homogeneous and consists of the wild type only. Then, one randomly chosen wild type individual is replaced by an advantageous mutant of either type, and some Δt generations later, another randomly chosen individual of the population (either wild type or mutant) is replaced by a mutant of the other type. Throughout this chapter, we will understand Δt to be the difference in generations between the appearance of mutant 2 and mutant 1, so that negative values of Δt indicate that mutant 2 appeared before mutant 1.

The population is finite of size N, and replication takes place in discrete generations, according to the Wright–Fisher model. All individuals in generation t are direct descendants of the individuals of the previous generation; the probability that an individual is the offspring of a particular parent is proportional to the parent's fitness. We introduce deleterious mutations into the offspring organisms with probability u. If a mutant of type 1 or 2 is hit by a mutation, then its fitness is set to 1 (that is, it reverts to the lower fitness wild type). We do not consider deleterious mutations in the wild type genotype.

We consider a genotype to be fixed if it has become the most-recent common ancestor of the whole population, regardless of whether some individuals in the population have a different genotype. This definition of fixation has been used recently to study fixation of beneficial mutations in a heterogeneous genetic background [1, 20], and also to investigate the process of fixation in a viral quasispecies [38].

3 Theoretical Analysis

3.1 Probability of Ultimate Fixation

In the Appendix, we derive equations for the probability of fixation $P(s, u)$ and time to fixation $T(s, u)$ of an individual mutant with selective advantage s and mutation rate u. For two mutants, we are mostly interested in the probability of ultimate fixation, that is, the probability that a mutant reaches fixation and is not subsequently replaced by the other mutant. In the following, we will denote the probability that mutant i reaches ultimate fixation under the condition that the two mutants are introduced Δt generations apart as $\pi_i(\Delta t)$.

We can calculate $\pi_i(\Delta t)$ for the two limiting cases $\Delta t \to -\infty$ and $\Delta t \to \infty$, and present a phenomenological description for intermediate Δt.

For $\Delta t \to -\infty$, that is, when mutant 2 arises much earlier than mutant 1, we have

$$\pi_1(-\infty) = [1 - P(s_2, u)]P(s_1, u), \tag{1}$$

$$\pi_2(-\infty) = P(s_2, u). \tag{2}$$

Since $s_2 > s_1$, mutant 1 can reach fixation only when mutant 2 has not reached fixation. For $\Delta t \to \infty$, we have to consider the possibility that mutant 1 reaches fixation first, but is later replaced by mutant 2. Therefore, we find

$$\pi_1(\infty) = P(s_1, u)\left[1 - P\left(\frac{1 + s_2}{(1 + s_1)(1 - u)} - 1, u\right)\right], \tag{3}$$

$$\pi_2(\infty) = P(s_1, u)P\left(\frac{1 + s_2}{(1 + s_1)(1 - u)} - 1, u\right) + [1 - P(s_1, u)]P(s_2, u), \tag{4}$$

where $P(\frac{1+s_2}{(1+s_1)(1-u)} - 1, u)$ is the probability that mutant 2 goes to fixation after mutant 1 has already reached fixation. Since $P(\frac{1+s_2}{(1+s_1)(1-u)} - 1, u)$ is always smaller than or equal to $P(s_2, u)$, we have $\pi_1(-\infty) \le \pi_1(\infty)$ and $\pi_2(-\infty) \ge \pi_2(\infty)$. In other words, both mutants have a higher probability of fixation when they are introduced first than when they are introduced second.

For intermediate values of Δt, there is no theory that enables us to derive a simple expression for $\pi_i(\Delta t)$ (but see [1]). We cannot use branching process theory, because it assumes that the presence of the invading mutants does not influence the mean fitness (which they do for intermediate Δt). Also, diffusion theory becomes unwieldy when there are more than two different sequence types. Nevertheless, we have found that we can develop a phenomenological description of the competition of two mutants as follows. We know that $\pi_i(t)$ must reach the two limiting values $\pi_i(\infty)$ and $\pi_i(-\infty)$ for sufficiently large positive or negative Δt. Moreover, as long as $\Delta t < 0$, we do not expect $\pi_i(\Delta t)$ to be very different from $\pi_i(-\infty)$, because mutant 1 has only a realistic chance of proliferating and going to fixation if mutant 2 is not present in the population. As long as $\Delta t \lesssim 0$, mutant 2 will either go to fixation relatively unscathed from the later appearance of mutant 1, or be lost to drift, in which case mutant 1 will have its turn. Likewise, for Δt larger than the time to fixation of mutant 1, T_1, we expect that $\pi_i(\Delta t) = \pi_i(\infty)$, because T_1 generations after the introduction of mutant 1, mutant 2 will find either a population in which mutant 1 has already gone to fixation, or one in which it has been lost to drift. For $0 \lesssim \Delta t \lesssim T_1$, we expect that $\pi_1(\Delta t)$ smoothly increases from $\pi_1(-\infty)$ to $\pi_1(\infty)$, while $\pi_2(\Delta t)$ smoothly decreases from $\pi_2(-\infty)$ to $\pi_2(\infty)$. These considerations suggest a sigmoidal form for $\pi_i(\Delta t)$. We use a logistic growth model to describe the smooth transition from $\pi_i(-\infty)$ to $\pi_i(\infty)$ within the range $0 \lesssim \Delta t \lesssim T_1$:

$$\pi_i(\Delta t) = \pi_i(-\infty) + \frac{\pi_i(\infty) - \pi_i(-\infty)}{1 + e^{-\gamma_1(\Delta t - T_1/2)}}, \tag{5}$$

where γ_1 is the fitness advantage of mutant 1 at finite mutation rate (see the definition following Eq. (8) below). This expression appears to be an acceptable description of the exact time dependence (see the numerical results) without free parameters.

3.2 Overall Probability of Fixation

Besides the individual probabilities $\pi_1(\Delta t)$ and $\pi_2(\Delta t)$, their sum $\pi = \pi_1(\Delta t) + \pi_2(\Delta t)$ is also of interest. This sum is the overall probability that at least one mutant goes to fixation. When we sum Eqs. (1) and (2), or Eqs. (3) and (4), we find that the overall fixation probability does not depend on the order in which the mutants are introduced, and has the form

$$\pi = P(s_1, u) + P(s_2, u) - P(s_1, u)P(s_2, u). \qquad (6)$$

Furthermore, if Eq. (5) is indeed a good description of $\pi_i(\Delta t)$ for arbitrary Δt, then π should not depend on Δt at all, because all time dependencies cancel when we use Eq. (5) to calculate $\pi_1(\Delta t) + \pi_2(\Delta t)$.

The invariance of the overall fixation probability π under the order by which the mutants appear is a general property. It holds also when more than two mutants arise. Assume that n advantageous mutants arise, with selective advantages $s_1, \ldots, s_n$ compared to the wild type. Further assume that the time intervals between the appearances of the mutants are large. Then, the probability that none of the mutants make it to fixation is $\prod_{i=1}^{n}[1 - P(s_i, u)]$, regardless of the order of their appearance. The probability that at least one mutant goes to fixation is therefore

$$\pi = 1 - \prod_{i=1}^{n}[1 - P(s_i, u)], \qquad (7)$$

which reduces to Eq. (6) in the case of $n = 2$. Using an extension of Kimura's well-known result for $P(s_i, u)$ derived in the Appendix, Eq. (18), we can simplify this expression even further. We find in the limit $N \to \infty$:

$$\pi = 1 - \exp\left[-2\left(\gamma_1 + \gamma_2 + \cdots + \gamma_n\right)\right], \qquad (8)$$

with $\gamma_i = (1 + s_i)(1 - u) - 1$ for $u < s_i/(1 + s_i)$, and $\gamma_i = 0$ otherwise. Moreover, in the absence of deleterious mutations, for $u = 0$, π becomes

$$\pi = 1 - \exp\left[-2\left(s_1 + s_2 + \cdots + s_n\right)\right]. \qquad (9)$$

In this limit, the overall probability of fixation is the same as the probability of fixation of a single mutant with selective advantage s equal to the sum of the selective advantages of all invading mutants, $s = \sum_i s_i$. Such a hypothetical mutant is extremely unlikely in clonal populations, because all beneficial mutations would have to hit the lineage sequentially, but could occur when beneficial mutations are shared via recombination. In the limit that all γ_i vanish (that is, for very large mutation rates or for vanishing s_i), Eq. (7) becomes

$$\pi = \frac{n}{N}\,. \tag{10}$$

3.3 Expected Fitness Increase and Expected Number of Fixed Mutants

Depending on which mutant goes to fixation, the average fitness of the final population is 1, $1 + \gamma_1$, or $1 + \gamma_2$, with γ as defined following Eq. (8). The expected fitness increase $\langle \gamma(\Delta t) \rangle$ after the introduction of the two mutants is therefore

$$\langle \gamma(\Delta t) \rangle = \gamma_1 \pi_1(\Delta t) + \gamma_2 \pi_2(\Delta t)\,. \tag{11}$$

Since we know that $\pi_1(\Delta t) + \pi_2(\Delta t)$ is constant, and $\gamma_2 \geq \gamma_1$ as a direct consequence of our assumption $s_2 > s_1$, it follows that $\langle \gamma(-\infty) \rangle \geq \langle \gamma(\infty) \rangle$. If we write $\pi_1(\infty) = \pi_1(-\infty) + \Delta\pi$, and $\pi_2(-\infty) = \pi_2(\infty) + \Delta\pi$, then we find

$$\langle \gamma(-\infty) \rangle - \langle \gamma(\infty) \rangle = (\gamma_2 - \gamma_1)\Delta\pi \geq 0\,. \tag{12}$$

Thus, the expected fitness increase is larger if we introduce the mutant with the higher selective advantage first.

Let n_{fix} denote the number of fixed mutants, that is, we have $n_{\text{fix}} = 0$ if none of the mutants reach fixation, $n_{\text{fix}} = 1$ if exactly one of the mutants reaches fixation, and $n_{\text{fix}} = 2$ if first mutant 1 reaches fixation and is later replaced by mutant 2. For $\Delta t \to -\infty$ and N large, n_{fix} can never be larger than one, because the probability that mutant 1 goes to fixation in the background of mutant 2 is zero. We find in this limit for the expected value of n_{fix}:

$$\langle n_{\text{fix}}(-\infty) \rangle = \pi_1(-\infty) + \pi_2(-\infty) = \pi\,. \tag{13}$$

In the limit $\Delta t \to \infty$, on the other hand, we find

$$\begin{aligned}
\langle n_{\text{fix}}(\infty) \rangle &= \pi_1(-\infty) + \pi_2(-\infty) + P(s_1, u)P\Big(\frac{1 + s_2}{(1 + s_1)(1 - u)} - 1, u\Big) \\
&= \pi + P(s_1, u)P\Big(\frac{1 + s_2}{(1 + s_1)(1 - u)} - 1, u\Big)\,.
\end{aligned} \tag{14}$$

Clearly, $\langle n_{\text{fix}}(\infty) \rangle \geq \langle n_{\text{fix}}(-\infty) \rangle$. This means that if the mutant with the smaller selective advantage appears before the mutant with the larger selective advantage, then the expected number of mutants that go to fixation is larger than if the mutant with the larger selective advantage appears first. However, at the same time the expected increase in average fitness is smaller, as we saw in the previous paragraph.

4 Numerical Simulation

In order to measure fixation probabilities, we carried out $100,000$ replicates of the simulation for each set of parameters, and recorded the final outcome (all individuals unmarked, or marked as descendants of either advantageous mutant). We studied a population of size $N = 1000$ and mutation rates $u = 0.0, 0.01, 0.02, 0.03, 0.05, 0.07, 0.1, 0.15, 0.20, 0.3, 0.4, 0.5, 0.6, 0.7, 0.8$. The selective advantages were $s_1 = 0.1$ and $s_2 = 0.2$, $s_1 = 0.1$ and $s_2 = 0.5$, $s_1 = 0.05$ and $s_2 = 0.2$. We also studied a population of size $N = 10,000$ for a subset of these parameters, in order to make sure that our results were robust against a change in population size. Because of the sizable amount of CPU time needed to carry out 100,000 replicates for $N = 10,000$, we could, however, not study this case exhaustively.

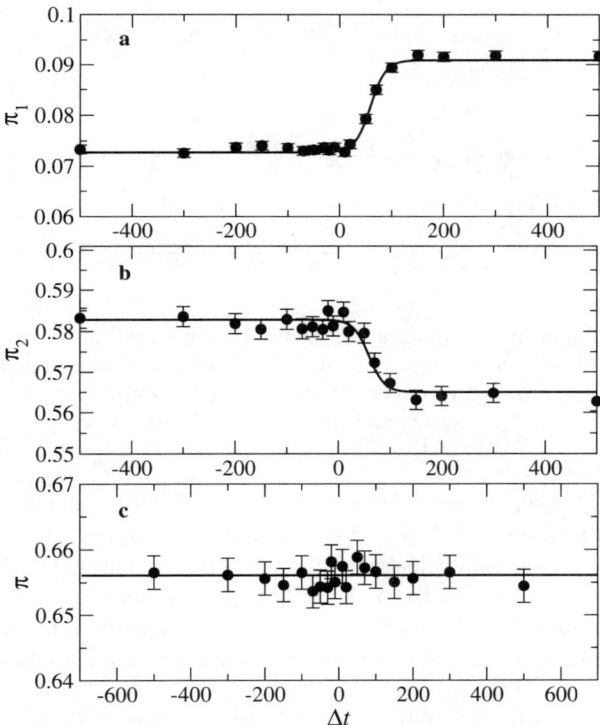

Fig. 1. Probability of fixation as a function of time interval Δt ($N = 1000$, $s_1 = 0.1$, $s_2 = 0.5$, $u = 0.0$). Solid lines represent the prediction according to the logistic growth model Eq. (5). a: Probability of fixation of genotype 1 (less beneficial mutation), π_1. b: Probability of fixation of genotype 2 (more beneficial mutation), π_2. c: Overall probability of fixation, π.

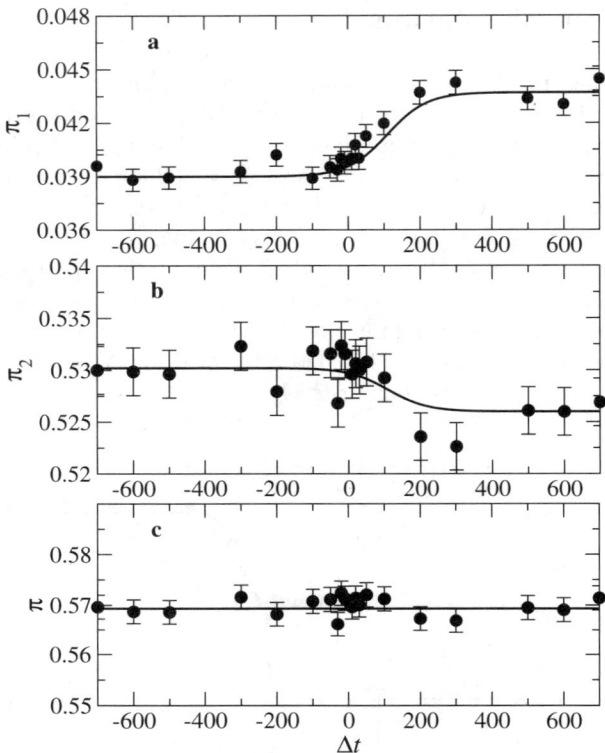

Fig. 2. Probability of fixation as a function of time interval Δt ($N = 1000$, $s_1 = 0.1$, $s_2 = 0.5$, $u = 0.05$). Solid lines represent the prediction according to the logistic growth model Eq. (5). a: Probability of fixation of genotype 1, π_1. b: Probability of fixation of genotype 2, π_2. c: Overall probability of fixation, π.

In order to keep track of the evolutionary history of each mutant, we marked the initial mutants 1 and 2 with two distinct inheritable neutral markers. In that way, we could distinguish wild type sequences that were descendants from mutants 1 or 2 from the wild type sequences that were originally present. We continued all simulations until all individuals in the population were unmarked, marked as descendants of mutant 1, or marked as descendants of mutant 2.

Figure 1 shows the probability of ultimate fixation of mutants 1 and 2, π_1 and π_2, and the overall probability of fixation $\pi = \pi_1 + \pi_2$, for $u = 0$. We see that the probability π_1 is constant for $\Delta t < 0$, and starts to increase as soon as Δt turns positive. Eventually, π_1 levels off again. We observe the opposite behaviour for the probability π_2. For $\Delta t < 0$, π_2 is constant, but decreases rapidly in the same range of Δt in which π_1 increases. Finally, π_2 levels off as well.

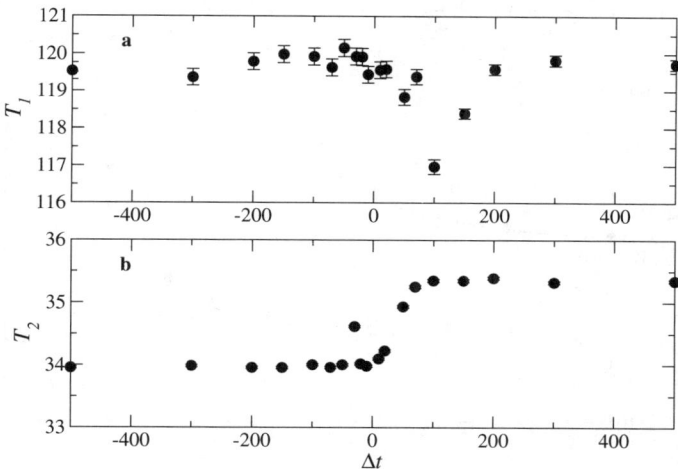

Fig. 3. Time to fixation as a function of the time interval between beneficial mutants Δt ($N = 1000$, $s_1 = 0.1$, $s_2 = 0.5$, $u = 0.0$). a: Time to fixation of genotype 1, T_1. b: Time to fixation of genotype 2, T_2.

The overall probability of fixation π is approximately constant for all values of Δt. The solid lines in Fig. 1a and b correspond to Eq. (5), and the solid line in Fig. 1c is $\pi_1(-\infty) + \pi_2(-\infty)$. We find that our phenomenological description Eq. (5) performs very well for intermediate Δt.

In Fig. 2 we plot the same quantities as those shown in Fig. 1, but now with a positive mutation rate $u = 0.05$. We observe that the two probabilities π_1 and π_2 have smaller values than in the absence of mutations. The phenomenological description still works well, and appears to correctly take into account the effects of mutation. The overall fixation probability π is again independent of Δt. In Fig. 3 we show the time to fixation of mutant 1, T_1, and the time to fixation of mutant 2, T_2, as functions of Δt. The parameter values are the same as those of Fig. 1. In Fig. 3a, we see that T_1 is approximately constant for all values of Δt except for the range $0 \leq \Delta t \leq 200$. For negative Δt the fixation of mutant 1 occurs only when mutant 2 has been eliminated. Therefore, T_1 corresponds to the result for the fixation time of a mutant with selective advantage s_1 in a homogeneous population with wild type individuals only. The same is true for large positive Δt because there mutant 1 has enough time to reach fixation without the interference of the second mutant. For small positive Δt, we observe a decrease of T_1. The decrease occurs for those Δt for which the two beneficial mutants coexist for several generations in the population. Fixation of mutant 1 in this regime occurs only when mutant 1 reaches fixation so quickly that mutant 2 has not had time to build up momentum. Otherwise, most likely mutant 1 will be displaced by mutant 2 before reaching fixation. For mutant 2, the time to fixation T_2 is shorter for negative Δt than for positive Δt. For positive Δt, in a fraction of cases

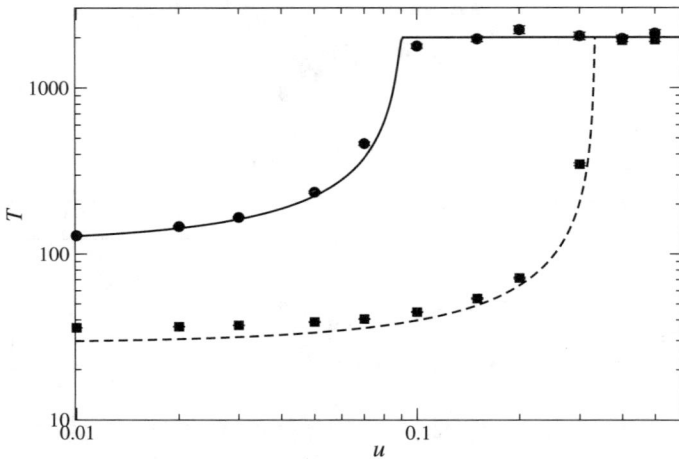

Fig. 4. Time to fixation as a function of the mutation probability u ($s_1 = 0.1$ and $s_2 = 0.5$, $\Delta t = 100$, $N = 1000$). Solid line: the time T_1 to fixation for the less beneficial mutant; dashed line: time to fixation of more beneficial mutant T_2.

mutant 2 has to go to fixation in a background of mutant 1, rather than in a background of wild type. In the background of mutant 1, the selective advantage of mutant 2 is smaller than in a wild type population, which explains the increased time to fixation. Interestingly, before T_2 starts to rise for increasing positive Δt, it quickly spikes at $\Delta t \approx 30$. This spike has the following explanation: if mutant 1 is introduced right before mutant 2 has reached fixation, then the time to fixation of mutant 2 is increased by the additional time it takes for mutant 1 to disappear again. This additional time will typically be one or two generations, which is in agreement with the height of the spike. (T_1 on the top of the spike is actually approximately half a generation larger than before the spike.) The limiting values for π_1 and π_2 when $\Delta t \to -\infty$ and $\Delta t \to -\infty$ were estimated by means of branching process theory which provides very good accuracy, especially for low mutation values.

In Fig. 4 we show the fixation times T_1 and T_2 as a function of mutation rate obtained from simulations, and compared to the prediction from diffusion theory, Eq. (19). As we increase the mutation rate u, the population moves from a strong selection regime, characterized by a short time to fixation, to a neutral regime, where $T \approx 2N$. We also observe that the transition between these two regimes occurs at different mutation rates for the two genotypes: T_1 shows an abrupt transition around the critical value $u = u_{1c} \approx 0.09$, whereas T_2 reaches the neutral regime around $u = u_{2c} \approx 0.30$. The mutation rate at which this transition occurs is known as the *error threshold* [8].

In Fig. 5 we can see the overall probability of fixation as a function of mutation rate for the case of two as well as three interfering mutants. Simulation results are again compared to the diffusion theory result Eq. (18), but also to

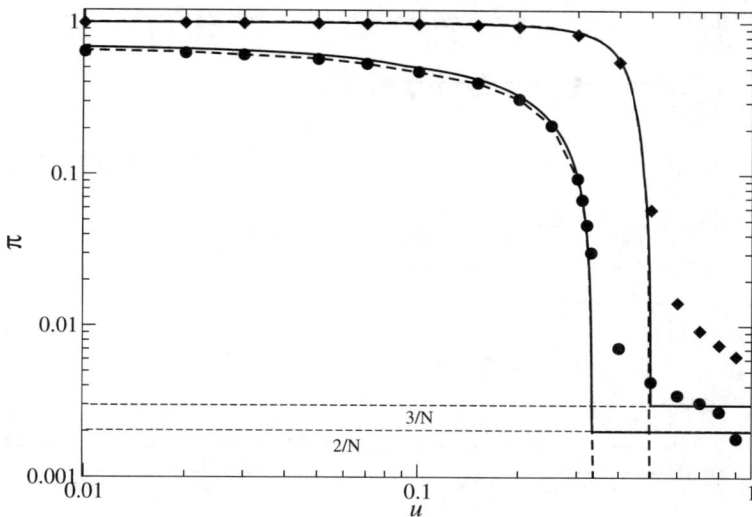

Fig. 5. Overall probability of fixation as a function of mutation rate u. Circles correspond to the situation of two invading mutants ($s_1 = 0.1$, $s_2 = 0.5$, $\Delta t = 100$, $N = 1000$), and squares correspond to the situation of three invading mutants ($s_1 = 0.8$, $s_2 = 0.9$, $s_3 = 1.0$, $\Delta t = 20$). The solid line is the prediction according to diffusion theory, while the long-dashed line corresponds to the branching process. The short dashes indicate the fixation probability levels according to the neutral theory Eq. (10), for two or three mutants, respectively.

a prediction from branching process theory, Eq. (28). For these simulations, we used $s_1 = 0.1$, $s_2 = 0.5$ and population size $N = 1000$. We also chose a time interval $\Delta t = 100$, which is smaller than the time required for fixation of mutant 1, to ensure that the dynamics takes place in the clonal interference regime. Above the error threshold, the probability of fixation is known to be $1/N$ for a single mutation ($2/N$ and $3/N$ for two or three mutants, respectively). Because the branching process description assumes infinite population size, it predicts zero probability of fixation above the error threshold. Diffusion theory, on the other hand, describes this regime adequately, while being less accurate at small mutation rates. The abrupt change between the ordered and disordered regime predicted by both theories is not present in the numerical data because the simulated system is finite.

Finally, we tested the prediction that the mean number of fixations in the population is larger when the mutant with the smaller selective advantage is introduced first, even though the mean fitness increase is smaller. Figure 6a shows the results for the expected increase in fitness as a function of the time interval Δt as defined in Eq. (11), as well as the solution of Eq. (11) using a sigmoidal *ansatz* for the probabilities π_1 and π_2, as defined in Eq. (5).

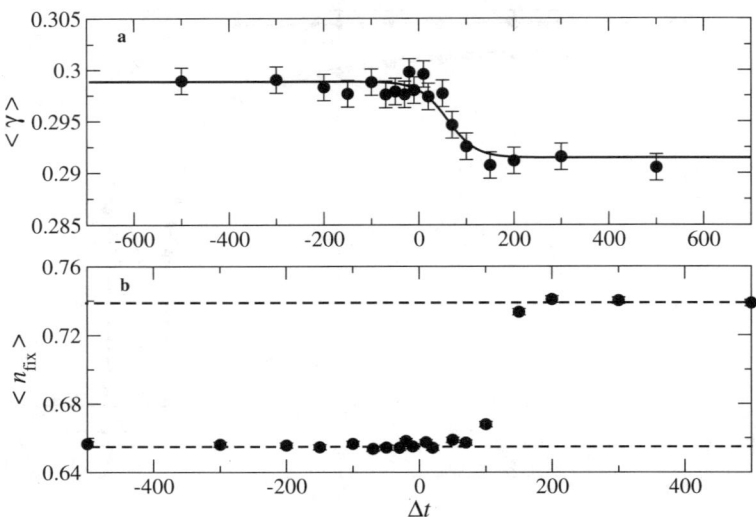

Fig. 6. a: Expected increase in fitness due to introduction of mutants as a function of time between their introduction. b: Expected number of fixed mutants (panel b) as a function of Δt ($s_1 = 0.1$, $s_2 = 0.5$, $u = 0.0$, $N = 1000$, for a and b). Solid line in panel (a) is expected result according to fixation probabilities given by Eq. (5).

As expected, the mean fitness gain decreases as Δt turns positive, while the expected number of fixed mutants, shown in Fig. 6b, increases.

5 Discussion

The rate of adaptation of a non-recombining chromosome or asexual population is a non-trivial quantity that depends on details of the fitness distribution of mutations. Beneficial mutations often take a long time to dominate in a population, and compete against each other in the background of deleterious mutations. The probability of fixation of any one mutation depends on the time of their introduction, stochastic drift, and the prevalent mutation rate. A good understanding of the rate of adaptation of asexuals is important in molecular phylogeny, because it influences the speed of the assumed molecular clock [25]. Since successive fixation events determine the branchings in the phylogenetic tree, a stochastic analysis of the fixation probability of competing mutations can provide insight into models of evolution used for tree reconstruction methods [31].

We studied the probability of fixation of beneficial mutations in the presence of other beneficial mutations that were introduced either earlier or later, and in the presence of other deleterious mutations, that is, we studied fixation in a non-equilibrium background. This problem has been addressed previously by [1] using a deterministic approach, which was unfortunately limited to small selective advantages. Similarly, Johnson and Barton [20] studied the probability of fixation in a changing background, but they confined themselves mostly to the case where deleterious mutations accumulate in the background after a selective sweep.

We found that the probability of fixation is easily understood from an analytic point of view as long as the time interval between the introduction of mutants is larger than the fixation time of the least beneficial mutation. In the interference regime, where the population consists of clones of the wild type as well as both beneficial mutants, a phenomenological approach based on logistic growth is successful at describing the competition. In general, we find that the probability of fixation of the most beneficial mutant is reduced when it competes against a less beneficial mutant, and there is a sizable probability that the less beneficial mutant will in fact survive (see Fig. 1). Yet, the probability that either one of the mutants survives does not depend on the timing of their introduction.

A number of authors noted the reduction of fixation probabilities caused by segregating deleterious mutations [3, 20, 26, 30, 32], and we observe the same qualitative behaviour here. We find that a background of deleterious mutations suppresses the fixation probability, and this suppression is more pronounced for the less beneficial mutant. In the model presented here, there is a simple relationship between beneficial and deleterious mutations: each mutation occurring on a clone created by a beneficial mutation is reduced to the wild type fitness. This simplified model corresponds to a single-peak landscape, exhaustively studied in the context of quasispecies theory [2,14,35, 36]. Consequently, we expect a more pronounced decrease of the probability of fixation for small s due to the existence of an error threshold, that is a point at which information is lost from the sequence due to a critically high mutation rate, see [8]. In the stochastic regime of evolution, which occurs for mutation rates above the error threshold, individuals replicate randomly, and each member of the population is expected to be the most recent ancestor of the population with the same probability $\pi = 1/N$ while the time to fixation is $T = 2N$ [7,22,24]. This prediction is confirmed in Fig. 4, where we see that for large u both T_1 and T_2 are approximately equal to $2N$. Accordingly, in the interval $u_{1c} < u < u_{2c}$, where u_{ic} denotes the error threshold for genotype of type i, we can describe the dynamics by ignoring the less beneficial mutant.

The explicit modelling of stochastic interference between beneficial mutation revealed an interesting observation concerning the mean number of fixation events and the expected fitness change. Naively one might surmise that, as the number of fixation events increases, so does the expected increase in fitness. Instead, we found the opposite dynamic: Fig. 6 clearly shows that if the mutant with the smaller selective advantage appears first, then we expect more fixation events because the second mutant then has a good chance to survive. Yet, this is not the best scenario if we are interested in maximizing fitness. In this case, it is better if the most beneficial mutant can establish itself first, even though this implies that there is little chance for another fixation event.

Finally, we observed that the probability for any beneficial mutant to go to fixation is equal to the probability that a genotype bearing *all* mutations will become the ancestor of the entire population. Moreover, this observation holds for an arbitrary number of competing mutations, even in the presence of deleterious mutations, independently of the order of appearance of the beneficial mutations. Naturally, while such a probability is very high if the sum of all benefits is large, the probability of a single sequence appearing with this combination of mutations is exponentially small in non-recombining populations. This result was confirmed using an extension of Kimura's result [21] to finite mutation rates (Eq. (18)) and also by means of a branching process formulation. Both approaches yield similar results and are in excellent agreement with our numerical simulations. We find that the branching process approach is more accurate at low mutation rates, away from the error threshold. As it is formulated in the $N \to \infty$ limit, it fails to predict the asymptotic result $\pi \approx n/N$ which is reached in the limit that all γ_i vanish.

The present analysis concerns the rate of adaptation as beneficial mutations compete against each other in the presence of deleterious mutations, in a non-equilibrium framework. This is a step toward the goal of characterizing the rate of adaptation of whole populations in arbitrary circumstances, and in the presence of recombination. We expect that these extensions are necessary before accurate predictions of optimal mutation rates in biological evolution can be made.

6 Appendix: Probability of Fixation and Time to Fixation for an Individual Mutant

We can calculate the probability of fixation of an individual mutant either from branching process theory [1,12,13,17,20,38] or from diffusion theory [21]. Branching process theory is applicable for very large population sizes, and arbitrary (but positive) selective advantages s. Diffusion theory is applicable to moderately large to large population sizes and small or even vanishing selective advantages s. Moreover, diffusion theory also gives an expression for the expected time to fixation.

6.1 Diffusion Theory

If the advantageous mutant suffers additional mutations while it goes to fixation, then we typically have to use multi-dimensional diffusion equations, which can be very unwieldy. However, in the simple case treating the advantageous mutant and the wild type only, and where the only effect of deleterious mutations is to revert the advantageous mutant back to wild type, one-dimensional diffusion theory, as developed by [21], and [23], is still applicable, and gives satisfying results [37,39].

The main quantities in diffusion theory are the mean $M_{\delta x}$ and the variance $V_{\delta x}$ in the rate of change per generation of the concentration x of the invading mutant. The derivation of these quantities is straightforward [11], and we find

$$M_{\delta x} = \gamma x(1 - x), \qquad (15)$$
$$V_{\delta x} = x(1 - x)/N \qquad (16)$$

to first order in $1/N$ and γ. Here, γ is the relative difference in mean fitness between a wild type population and a population with fixed advantageous mutant. The only difference between these expressions and those of [21] is that γ replaces the selective advantage s in $M_{\delta x}$. Without mutations ($u = 0$), γ is equal to s, and we recover the standard results. For positive u, γ has the form

$$\gamma = \begin{cases} (1 + s)(1 - u) - 1 & \text{for } u < s/(1 + s), \\ 0 & \text{for } u \geq s/(1 + s). \end{cases} \qquad (17)$$

The mutation rate u_c at which γ reaches 0, $u_c = s/(1 + s)$, corresponds to the error threshold of quasispecies theory [9,10,35]. For mutation rates larger than u_c, the invading mutant does not have an advantage over the wild type, and the fixation process corresponds to that of a neutral mutant.

The probability of fixation follows now from [21] as

$$P(s, u) = \frac{1 - e^{-2\gamma}}{1 - e^{-2\gamma N}}. \qquad (18)$$

Likewise, the expected time to fixation follows from [23] as

$$T(s, u) = J_1 + \frac{1 - P(s, u)}{P(s, u)} J_2, \qquad (19)$$

where

$$J_1 = \frac{1}{\gamma(1 - e^{-2\gamma N})} \int_{1/N}^{1} \frac{(e^{2\gamma N x} - 1)(e^{-2\gamma N x} - e^{-2\gamma N})}{x(1 - x)} dx, \qquad (20)$$

$$J_2 = \frac{1}{\gamma(1 - e^{-2\gamma N})} \int_{0}^{1/N} \frac{(e^{2\gamma N x} - 1)(1 - e^{-2\gamma N x})}{x(1 - x)} dx. \qquad (21)$$

6.2 Branching Process Theory

According to [1], the probability that a beneficial mutation reaches fixation in a genotype with genetic background i follows from iterating the following set of equations:

$$(1 - P_{i,t-1}) = \sum_{j=0}^{\infty} W_{i,j}(1 - P_{i,t}^*)^j, \qquad (22)$$

where $P_{i,t}$ is the probability of fixation of an allele that is present in a single copy in site i in generation t, $W_{i,j}$ denotes the probability that an allele in site i contributes with j offspring to the next generation, and

$$P_{i,t}^* = \sum_k M_{i,k} P_{k,t} \qquad (23)$$

is the probability that an allele in background i at time $t-1$ would be fixed, given that at time t it is passed to one offspring. The quantity $M_{j,k}$ gives the probability that an offspring from a parent at background i will be at background k. If the distribution of offspring is given by a Poisson distribution with mean $(1+s_i)$, that is,

$$W_{i,j} = \frac{(1+s_i)^j}{j!} e^{-(1+s_i)}, \qquad (24)$$

then Eq. (22) becomes

$$(1 - P_{i,t-1}) = \exp\left[-(1+s_i) P_{i,t}^*\right]. \qquad (25)$$

The fixation probabilities correspond to the solution of Eq. (22) obtained in the limit $t \to \infty$.

In our model, we consider only two distinct classes of genotypes, and mutations occur only from advantageous mutant to wild type. Because the wild type has a mean number of offspring per generation exactly equal to one, the probability of fixation of wild type sequences in the branching process formulation vanishes [18]. Therefore, the probability of fixation $P(s, u)$ is given by the solution of the equation

$$P(s, u) = 1 - \exp\left[-(1+s)(1-u) P(s, u)\right]. \qquad (26)$$

Unfortunately, it is not possible to derive a closed-form solution to this equation. $P(s, u)$ has to be determined numerically from iterating Eq. (26).

Since the relevant parameter seems to be $\gamma = (1+s)(1-u) - 1$, the above equation can be written as

$$P(s, u) = 1 - \exp\left[-(\gamma + 1) P(s, u)\right]. \qquad (27)$$

Although a derivation cannot be performed when n beneficial mutants are considered, we found empirically that in this case

$$P(s, u) = 1 - \exp\left[-\left[\prod_{i=1}^n (\gamma_i + 1)\right] P(s, u)\right]. \qquad (28)$$

This expression is used to compare numerical results to the branching process theory prediction in Fig. 5.

Acknowledgments

P. R. A. Campos is supported by Fundação de Amparo à Pesquisa do Estado de São Paulo, Proj. No. 99/09644-9. This work is supported by the NSF under contract No. DEB-9981397. The work of C.A. was carried out in part at the Jet Propulsion Laboratory (California Institute of Technology), under a contract with the National Aeronautics and Space Administration.

References

1. Barton, N. H. (1995). Linkage and the limits to natural selection. *Genetics 140*, 821–841.
2. Campos, P. R. A. and J. F. Fontanari (1998). Finite-size scaling of the quasi-species model. *Phys. Rev. E 58*, 2664–2667.
3. Charlesworth, B. (1994). The effect of background selection against deleterious mutations on weakly selected, linked variants. *Genet. Res. Camb. 63*, 213–227.
4. Cuevas, J. M., S. F. Elena, and A. Moya (2002). Molecular basis of adaptive convergence in experimental populations of RNA viruses. *Genetics 162*, 533–542.
5. de Visser, J. A. G. M., C. W. Zeyl, P. J. Gerrish, J. L. Blanchard, and R. E. Lenski (1999). Diminishing returns from mutation supply rate in asexual populations. *Science 283*, 404–406.
6. Dobzhansky, T. (1973). Nothing in biology makes sense except in the light of evolution. *Am. Biol. Teach. 35*, 125–129.
7. Donnelly, P. and S. Tavaré (1995). Coalescents and genealogical structure under neutrality. *Annu. Rev. Genet. 29*, 401–421.
8. Eigen, M. (1971). Selforganization of matter and evolution of biological macromolecules. *Naturwissenschaften 58*, 465–429.
9. Eigen, M., J. McCaskill, and P. Schuster (1988). Molecular quasi-species. *J. Phys. Chem. 92*, 6881–6891.
10. Eigen, M., J. McCaskill, and P. Schuster (1989). The molecular quasi-species. *Adv. Chem. Phys. 75*, 149–263.
11. Ewens, W. J. (1979). *Mathematical Population Genetics*. Berlin: Springer.
12. Fisher, R. A. (1922). On the dominance ratio. *Proc. Roy. Soc. Edinb. Sect. B Biol. Sci. 42*, 321–341.
13. Fisher, R. A. (1930). *The Genetical Theory of Natural Selection*. Oxford: Clarendon Press.
14. Galluccio, S. (1997). Exact solution of the quasispecies model in a sharply peaked fitness landscape. *Phys. Rev. E 56*, 4526–4539.
15. Gerrish, P. (2001). The rhythm of microbial adaptation. *Nature 413*, 299–302.
16. Gerrish, P. J. and R. E. Lenski (1998). The fate of competing beneficial mutations in an asexual population. *Genetica 102/103*, 127–144.
17. Haldane, J. B. S. (1927). A mathematical theory of natural and artificial selection. Part V: Selection and mutation. *Proc. Camb. Phil. Soc. 26*, 220–230.
18. Harris, T. E. (1963). *The Theory of Branching Processes*. Berlin Heidelberg New York: Springer.
19. Hill, W. G. and A. Robertson (1966). The effect of linkage on the limits to artificial selection. *Genet. Res. 8*, 269–294.

20. Johnson, T. and N. H. Barton (2002). The effect of deleterious alleles on adaptation in asexual organisms. *Genetics 162*, 395–411.
21. Kimura, M. (1962). On the probability of fixation of mutant genes in a population. *Genetics 47*, 713–719.
22. Kimura, M. (1968). Evolutionary rate at the molecular level. *Nature 217*, 624–626.
23. Kimura, M. and T. Ohta (1969). The average number of generations until fixation of a mutant gene in a finite population. *Genetics 61*, 763–771.
24. Kingman, J. F. C. (1982). On the genealogies of large populations. *J. Appl. Prob. 19A*, 27–43.
25. Kuhner, M. and J. Felsenstein (1994). A simulation comparison of phylogeny algorithms under equal and unequal evolutionary rates. *Mol. Biol. Evol. 11*, 459–68.
26. Manning, J. T. and D. J. Thompson (1984). Muller's ratchet and the accumulation of favourable mutations. *Acta Biotheor. 33*, 219–225.
27. McVean, G. A. T. and B. Charlesworth (2000). The effects of Hill-Robertson interference between weakly selected mutations on patterns of molecular evolution and variation. *Genetics 155*, 929–944.
28. Miralles, R., P. J. Gerrish, A. Moya, and S. F. Elena (1999). Clonal interference and the evolution of RNA viruses. *Science 285*, 1745–1747.
29. Muller, H. J. (1964). The relation of recombination to mutational advance. *Mutat. Res. 1*, 2–9.
30. Orr, H. A. (2000). The rate of adaptation in asexuals. *Genetics 155*, 961–968.
31. Page, R. and L. Holmes (1998). *Molecular Evolution: A Phylogenetic Approach.* Oxford: Blackwell Science.
32. Peck, J. R. (1994). A ruby in the rubbish: Beneficial mutations, deleterious mutations and the evolution of sex. *Genetics 137*, 597–606.
33. Rozen, D. E., J. A. G. M. de Visser, and P. J. Gerrish (2002). Fitness effects of fixed beneficial mutations in microbial populations. *Curr. Biol. 12*, 1040–1045.
34. Shaver, A. C., P. G. Dombrowski, J. Y. Sweeney, T. Treis, R. M. Zappala, and P. D. Sniegowski (2002). Fitness evolution and the rise of mutator alleles in experimental *Escherichia coli* populations. *Genetics 162*, 557–566.
35. Swetina, J. and P. Schuster (1982). Self-replication with errors: A model for polynucleotide replication. *Biophys. Chem. 16*, 329–345.
36. Tarazona, P. (1992). Error thresholds for molecular quasi-species as phase-transitions - from simple landscapes to spin-glass models. *Phys. Rev. A 45*, 6038–6050.
37. van Nimwegen, E., J. P. Crutchfield, and M. Mitchell (1999). Statistical dynamics of the Royal Road genetic algorithm. *Theor. Comput. Sci. 229*, 41–102.
38. Wilke, C. O. (2003). Probability of fixation of an advantageous mutant in a viral quasispecies. *Genetics 162*, 467–474.
39. Wilke, C. O., C. Ronnewinkel, and T. Martinetz (2001). Dynamic fitness landscapes in molecular evolution. *Phys. Rep. 349*, 395–446.

Software Verification of Biomolecular Systems

Gabriel Ciobanu

Romanian Academy, Institute of Computer Science, Iaşi
gabriel@iit.tuiasi.ro

Summary. This chapter describes the kinetics of the sodium–potassium exchange pump in terms of the π-calculus process algebra. The π-calculus has the advantage of a software verification tool. We emphasize that this software tool is able to check various properties and to provide confidence in the formal description of the pump. This model checker is used to verify that our model of the pump is deadlock free. It is also used to prove that a detailed description with a large number of states has the same behaviour with a model with a smaller number of states. This simpler model can become a part of a larger system, and in this way we get a scalable and compositional abstraction for biomolecular systems.

Keywords: process algebra, π-calculus, model checking, temporal logic.

1 Introduction

One of the goals of the new field of "computational methods in system biology" is to model and simulate various biological and biochemical networks (metabolic networks, molecular networks, gene networks) that are so complex that they require a formal framework for an accurate representation. Recent work by Cardelli, Harel, Pnueli, Regev, Shapiro and others suggests that process algebras, in particular the π-calculus and mobile ambients, may become valuable tools in modelling and simulation of the biological systems where the interaction and mobility are important features. The field may have an important impact in understanding how the biological systems work, giving at the same time a way to describe, manipulate, and analyze them.

A real challenge for computer science is to understand what paradigm of computation the cell is using, and what are the appropriate tools to study it. An essential step is to find a good and appropriate abstraction. Such an appropriate abstraction for biological systems should be able to highlight the main properties of a system, ignoring unimportant aspects. The model should be relevant and understandable, providing a conceptual framework to express desirable features of a system, and then to prove some properties. The molecular biology community is looking for a unifying abstraction for describing the dynamics of molecular systems and able to describe faithfully the interaction of various components, qualitative and quantitative reasoning, as well as similar behaviours of two related systems, providing a scalable approach to systems of higher levels.

It is reasonable to expect that, in order to model biological systems, we can adapt and apply the range of tools developed in concurrency theory from the formalisms of process algebras to the accompanying verification techniques for temporal logics. This is what we do in this chapter. We present a system of interacting entities as a system of computational interacting entities, and then use a process algebra to describe, simulate, and, more important, verify automatically various properties of molecular systems. We insist on the model checking aspects related to this representation of biomolecular systems.

The biomolecular system used in this chapter is the sodium–potassium exchange pump. The main function of the pump is to move sodium and potassium ions across a cell membrane, namely Na ions from inside to outside, and K ions from outside to inside. We use the π-calculus [10, 14] to describe the changing configurations and movements of the pump. We describe the molecular components as computational processes of the π-calculus, individual domains as communication channels, and molecular interaction as channel transmission according to the π-calculus rules. This abstraction helps us to reason about complex biological systems. The π-calculus has a software verification tool, and we emphasize that this software tool is able to check properties and to provide confidence in the formal description of the pump. The properties are checked upon their models, and their verification is based on the exhaustive search of the state space generated by the π-calculus model. The verification tool is able to prove that the behaviour of the π-calculus processes are faithful to the biological components.

2 Sodium–Potassium Exchange Pump

Cell membranes are crucial to the life of the cell. Defining the boundary of the living cells, membranes have various functions and participate in many essential cell activities including barrier functions, transmembrane signalling, and intercellular recognition. A substantial fraction of the energy available for life processes is provided across biological membranes, and it depends on the corresponding gradients. We refer to the sodium–potassium exchange pump which is a membrane-bound protein that establishes and maintains the high internal K^+ and low internal Na^+ concentrations in cells. It is an important physiologic process present in all animal cells. By using the energy from the hydrolysis of one molecule of ATP, the pump transports three Na^+ outside the cell, in exchange for two K^+ that are taken inside the cell. This exchange is critical in maintaining the osmotic balance of the cell, the resting membrane potential of most tissues, and the excitable properties of muscle and nerve cells.

This molecular process concerns phenomena related to distribution, cooperation, but with mobility and adaptability as well. We describe the molecular interactions and conformational transformations of the sodium–potassium exchange pump in an explicit way. We manipulate formally the changing conformations and describe the corresponding dynamic systems using discrete mathematics instead of the usual partial differential equations. The transfer mechanisms are described in more detail, step by step.

The sodium–potassium pump is a primary active transport system driven by a cell membrane ATPase carrying sodium ions out and potassium ions in (Fig. 1). An animated representation of the pump is available on the web at http://arbl.cvmbs.colostate.edu/hbooks/molecules/sodium_pump.html.

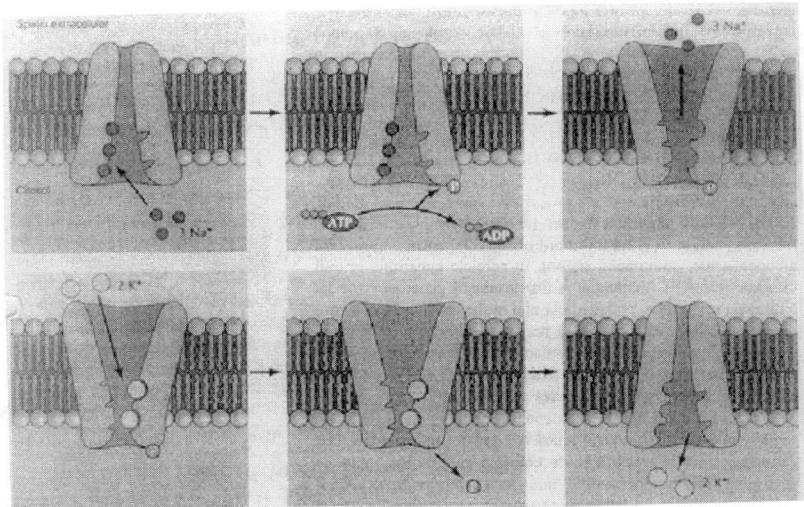

Fig. 1. The sodium–potassium exchange pump

The description given in Table 1 is known as the Albers–Post model. According to the Albers–Post cycle, the Na–K pump has essentially two conformations E_1 and E_2, which both may be phosphorylated or dephosphorylated. Ion transport is mediated by transitions between these conformations. In the table, $A + B$ means that A and B are present together (e.g., in a test tube). $A \cdot B$ means that A and B are bound to each other noncovalently. $E_2 \sim P$ indicates that the phosphoryl group is covalently bound to E_2. P_i is the inorganic phosphate group. $\rightleftharpoons$ indicates that the process can go either way, i.e., can proceed reversibly.

E_1 binds Na^+ to a high-affinity site available only from the inside (1). The binding of the sodium stimulates the enzyme to hydrolyze ATP (2), forming

Table 1. The Albers–Post model

$$E_1 + Na_{in}^+ \rightleftharpoons Na^+ \cdot E_1 \tag{1}$$

$$Na^+ \cdot E_1 + ATP \rightleftharpoons Na^+ \cdot E_1 \sim P + ADP \tag{2}$$

$$Na^+ \cdot E_1 \sim P \rightleftharpoons Na^+ \cdot E_2 \sim P \tag{3}$$

$$Na^+ \cdot E_2 \sim P \rightleftharpoons E_2 \sim P + Na_{out}^+ \tag{4}$$

$$E_2 \sim P + K_{out}^+ \rightleftharpoons K^+ \cdot E_2 \sim P \tag{5}$$

$$K^+ \cdot E_2 \sim P \rightleftharpoons K^+ \cdot E_2 + P_i \tag{6}$$

$$K^+ \cdot E_2 \rightleftharpoons K^+ \cdot E_1 \tag{7}$$

$$K^+ \cdot E_1 \rightleftharpoons K_{in}^+ + E_1 \tag{8}$$

a phosphorylated enzyme intermediate. Then conformation E_1 changes to E_2 (3): Na^+ is exposed to the outside surface and Na^+ binding is of a low-affinity type. Na^+ is then released to the outside (4). On the outside surface is a potassium binding site exposed by the E_2 phosphorylated enzyme. When K^+ binds (5), the phosphoenzyme P is hydrolyzed (6). This stimulates the enzyme to expose the potassium binding site to the inside surface of the membrane, changing its conformation from E_2 to E_1 (7). K^+ binding becomes of low affinity and we have the release of the potassium ions to the inside (8). The ATPase is now ready to bind Na^+ once more. Inside and outside in this mechanism refer to the inside and the outside of the cell plasma membrane in which the Na^+/K^+–ATPase resides.

Regarding the relationship between the kinetic parameters of the transport process and the efficiency of the pump, we can mention that the rate constants of competing steps (that would decrease the efficiency) are small. This ensures that the binding and the release of substrate occur at the proper point in the cycle. For example, the reaction $E_1 + ATP \rightleftharpoons E_1 \sim P + ADP$ of (2) is slower than the reaction described by (1). As a consequence, E_1 has enough time to bind sodium ions before undergoing the transition to E_2. Similar relationships among rate constants ensure that ions are released from the enzyme before they come back to the side at which they were initially bound. In other words, the slow rate constants channel the enzyme along a reaction path in which the hydrolysis of ATP is tightly coupled to the transport process.

3 The π-calculus

In computer science there exist many formalisms to describe mobile, concurrent, and distributed systems. Among them, the π-calculus is a widely accepted theory of interacting systems with dynamically evolving communication topology [9, 10, 14]. The π-calculus is a general model of computation which takes interaction as a primitive. It has a simple semantics and a tractable algebraic theory. The π-calculus computation is given by interaction matchings and their appropriate reduction rules. Both the sender and receiver offer their availability for interaction. Similar mechanisms work in computation and in biology [2, 4].

The π-calculus was introduced by Milner, Parrow, and Walker [11] as an attempt to describe concurrent communicating processes. It models networks in which messages are sent from one site to another site, possibly containing links to active processes or to other sites, and allowing dynamic reconfiguration among processes. It is able to describe complex systems, providing a conceptual framework and mathematical tools. The computational world of the π-calculus contains processes (also called agents) and channels (also called names or ports). The processes are denoted by $P, Q, \ldots$, and the channels by $x, y, \ldots$. There are two types of prefixes (or guards): the input prefix $x(y)$ to receive a name for y along the channel x, and the output prefix $\overline{x}\langle z \rangle$ to send the name z along the channel x. Interaction is established by a nondeterministic matching which dynamically binds "senders" to eligible "receivers". Even though there are many pairs which can satisfy the matching condition, only a single receiver gets the commitment of the sender. Thus processes can interact by using names they share. A name received in one interaction can be used in another; by receiving a name, a process can interact with processes which are unknown to it, but now they share the same channel name. The π-calculus mobility comes from its scoping of names and extrusion of names from their scopes.

Starting with atomic actions and simpler processes, complex processes can be constructed in many ways. The process expressions are defined by guarded processes, parallel composition $P|Q$, nondeterministic choice $P + Q$, replication $!P$, and a restriction operator $(\nu x)P$ creating a local fresh channel x for a process P. The evolution of a process is described in the π-calculus by a reduction relation over processes, also called reaction. This reaction relation contains those transitions that can be inferred from a set of rules.

Without loss of generality, we present in this section the monadic version of the π-calculus: this means that a message between two processes consists of exactly one name. The polyadic version allows messages of 0 or more names, and it can be expressed in terms of the monadic version [10]; as a consequence, we may use the polyadic version as well in our description of the biomolecular systems.

Let $\mathcal{X} \subset N$ be an infinite countable set of *names*. The elements of $\mathcal{X}$ are denoted by $x, y, z \ldots$. The terms of this formalism are called processes and processes are denoted by $P, Q, R \ldots$.

Definition 1. *The* **processes** *over the set $\mathcal{X}$ of names and using the prefixes* $\pi ::= \overline{x}\langle z \rangle \mid x(y) \mid \tau \mid [x = y]\pi$ *are defined by*

$$P ::= 0 \mid \pi.P \mid P + Q \mid P \mid Q \mid !P \mid (\nu x)P$$

Processes evolve by performing interactions, and these interactions are given by their prefixes π. The output prefix $\overline{x}\langle z \rangle$ sends z along x; an input prefix $x(y)$ waits until a name is received along x and substitutes it for the bound variable y. τ is an unobservable action; τ can be thought of as expressing an internal action of a system. The match prefix $[x = y]\pi.P$ can evolve as $\pi.P$ if x and y are the same, and do nothing otherwise; 0 is the empty (do nothing) process. $P + Q$ represents a nondeterministic choice of P or Q. $P \mid Q$ represents the parallel composition of P and Q. A replicated process $!P$ denotes a process that allows us to generate arbitrary instances of P in parallel. The replication $!P$ can be expressed by recursive equations of parametric processes as well. The informal meaning of the restriction $(\nu x)P$ is that x is a local fresh channel for P.

The parallel composition $\overline{x}\langle z \rangle.P \mid x(y).Q$ describes the interaction along the channel x. An interaction is actually defined by a "sender" $\overline{x}\langle z \rangle.P$ and a "receiver" $x(y).Q$, and it can be represented by the transition

$$\overline{x}\langle z \rangle.P \mid x(y).Q \xrightarrow{\tau} P \mid Q\{z/y\}$$

This is a synchronous interaction, where the send operation is blocking: an output action cannot be passed without the simultaneous occurrence of its corresponding input action.

The prefix $x(y)$ binds the name y, and (νx) binds the name x. The definitions of free and bound names are standard. We denote by $fn(P)$ the set of the names with free occurrences in P, by $bn(P)$ the set of bound names of P, and by $n(P)$ the names of P. The same notations are used whenever we consider the input and output actions α. For instance, if $\alpha = x(y)$ then $bn(\alpha) = \{y\}$, and if $\alpha = \overline{x}\langle z \rangle$, then $n(\alpha) = \{x, z\}$ and $bn(\alpha) = \emptyset$. We denote by $P\{v/u\}$ the result of simultaneous substitution in P of all free occurrences of the name u by the name v, using α-conversion wherever necessary to avoid name capture. We denote by $=_\alpha$ the standard α-conversion.

A structural congruence relation is defined over the set of processes; this relation provides a static semantics of some formal constructions. The structural congruence deals with the aspects related to the structure and the names of the processes.

Definition 2. *The relation $\equiv$ over processes is called structural congruence and it is defined as the smallest congruence over processes which satisfies*

- $[x = x]\pi.P \equiv \pi.P$
- $P \equiv Q$ *if* $P =_\alpha Q$
- $P + 0 \equiv P,\ P + Q \equiv Q + P,\ (P + Q) + R \equiv P + (Q + R),$
- $P \mid 0 \equiv P,\ P \mid Q \equiv Q \mid P,\ (P \mid Q) \mid R \equiv P \mid (Q \mid R),$
- $!P \equiv P \mid !P$
- $\nu x 0 \equiv 0,\ \nu x \nu y P \equiv \nu y \nu x P,$
  $\nu x(P \mid Q) \equiv P \mid \nu x Q$ *if* $x \notin fn(P)$.

The evolution of a process is described in the π-calculus by a reaction relation over processes. This reaction relation contains those transitions which can be inferred from a set of rules. We can emphasize the use of specific prefixes by labelling the corresponding reaction step (see the corresponding rule of the definition). Accordingly, the interaction rule *com* is changed. We have the same meaning for τ, namely an interaction between a "sender" $\bar{x}\langle z\rangle$ and a "receiver" $x(y)$.

Definition 3. *The reaction relation over processes is defined as the smallest relation $\rightarrow$ satisfying the following rules, where α denotes an input or output prefix.*

$$\text{struct: } \frac{P \equiv P' \; P \xrightarrow{\alpha} Q \; Q' \equiv Q}{P' \xrightarrow{\alpha} Q'} \qquad\qquad \text{pre: } \alpha.P \xrightarrow{\alpha} P$$

$$\text{par: } \frac{P \xrightarrow{\alpha} P'}{P|Q \xrightarrow{\alpha} P'|Q} \; bn(\alpha) \cap fn(Q) = \emptyset \qquad \text{sum: } \frac{P \xrightarrow{\alpha} P'}{P+Q \xrightarrow{\alpha} P'}$$

$$\text{com: } \frac{P \xrightarrow{\bar{x}\langle z\rangle} P' \; Q \xrightarrow{x(y)} Q'}{P|Q \xrightarrow{\tau} P'|Q'\{z/y\}} \qquad\qquad \text{match: } \frac{P \xrightarrow{\alpha} P'}{[x=x]P \xrightarrow{\alpha} P'}$$

$$\text{res: } \frac{P \xrightarrow{\alpha} P'}{(\nu x)P \xrightarrow{\alpha} (\nu x)P'} \; x \notin n(\alpha) \qquad \text{rep: } \frac{P|!P \xrightarrow{\alpha} P'}{!P \xrightarrow{\alpha} P'}$$

The most studied forms of behavioural equivalence in process algebras are based on the notion of bisimulation. Several definitions have been given in the literature for bisimilarity; we choose the notion of *open bisimilarity*, which is finer than many other equivalences. Its strong version is a congruence and has a simple axiomatization. More importantly, open bisimulation has an efficient characterization providing a technique for deciding bisimilarity of finite-state systems. The definition of open bisimilarity is given by using the labelled transition system defined by its reaction rules. Here we use here the so-called late-style transition system.

Definition 4. *A relation $\mathcal{S}$ defined over processes is called an open simulation if for all P, Q whenever $P \mathcal{S} Q$ then for all substitutions σ the following holds:*
if $P\sigma \xrightarrow{\alpha} P'$, there exists Q' such that $Q\sigma \xrightarrow{\alpha} Q'$ and $P' \mathcal{S} Q'$.
$\mathcal{S}$ *is an open bisimulation if both $\mathcal{S}$ and $\mathcal{S}^{-1}$ are open simulations. Two processes P and Q are open bisimilar $P \sim Q$ if there exists an open bisimulation $\mathcal{S}$ that relates them, i.e., $P \mathcal{S} Q$.*

The bisimilarity between two processes can be verified automatically by a software program called Mobility Workbench. The bisimulation used in the verification process is called *weak open bisimilarity*. It allows the basic verification technique for proving properties of the concurrent communicating systems modelled in the π-calculus[1]. Let $\Longrightarrow \stackrel{def}{=} \xrightarrow{\tau}^{*}$ be the transitive and

---

[1] Note, however, that the usual model checking techniques are still applicable where the state space of a process is finite.

reflexive closure of the $\xrightarrow{\tau}$ relation and let $\xRightarrow{\alpha} \overset{def}{=} \Longrightarrow$ whenever $\alpha = \tau$ and $\Longrightarrow \xrightarrow{\alpha} \Longrightarrow$ if $\alpha \neq \tau$. The *weak open bisimulation* denoted by $\approx$ is defined exactly as the open bisimulation, replacing $Q\sigma \xrightarrow{\alpha} Q'$ with $Q\sigma \xRightarrow{\alpha} Q'$.

4 Formal Description of the Sodium–Potassium Pump

We have briefly presented a computational model of the Na–K exchange pump in [5]. The equations of the Albers–Post model are translated into an appropriate operational semantics which can describe both protein interactions (conformational transformations) and membrane transportation occurring in the pump mechanism. Here we refine the model and emphasize the automated verification associated with the computational model. In this way we show how the properties of the Na–K pump can be automatically checked.

Table 2. The π-calculus description

$Inside(side1, Na) = \overline{side1}\langle Na \rangle.side1(j).Inside(side1, Na)$

$Phase1(side1, side2, ATP) = side1(i).ATP.\overline{side2}\langle i \rangle.side2(j).\overline{side1}\langle j \rangle.$
$\qquad Phase1(side1, side2, ATP)$

$Phase2(side2, side3, P) = side2(i).\overline{side3}\langle i \rangle.side3(j).\overline{P}.\overline{side2}\langle j \rangle.$
$\qquad Phase2(side2, side3, P)$

$Outside(side3, K) = side3(i).\overline{side3}\langle K \rangle.Outside(side3, K)$

$Energy(ATP, P) = \overline{ATP}.P.Energy(ATP, P)$

$System(side1, side2, side3, Na, K, ATP, P) = (\nu \ side1 \ side2 \ side3 \ ATP \ P)$
$\qquad (Inside(side1, Na) \mid Phase1(side1, side2, ATP) \mid Phase2(side2, side3, P) \mid$
$\qquad Outside(side3, K) \mid Energy(ATP, P))$

$PUMP(side1, side3, Na, K) =$
$\qquad (\nu \ ATP \ P \ side2) \ (System(side1, side2, side3, Na, K, ATP, P))$

Generally speaking, the molecular components could be treated as computational processes where their individual domains correspond to communication channels. The complementary molecular domains that allow their interaction can be modelled as the ends of a channel (one end for input, and another for output). In this way, molecular interaction coincides with communication and channel transmission. The membrane transport system involves both information and energy. Consequently, we assume that the π-calculus can model the interactions of the Na–K exchange pump. These interactions are defined syntactically and they have a clear operational semantics given by the π-calculus reaction relation. In this way it is possible to define and study rigorously the behaviour of the pump. The molecular conformational shapes are explicitly modified, and the capabilities of the interacting components are dynamically changed in this model.

Table 2 presents the computational model of the Albers–Post mechanism. Using the reduction rules of the π-calculus, we can describe the dynamics of the pump, step by step. Figure 2 can help to understand these steps. In this figure, s1 is side1, s2 is side2, and s3 is side3 of the previous π-calculus description.

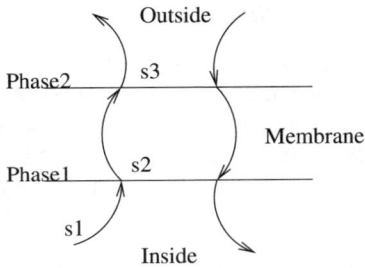

Fig. 2.

First the Na ions bind to the ATPase in conformation E_1. The Na ions are transmitted by the agent $Inside$ along channel $side1$. According to the reduction rules and the transition system of the π-calculus, the system evolution is given by the fact that

$$Inside \xrightarrow{\overline{side1}\langle Na \rangle} side1(j).Inside$$

and

$$Phase1 \xrightarrow{side1(i)} ATP.\overline{side2}\langle i \rangle.side2(j).\overline{side1}\langle j \rangle.Phase1$$

inferred with rule pre. Consequently,

$$Inside \mid Phase1 \mid Phase2 \mid Outside \mid Energy \xrightarrow{\tau} side1(j).Inside \mid$$

$$ATP.\overline{side2}\langle Na \rangle.side2(j).\overline{side1}\langle j \rangle.Phase1 \mid Phase2 \mid Outside \mid Energy$$

by rules com and par applied to the above transitions.

The pump requires some energy to proceed any further. This step corresponds to the second equation of the Albers–Post model (Table 1). We have an interaction between the complementary $\overline{ATP}$ of $Energy$ and ATP of $Phase1$. Using the rules pre, com, and par, the whole system evolves to

$$\xrightarrow{\tau} side1(j).Inside \mid \overline{side2}\langle Na \rangle.side2(j).\overline{side1}\langle j \rangle.Phase1 \mid$$

$$Phase2 \mid Outside \mid P.Energy$$

The pump suffers a conformation change. Now $\overline{side2}\langle Na \rangle$ of $Phase1$ interacts with $side2(i)$ of $Phase2$:

$$\xrightarrow{\tau} side1(j).Inside \mid side2(j).\overline{side1}\langle j \rangle.Phase1 \mid$$

$$\overline{side3}\langle Na \rangle.side3(j).\overline{P}.\overline{side2}\langle j \rangle.Phase2 \mid Outside \mid P.Energy$$

$\overline{side3}\langle Na \rangle$ of $Phase2$ interacts with $side3(i)$ of $Outside$, and the Na ions finish their movement across the membrane and are released outside:

$$\xrightarrow{\tau} side1(j).Inside \mid side2(j).\overline{side1}\langle j \rangle.Phase1 \mid$$

$$side3(j).\overline{P}.\overline{side2}\langle j \rangle.Phase2 \mid \overline{side3}\langle K \rangle.Outside \mid P.Energy$$

The pump transports the Na ions out of the cell in exchange for K ions entering the cell. We are now in the second phase of the process when the pump is open to the outside and it has a high affinity for the K ions. As a consequence it accepts any K ion available from the outside of the cell. $\overline{side3}\langle K \rangle$ sends K ions from $Outside$ to $Phase2$ (interacting with $side3(j)$ of $Phase2$):

$$\xrightarrow{\tau} side1(j).Inside \mid side2(j).\overline{side1}\langle j \rangle.Phase1 \mid \overline{P}.side2\langle K \rangle.Phase2 \mid$$

$$Outside \mid P.Energy$$

At the previous stage of the conformational cycle, P is transferred from ATP to the carboxyl of a glutamate or aspartate residue, forming a "high energy" anhydride linkage given by P. Now the phosphate P is released by hydrolysis, and the conformation changes opening to inside. This process is reflected by the interaction between P of the $Energy$ process and $\overline{P}$ of $Phase2$:

$$\xrightarrow{\tau} side1(j).Inside \mid side2(j).\overline{side1}\langle j \rangle.Phase1 \mid \overline{side2}\langle K \rangle.Phase2 \mid$$

$$Outside \mid Energy$$

$\overline{side2}\langle K \rangle$ sends the K ions to $side2(j)$ of $Phase1$:

$$\xrightarrow{\tau} side1(j).Inside \mid \overline{side1}\langle K \rangle.Phase1 \mid Phase2 \mid Outside \mid Energy$$

Finally, $\overline{side1}\langle K \rangle$ of $Phase1$ interacts with $side1(j)$ of $Inside$. The K ions are inside the cell, and the pump is now ready to start a new cycle:

$$\xrightarrow{\tau} Inside \mid Phase1 \mid Phase2 \mid Outside \mid Energy$$

5 Model Checking

In this section we focus on a model checking tool for the π-calculus, emphasizing the bisimulation equivalences and properties as deadlocks. We present the Mobility Workbench code of the Na–K pump, providing a Mobility Workbench session to verify and analyze the π-calculus description.

The model checking approach to verification [7, 12] is to abstract out key elements of the software and to verify just these elements. These key abstractions are binary predicates, and various techniques and structures have been developed to automatically and efficiently check the abstract elements against specified properties. Given the underlying reliance on binary abstractions, it is no surprise that model checking is being used in the analysis of digital electronic circuits, but it has also proved effective in the software domain, particularly in the areas of protocol analysis, the behaviour of reactive systems, and for checking concurrent systems. We intend to apply this approach to the π-calculus model of the Na–K pump.

An extensive theory was developed for the π-calculus many years before its use as an abstraction for biomolecular systems. Based on the theory, methods and tools for formal verification have been developed, and they were used to implement a software tool able to verify the properties of complex concurrent systems described in the π-calculus. Mobility Workbench (MWB) is a model checking tool for manipulating and analysing mobile concurrent systems described in the π-calculus [13, 17, 18].

The π-calculus processes are called agents in MWB. An important functionality of the MWB is to decide the strong and weak open bisimilarity between two systems described as agents, as well as checking deadlocks and other properties expressed as μ-calculus formulas. The syntax of the μ-calculus formulas contains the truth values, conjunction, disjunction and negation, universal and existential quantifiers, temporal operators such as "eventually" and "always", the least fixpoint and the greatest fixpoint operators. More details of the description of the syntax and the semantics of the μ-calculus formulas are given in [8, 17], and at http://www.it.uu.se/research/docs/fm/mobility/mwb. The propositional μ-calculus is a powerful temporal logic capable of embedding CTL* (and therefore CTL and LTL) [7]. Its formulas correspond either to properties of states which can be statically checked for each state, or to temporal properties which are described in terms of computation paths. The infinite computations are expressed by using the least and the greatest fixpoints. The least fixpoints correspond to eventuality properties, and the greatest fixpoints to global properties. However, it is not easy to express various properties of a (biomolecular) system as μ-calculus formulas. It would be very useful to have a friendly interface to specify the properties. We are working on such a software interface. Here we utilize the MWB commands that are easy to use.

Therefore the π-calculus description of the Na–K pump has the advantage of an automated verification tool. Once we have used the π-calculus to describe a biomolecular system, we may use a computer program to verify various properties, using open bisimulation and propositional μ-calculus. In the previous section we have presented the π-calculus description of the Na–K pump, together with the dynamics of the pump according to the π-calculus reaction rules. We now present the MWB descriptions of the pump (Table 3), and then verify some properties. As a matter of notation, the ν operator is denoted by $^\wedge$, and $\overline{side1}\langle Na \rangle$ is written as 'side1<Na>. Note that we use the polyadic π-calculus; any information is sent along the channels ATP and P, i.e., the pairs 'ATP, ATP and 'P, P are used for the synchronization of two processes (in our case, the processes Energy and Phase1 along ATP, and Energy and Phase2 along P).

Table 3. The MWB code of the Na–K pump

```
agent Inside(side1,Na) = 'side1<Na>.side1(j).Inside(side1,Na)
agent Phase1(side1,side2,ATP) = side1(i).ATP.'side2<i>.side2(j).'side1<j>.
        Phase1(side1,side2,ATP)
agent Phase2(side2,side3,P) = side2(i).'side3<i>.side3(j).'P.'side2<j>.
        Phase2(side2,side3,P)
agent Outside(side3,K) = side3(i).'side3<K>.Outside(side3,K)
agent Energy(ATP,P) = 'ATP.P.Energy(ATP,P)
agent System(side1,side2,side3,Na,K,ATP,P) = (^side1,side2,side3,ATP,P)
    (Inside(side1,Na) | Phase1(side1,side2,ATP) | Phase2(side2,side3,P) |
        Outside(side3,K) | Energy(ATP,P))
agent PUMP(side1,side3,Na,K) =
        (^ATP,P,side2)(System(side1,side2,side3,Na,K,ATP,P))
```

MWB may check whether the components of the system are deadlock-free (i.e., they are working well and they do not stop accidentally), or may check the bisimilarity of two systems. In the MWB session of Table 4 the reader can see that the whole system is deadlock-free. It is useful to know that the size of the whole system reduces drastically when we make the variables local (by using the ν operator) to the system, and in this way restrict the interaction with other components. In the definition of the system, the variables side1, side2, side3, and also ATP and P are local, and the size reduces to 8 states. If no variable is local, then the size is 3864. If only side1 is local, the size reduces to 220. If side1 and side2 are local, the size is 40. If side1, side2, and side3 are local, the size is 16. Finally, if side1, side2, side3, atp, and p are local, the size is 8.

The bisimulation equivalence is a very useful mechanism, able to provide a scalable approach. Bisimulation could be checked with MWB. Many other specific properties can also be verified with MWB. A user can formulate various properties or questions regarding the described system, translate them into corresponding formulas, and then verify them by using MWB.

To conclude, MWB provides a software alternative of lab experiments whenever we have a faithful description of a biomolecular systems. The π-

calculus supports both qualitative and quantitative properties. We can verify certain qualitative assertions on biomolecular systems by verifying a corresponding formula using MWB.

Table 4. An MWB session

```
 The Mobility Workbench
(MWB'99, version 4.134, built Fri Apr 11 2003)
MWB: input "biospec.mwb"
MWB: env
agent Energy(atp,p) = 'atp.p.Energy<atp,p>
agent Inside(side1,na) = 'side1<na>.side1(j).Inside<side1,na>
agent Outside(side3,k) = side3(i).'side3<k>.Outside<side3,k>
agent Phase1(side1,side2,atp) = side1(i).atp.'side2<i>.side2(j).
'side1<i>.Phase1<side1,side2,atp>
agent Phase2(side2,side3,p) = side2(i).'side3<i>.side3(j).'p.
'side2<j>.Phase2<side2,side3,p>
agent System(side1,side2,side3,na,k,atp,p) =
(^side1,side2,side3,atp,p) (Inside<side1,na>
| Phase1<side1,side2,atp> | Phase2<side2,side3,p> | Outside<side3,k>
| Energy<atp,p>)
agent PUMP(side1,side3,na,k) = (^atp,p,side2)
System <side1,side2,side3,na,k,atp,p>
agent TempPhase1(side1,side2,atp,na) = (^side1)(Inside<side1,na> |
Phase1<side1,side2,atp>)
agent TempPhase2(side1,side2,atp,na,side3,p) = (^side2)
(TempPhase1<side1,side2,atp,na> | Phase2<side2,side3,p>)
agent TempPhase3(side1,side2,atp,na,side3,p,k) = (^side3)
(TempPhase2<side1,side2,atp,na,side3,p> | Outside<side3,k>)
agent TempPhase4(side1,side2,atp,na,side3,p,k) = (^atp,p)
(TempPhase3<side1,side2,atp,na,side3,p,k> | Energy<atp,p>)

MWB: weq System TempPhase4
The two agents are equal.
Bisimulation relation size = 16.
MWB: eq System TempPhase4
The two agents are equal.
Bisimulation relation size = 8.
MWB: weq System PUMP
The two agents are NOT equal.

MWB: deadlocks System
No deadlocks found.
MWB: deadlocks PUMP
No deadlocks found.
```

Moreover, we can compare two π-calculus descriptions to determine the degree of similarity of their behaviour. This allows scalability without losing transparency and useful details. We can describe components taking into consideration many details and verifying various properties. Then we use bisimulation to confirm that the detailed description has the same behaviour with a much simpler description. This simpler description is integrated as a small part of a larger system, and so on.

5.1 Methodology

We present here some elements related to the methodology of using MWB and the process of verification. Let us suppose that we have a large, complex biomolecular system. Using the π-calculus, we start by describing small components in detail. The components are then considered together, and we refer to this faithful description as *the model*. At this point we have two possibilities.

Generally speaking, a system interacts with its environment. We may characterize the system from an external observer's viewpoint, i.e., considering only the observable interaction of the system with its environment. Then we can check for weak open bisimulation between the model and this simplified description. If they are bisimilar, we may include the simpler description as a part of a larger description where the details concern the interaction between the components already described. In this way we get a very useful scalable formalism able to describe faithfully large biological systems, component by component. For instance, considering our description of the Na–K pump, the specification $PUMP$ is bisimilar to $System$, i.e., $PUMP \approx System$. This means that a pump with a never-ending source of energy behaves externally, with regard to the flux of transported ions through the membrane as an agent whose carrier behaviour is desirable, and reflects the general understanding of how the pump works.

Another possibility is to check certain properties of the system. These properties are expressed using a logic. We refer to these properties as the *specifications*, and we can check if the specifications satisfy the model. This verification is automatic, i.e., done by a computer. Moreover, when a formula does not hold, the verification tool MWB can provide an error trace or a counterexample. For instance, considering our description of the Na–K pump, we may verify that System works properly forever, and it does not deadlock. MWB has a predefined name *deadlocks* for the corresponding μ-calculus formula expressing that the transition system of the described system has successors for each of its states (a process deadlocks if its transition system has states with no successors). Whenever we want to check that a system is deadlock-free, we can run the MWB program and simply write *deadlocks System*, expecting the answer *No deadlocks found*.

6 A More Detailed Description of the Pump

In this section we provide a more detailed description of the Na–K exchange pump. The new feature is given by the existence of occluded states $E_1^P \cdot (3Na^+)_{occ}$ and $E_2 \cdot (2K^+)_{occ}$ (Fig. 3 and Fig. 4). This description refines the activity of the pump, opening new perspectives.

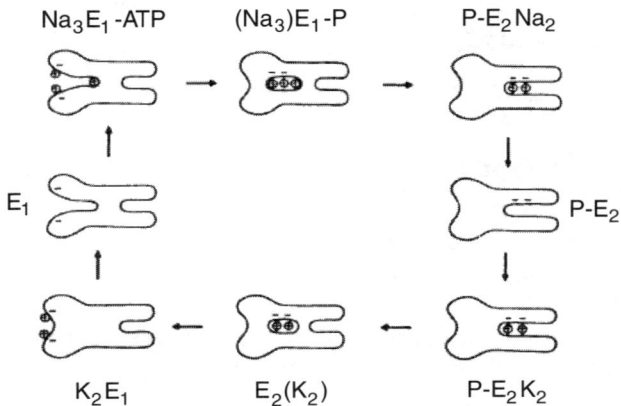

Fig. 3. The sodium–potassium pump with occluded states

We refine the Albers–Post model, adding the occluded states in which the the pump is unable to exchange ions in the surrounding media. We have the same two conformations E_1 and E_2; these conformations correspond to the mutually exclusive states in which the pump exposes ion binding sites alternatively on the cytoplasmic (E_1) and extracellular (E_2) sides of the membrane. Ion transport is mediated by transitions between these conformations.

Let us consider an initial state following the release of K ions to the cytosol (Fig. 4, left middle). The pump is in the conformation E_1, it is associated with ATP, and it has its cation binding sites empty and open to the intracellular space. In this situation, the affinity is high for Na^+ and low for K^+. Consequently, three Na^+ ions binds to the intracellular cation sites; this corresponds to the first equation of Table 5.

The binding of Na^+ is followed by the transfer of the γ phosphate of ATP to the aspartate residue of the pump structure. During this process Na ions are occluded (Fig. 4, right up). Thereafter the pump undergoes a conformational change to the E_2 state and loses its affinity for Na^+. The Na ions are subsequently released; first one Na ion is released during the conformational change from E_1 to E_2 when the cation binding sites are oriented toward the extracellular side. The pump is in the E_2^P state, and the affinity for Na ions is very

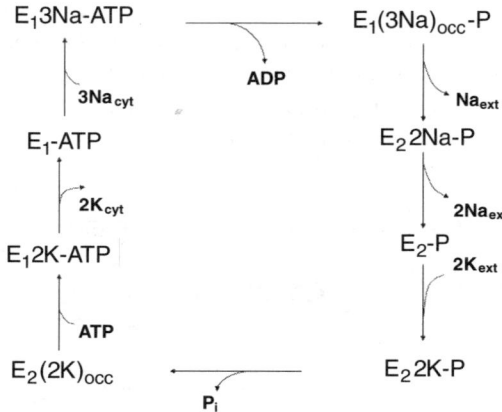

Fig. 4. The steps in the evolution of the pump

Table 5. The Albers–Post cycle with occluded states

$$E_1 \cdot ATP + 3Na_{cyt}^+ \rightleftharpoons E_1 \cdot ATP \cdot 3Na^+ \tag{9}$$

$$E_1 \cdot ATP \cdot 3Na^+ \rightleftharpoons E_1^P \cdot (3Na^+)_{occ} + ADP \tag{10}$$

$$E_1^P \cdot (3Na^+)_{occ} \rightleftharpoons E_2^P \cdot 2Na^+ + Na_{ext}^+ \tag{11}$$

$$E_2^P \cdot 2Na^+ \rightleftharpoons E_2^P + 2Na_{ext}^+ \tag{12}$$

$$E_2^P + 2K_{ext}^+ \rightleftharpoons E_2^P \cdot 2K^+ \tag{13}$$

$$E_2^P \cdot 2K^+ \rightleftharpoons E_2 \cdot (2K^+)_{occ} + P_i \tag{14}$$

$$E_2 \cdot (2K^+)_{occ} + ATP \rightleftharpoons E_1 \cdot ATP \cdot 2K^+ \tag{15}$$

$$E_1 \cdot ATP \cdot 2K^+ \rightleftharpoons E_1 \cdot ATP + 2K_{cyt}^+ \tag{16}$$

low; the two remaining Na ions are released into the extracellular medium. The binding sites now have a high affinity for K^+. Two external K^+ ions can bind; this corresponds to (13) of Table 5, and to the right down corner of Fig. 4. The binding of K^+ induces the dephosphorylation of the E_2^P conformation. The release of the inorganic phosphate into the intracellular medium is accompanied by the occlusion of the K^+ ions (14). ATP is then bound and this allows a conformational change to E_1 and K ions are deoccluded. The affinity for K ions reduces and they are released into the intracellular medium (cytosol). The pump protein is now ready to initiate a new cycle.

The π-calculus description of the Albers–Post cycle with occluded states is given in Table 6. This description allows biologists to follow the behaviour of the system step by step; this is the way biologists think of the systems they study. We can say that the models based on differential equations are not so intuitive as the π-calculus descriptions.

Table 6. The MWB code of the pump with occluded states

agent To1(s,i) = s<i>. To1(s,i)
agent To2(s,i,j) = To1(s,i) | To1(s,j)
agent From1(s,i) = s(i). From1(s,i)
agent From2(s,i,j) = From1(s,i) | From1(s,j)
agent Tos1(s1a,s1b,na,naplus,no) = To1(s1b,na) | To2(s1a,naplus,no)
agent Inside(s1a,s1b,na,naplus,u,v,no) = 's1b<na>.('s1a<naplus> |'s1a<no>).
    From2(s1a,u,v).Inside(s1a,s1b,na,naplus,u,v,no)
agent Phase1(s1a,s1b,s2a,s2b,atp,i,j,u,v,t1) = s1b(t1).From2(s1a,i,j).atp.'s2b<t1>.
    To2(s2a,i,j).From2(s2a,u,v).To2(s1a,u,v).Phase1(s1a,s1b,s2a,s2b,atp,i,j,u,v,t1)
agent Phase2(s2a,s2b,s3a,s3b,p,i,j,u,v,t1) = s2b(t1).From2(s2a,i,j).'s3b<t1>.
    To2(s3a,i,j).From2(s3a,u,v).'p.To2(s2a,u,v).Phase2(s2a,s2b,s3a,s3b,p,i,j,u,v,t1)
agent Outside(s3a,s3b,kplus,i,j,k) = s3b(t1).From2(s3a,i,j).To2(s3a,kplus,k).
    Outside(s3a,s3b,kplus,i,j,k)
agent Energy(atp,p) = 'atp.p.Energy(atp,p)
agent System(s1a,s1b,s2a,s2b,s3a,s3b,na,naplus,kplus,atp,p,u,v,no,k,i,j,t1) =
    (^s1a,s1b,s2a,s2b,s3a,s3b,atp,p)   (Inside(s1a,s1b,na,naplus,u,v,no) |
    Phase1(s1a,s1b,s2a,s2b,atp,i,j,u,v,t1) | Phase2(s2a,s2b,s3a,s3b,p,i,j,u,v,t1) |
    Outside(s3a,s3b,kplus,k) | Energy(atp,p))
agent Pump(s1a,s1b,s3a,s3b,na,naplus,kplus,no,k,i,j,t1) = (^s2a,s2b,atp,p) Sys-
    tem(s1a,s1b,s2a,s2b,s3a,s3b,na,naplus,kplus,atp,p,no,k,i,j,t1)

Figure 5 can help to understand the π-calculus description of the Na–K pump with occluded states. It is possible to describe the dynamics of the pump according to the reaction rules of the π-calculus, as in Sect. 4. The verification of properties for the new description is done by using MWB, similarly to Sect. 5.

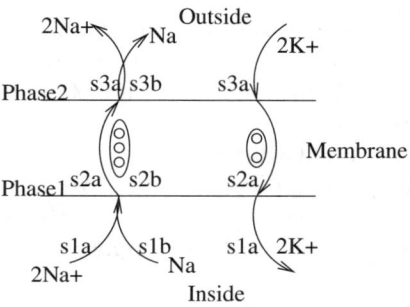

Fig. 5.

7 Conclusions

Various approaches from mathematics and computer science have been used for the description of molecular processes. The use of the π-calculus or other process algebras to model the molecular interaction is quite new. In computer science, the π-calculus is a widely accepted model of interacting systems with dynamically evolving communication topology. The π-calculus has a well-defined semantics and an appropriate algebraic theory. We think the π-calculus is a formalism capable to describe many biomolecular processes.

As far as we know, the first papers using the π-calculus to describe molecular processes were [1] and [3], followed by the successful papers [15, 16] that represent and simulate biomolecular processes, giving a set of steps taken to adapt, extend and implement a core language to describe the requirements of biochemical systems, first on a qualitative and then on a quantitative, stochastic scale. In [1], the π-calculus is used to describe DNA methylation. In [3] the so-called molecular structures are defined, and it is proved that they have the same expressive power as the π-calculus (which has the same computational power as Turing machines). A more detailed approach is presented in [4].

In this chapter we motivate the use of the π-calculus as an adequate formalism for automated verification of biomolecular systems. We describe the dynamics of the sodium–potassium exchange pump, an important physiologic process present in all animal cells. We present a computational model based on molecular interaction, which can cope with phenomena beyond the classical approach, including a new scalability providing by bisimulation. We manipulate formally the changing conformations and describe explicitly the corresponding dynamic systems using discrete mathematics instead of the (usual) partial differential equations. The transfer mechanisms are described step by step. Moreover, we can use some software tools of verification developed for the π-calculus. This means that it would be possible to verify properties of the described systems by using a computer program. The main consequence of this approach is the use of the verification software as a substitute for expensive lab experiments.

Biologists are of course interested in more detail. There are various ways to extend the approach based on the π-calculus. We have used in [5] a version of the π-calculus called stochastic π-calculus to describe the efficiency of the Na–K pump. In the stochastic π-calculus, the prefix $\pi.P$ is replaced by $(\pi, d).P$, where d is a probability distribution that characterizes the stochastic behaviour of the activity corresponding to the prefix π. This version of the π-calculus allows us to describe the complexity of the molecular interactions involving the dynamic efficiency of the pump and other quantitative aspects (e.g., kinetics rates, energy, pump failures).

Another attempt is presented in [6]. The idea is to descend from the π-calculus to a lower level of abstraction given by the communicating automata. This step allows us to add more details to the description of the system. Moreover, it is possible to get new results regarding the computational power of the communicating automata, and to obtain a different software implementation able to model and simulate molecular networks.

Acknowledgments

Thanks to my former students Vlad Ciubotariu and Bogdan Tanasă for their collaboration. This work was partially supported by an academic research grant of the National University of Singapore.

References

1. G. Ciobanu. Formal Description of the Molecular Processes. In C. Martin-Vide, Gh. Păun (Eds.): *Recent Topics in Mathematical and Computational Linguistics*, 82-96, Ed. Academiei, Bucharest, 2000.
2. G. Ciobanu, M. Rotaru. A π-calculus machine. *Journal of Universal Computer Science*, vol.6(1), 39-59, 2000.
3. G. Ciobanu. Molecular Structures. In C. Martin-Vide, V. Mitrana (Eds.): *Words, Sequences, Languages*, 299-317, Kluwer, Dordrecht, 2001.
4. G. Ciobanu, M. Rotaru. Molecular Interaction. *Journal of Theoretical Computer Science*, vol.289(1), 801-827, 2002.
5. G. Ciobanu, V. Ciubotariu, B. Tanasă. A pi-calculus model of the Na pump, *Genome Informatics*, 469-472, Universal Academy Press, Tokyo, 2002.
6. G. Ciobanu, B. Tanasă, D. Dumitriu, D. Huzum, G. Moruz. Molecular networks as communicating membranes. In Gh. Păun, C. Zandron (Eds.), Proceedings of Workshop On Membrane Computing, MolCoNet vol.1, 163-175, 2002.
7. E. Clarke, O. Grumberg, D.Peled. *Model Checking*. MIT Press, Cambridge, MA, 1999.
8. M. Dam. Model Checking Mobile Processes. *Information and Computation*, vol.129(1), 35-51, 1996.
9. R. Milner. Elements of Interaction. Turing Award Lecture, *CACM*, vol.36(1), 78-89, 1993.
10. R. Milner. *Communicating and mobile systems : the π-calculus*. Cambridge University Press, Cambridge, 1999.
11. R. Milner, J. Parrow, D. Walker. A calculus of mobile processes. *Journal of Information and Computation*, vol.100, 1-77, 1992.
12. K.L. McMillan. *Symbolic Model Checking*, Kluwer, Dordrecht, 1993.
13. F. Orava, J. Parrow. An Algebraic Verification of a Mobile Network. *Formal Aspects of Computing*, vol.4, 497-543, 1992.
14. J. Parrow. An Introduction to the π-calculus, J.Bergstra, A.Ponse, S.Smolka (Eds.): *Handbook of Process Algebra*, Elsevier, Amsterdam, 2000.
15. A. Regev, E. Shapiro. Cellular abstractions. *Nature*, vol. 419, 343, 2002.
16. A. Regev, W. Silverman, E. Shapiro. Representation and simulation of biochemical processes using the pi-calculus process algebra. In *Pacific Symposium on Biocomputing*, vol.6, 459-470, World Scientific Press, Singapore, 2001.
17. B. Victor. A Verification Tool for the Polyadic π-Calculus, Licentiate Thesis, Uppsala University, 1994.
18. B. Victor, F. Moeller. The Mobility Workbench: A tool for the π-calculus, In *CAV'94: Computer-Aided Verification*, LNCS vol.818, 428-440, Springer-Verlag, Berlin Heidelberg New York, 1994.

Kinetic Modelling as a Modern Technology to Explore and Modify Living Cells

Oleg V. Demin[1], Galina V. Lebedeva[2], Alex G. Kolupaev[1], E.A. Zobova[1],
T.Yu. Plyusnina[2], A.I. Lavrova[2], A.Dubinsky[3],
E.A. Goryacheva[4], Frank Tobin[5], and Igor I. Goryanin[6]

[1] A.N. Belozersky Institute of Physico-Chemical Biology, Moscow State
University, Russia
[2] Biophysics Department, Faculty of Biology, Moscow State University, Russia
[3] Moscow Physico-Technical Institute, Russia
[4] Institute of Bioorganic Chemistry, Russian Academy of Sciences, Moscow, Russia
[5] GlaxoSmithKline, Scientific Computing & Mathematical Modelling, Upper
Merion, PA 19406, USA
[6] GlaxoSmithKline, Scientific Computing & Mathematical Modelling, Stevenage,
SG2 8PU, UK

Summary. We describe a general strategy that enables us to develop kinetic models
of large-scale metabolic systems by collecting and using all available metabolic experimental data. Our approach can be used to explore the local and global regulatory
properties of selected metabolic pathways, and to predict how cell genome modifications can meet selected biotechnological and biomedical criteria. We have applied
the strategy for the development and use of detailed kinetic models of catabolic and
anabolic pathways of *E. coli* and mitochondrial energy metabolism.

1 Introduction

The last few years have seen substantial progress in molecular biology and
genetic research. Genomes of *Escherichia coli, Saccharomyces cerevisia, Mus
musculus, Home sapience*, and more then 100 other organisms have been sequenced [1–3]. That research has stimulated the emergence of functional genomics, a discipline that sets out to understand the meaning of sequenced
data using high-throughput gene and protein expression data. Life scientists
have transformed old style protein chemistry to proteomics, traditional biochemistry to metabolomics. These new fields provide essential clues to the
underlying metabolic, gene regulation, and signalling networks that operate
in different organisms under different conditions.

All these achievements can create the illusion that this voluminous knowledge will enable us to predict whole cell behaviours for the purposes of mechanistic understanding and bioengineering control. It is an easy trap to fall into

when one has knowledge of most, or all, of the major biological entities and their static interactions. But will this knowledge suffice to handle problems appearing in the medicine, biotechnology, and bioengineering areas? Will we be able to find remedies for modern diseases and construct new versions of a cell to produce any desired metabolites? In some specific cases, it is possible to make plausible predictions based on "static" information without relying upon kinetic data. Unfortunately, this is not the general case, and the knowledge of a cell's genome or proteome will not allow us to predict correctly the cell behaviour, particularly in non-normal, or stressful, situations. Overall cellular behaviour is determined not only by what biological entities are available, but mainly by their dynamic interactions and their individual properties. Such interactions ebb and flow with a changing external environment and it is the overall dynamic nature of the cell that determines not only its present properties, but its future ones as well. It has long been recognized in other fields of science that such complex networks can produce a wide range of possible cellular states. Such emergent behaviours are extremely hard to predict, are often non-intuitive, and depend upon a full kinetic knowledge of the dynamics of the system.

The main purpose of this contribution is to describe a new approach for constructing large-scale kinetic network models. In the framework of this approach we suggest a novel way to collect and mine large-scale experimental data, and use them to build and verify kinetic models. We show how to apply models for various practical problems in biotechnology, bioengineering, and biomedicine.

First, we describe how a kinetic model of aromatic acid biosynthesis is used to optimize *in vitro* experiments for drug development. Then, we show how a kinetic model of energy metabolism in hepatic mitochondria enables us to hypothesize the ways to relieve side-effects of medicines such as aspirin. And finally, we analyze the kinetic model of the phosphotransferase glycolysis system in *E. coli* and study the regulatory properties of the cell. From these predictions from the kinetic models, we generate new functional biological knowledge about the phenotypes of the cell based upon the expression levels of the corresponding enzymes.

2 Basic Principles of Kinetic Model Construction

The term "kinetic model" is used in two senses, one biological and one mathematical. In the biological sense, it is used to indicate that the network interactions between the different biological entities are always changing and the network always includes them, even if the fluxes may become zero. It is important to consider two entities as connected, if, for any time, there is ever a concentration flux connecting them. Thus, there is only a static network with temporally changing concentrations as the cell undergoes genotypic and phenotypic changes. In the mathematical sense, it refers to a system of mechanistic ordinary differential equations that determine the temporal state of the corresponding system of biochemical reactions. In these equations there is mass conservation between the production and consumption of each species:

$$dX/dt = V_{production} - V_{consumption},$$

where $V_{production}$ and $V_{consumption}$ are the respective rates of production and consumption of species X.

The development of kinetic models for metabolic systems, gene regulation, and signalling networks is accomplished in several steps. The first step is to elucidate a static model of the system, i.e., to find out all cellular players, intermediates, enzymes, small molecules, co-factors, and all non-enzymatic processes in the cellular network. The result is a network (i.e., a directed bond graph) of all interactions connecting all the species. For the network to be proper, each species must exist in at least one reaction or behave as a co-factor. Disconnected fragments, resulting from incomplete knowledge, can, optionally, be considered part of the network, although for all practical purposes they will be treated as separate networks.

Once an appropriate static network has been chosen, the second stage is to generate rate equations to describe the dependence of each reaction rate against concentrations of intermediates involved in the selected pathway. To make the models scalable and comparable with different kinds of experimental data we have developed both *detailed* and *reduced* descriptions for every biochemical process in the model. The *detailed* reaction description includes the exact molecular mechanism of the biomolecular reaction (i.e., enzyme catalytic cycle) and takes into account all possible states of the protein, including possible non-active states (i.e., phosphorylated) or dead-end inhibitor complexes. Usually, the *detailed* description comprises a set of differential algebraic equations from the ordinary differential flux equations and non-linear algebraic equations (if steady-state or conservation constraint assumptions are made) simultaneously. For determining the proper rate equations, the steady-state assumptions may be made. However, in some situations, such simplifications can or must be relaxed and the full differential equation approach used.

The *reduced* description represents the reaction rates as an explicit analytic function of the substrates and products. We identify from the literature or hypothesize the catalytic cycle based on 3D structures and other relative biological information for each active protein involved in the model (i.e., enzyme with catalytic function). To derive the corresponding rate equations from the catalytic cycle, we have used quasi-steady state and rapid equilibrium approaches. The catalytic cycle of each enzyme is described by non-linear differential equations. Initially, concentrations of substrates, products, and effectors (inhibitors and activators) are assumed to be buffered, i.e., do not change with time. At this stage we could include static constraints derived from static network structural analyses such as mass and flux balance linear dependencies.

The quasi-steady state of the system is calculated as a function of substrates, products, inhibitors, activators, total protein concentrations, and all kinetic constants of the processes. The rate law for every process is derived as a flux from the catalytic cycle for this quasi-steady state. Finally, the rate law depends on temporal changes of the total concentration of the protein, concentrations of the effectors (activators, inhibitors, agonists, and antagonists), substrates, products, and the values of the kinetic parameters (K_m, K_i, K_d, and elementary rate constants). While derived from a quasi-steady state approach, these rate laws will be used in simulations based upon the full differential equations with any such simplifications made.

Parameter estimation is the third stage of model development. To estimate the kinetic parameter values we use the following sources:

1. literature data on values of K_m, K_i, K_d, and rate constants
2. electronic databases; only a few databases with specific kinetic content are available at the moment, in particular EMP [4], and BRENDA [5]
3. experimentally measured dependencies of the initial reaction rates on concentrations of substrates, products, inhibitors, and activators
4. time series data from enzyme kinetics

However, many processes (i.e., enzyme reactions) have not been studied kinetically. Many kinetic parameters cannot be estimated from the literature or databases due to a lack of available experimental data. One remedy is to express these unknown or "free" parameters in terms of other measured kinetic parameters. The result is the establishment of functional relationships between "free" parameters and measured kinetic parameters. Each parameter value, of course, is constrained by physico-chemical properties and any other information available from other organisms or related processes. The more constraints available, the more dimensionality reduction can occur.

For example, the maximal enzyme velocity ($V_{\max}$) *in vivo* can be estimated from the value of specific activity (SA) of the enzyme in cell lysate/crude extracts using a relationship between the intracellular volume and dry cell weight (d.c.w.) or protein weight. In *Escherichia coli* [6] proteins constitute 55% of the dry cell weight, 70% of the protoplasm is water, and the density of protoplasm is 1.1. Taking all this into account yields:

1 mg d.c.w. corresponds approximately to $3\,\mu l$ of intracellular volume

1 mg total cellular protein corresponds to $5.5\,\mu l$ of intracellular volume

and $V_{\max}$ can be approximated in terms of the specific activity as:

$$
\begin{aligned}
V_{\max} &= SA \text{ mmol/min/(mg intracellular protein)} \\
&= SA \text{ mmol/min/}(5.5\ \mu l \text{ of intracellular volume}) \quad (1) \\
&= (SA/5.5)\text{M/min} = 182 \cdot SA \text{ mM/min}
\end{aligned}
$$

The next stage of modelling is to generate the corresponding differential equations for the pathway under investigation. These can be put into the following form:

$$dx/dt = \mathbf{N} \cdot \mathbf{v}(\mathbf{x}; \mathbf{e}, \mathbf{K}) = \mathbf{v}^{prod}(\mathbf{x}; \mathbf{e}, \mathbf{K}) - \mathbf{v}^{cons}(\mathbf{x}; \mathbf{e}, \mathbf{K}) \qquad (2)$$

where $\mathbf{x} = [x_1, \ldots, x_m]$ is a vector of the concentrations of all biological entities in the model (protein, metabolites, small molecules, etc.); $\mathbf{v} = [v_1, \ldots, v_n]$ is a vector of the n rate equations; $\mathbf{e} = [e_1, \ldots, e_n]$ is a vector of total enzyme concentrations; $\mathbf{K} = [K_1, \ldots, K_p]$ is the vector of the kinetic parameters. $\mathbf{N}$ is an m-by-n stoichiometric matrix corresponding to the kinetic scheme constructed in the first stage of the model development process. Additionally, the formal fluxes can be thought as being divided into production fluxes, $\mathbf{v}^{prod}$, and consumption fluxes, $\mathbf{v}^{cons}$. Once the equations are generated, numerical integration can be accomplished to simulate a sequence of different biological scenarios and a variety of mathematical analyses are possible. Such additional analyses can include asymptotic stability, parameter sensitivity, and linear stability. By examining the simulation results, extensive analysis can look for "interesting" phenomena that may not have been previously observed. To develop and to analyze kinetic models we use the software package DBSolve [7].

3 Kinetic Model of PTS and Glycolysis in *E. coli*. Expression Level of Glucokinase Controls the *E. coli* Metabolism

The phosphoenolpyruvate-dependent phosphotransferase system (PTS) is one of the main mechanisms responsible for substrate transport in *E. coli* [8]. Glucose PTS consists of four proteins localized in both the cytoplasm and the inner bacterial membrane [8]. PTS is responsible for the transport and phosphorylation of glucose. Glucose PTS is tightly coupled with glycolysis and, therefore, the PTS glycolysis in *E. coli* is a cyclic metabolic pathway. Utilizing PEP to phosphorylate glucose, PTS couples upstream glycolytic stages (activation of glucose via its phosphorylation) with downstream reactions of substrate phosphorylation. Indeed, both glucokinase and PTS can catalyze phosphorylation of glucose to form glucose-6-phosphate (G6P). Similarly, both pyruvate kinase and PTS can dephosphorylate PEP. One can question what the implications of this redundancy are, and how the cell can regulate its metabolism under different external conditions by utilizing this particular pathway organization.

3.1 The Kinetic Model

As a preliminary step to study the regulatory properties of carbohydrate catabolic mechanisms and to understand possible control schemes, we have developed kinetic models of the following metabolic pathways:

1. glucose PTS
2. glucokinase
3. pyruvate kinase
4. reaction of the production of two molecules of PEP and one molecule of ATP from one molecule of G6P and one molecule of ADP. This reaction mimics the part of glycolysis from G6P isomerization up to PEP formation.
5. ATP consumption process

Glucose PTS consists of the following proteins:

cytoplasmic protein Enzyme I (EI)
cytoplasmic protein HPr
cytoplasmic protein Enzyme IIA^{glc} (EIIA^{glc})
protein of inner bacterial membrane Enzyme IICB^{glc} (EIICB^{glc})

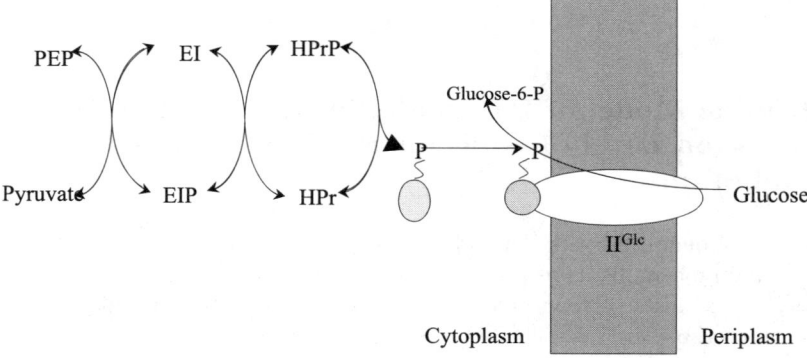

Fig. 1. Structural and functional organization of glucose PTS of *Escherichia coli*

The transfer (Fig. 1) of the phosphate group from PEP to glucose is coupled with the transport of glucose molecules inside the cell. First, EI is auto-phosphorylated by PEP as a phosphate donor, then phosphorylated Enzyme I (EIP) donates a phosphate group to HPr and forms HPrP which, in turn, phosphorylates EIIA^{glc}. Phosphorylated Enzyme IIA^{glc}, EIIA^{glc}P, associates with the membrane protein EIICB^{glc} and phosphorylates it. Finally, complex EIIA^{glc} and EIICB^{glc}P catalyze the transfer and phosphorylation of glucose [9].

Enzyme I

Enzyme I catalyzes the transfer of a phosphate group from PEP to HPr. The catalytic cycle in Fig. 2 reflects the peculiarities of the way Enzyme I functions. Monomers of Enzyme I can associate to form a dimer, EI$_2$ (reaction 1), which is the catalytically active form. The level of dimerization strongly depends on Mg^{2+} concentrations and, as was shown in [8], approaches 90% at physiological concentrations of Mg^{2+}. Therefore, we assume that all the Enzyme I proteins are present in the dimeric form. Upon dimer formation each monomer is phosphorylated on histidine 189 (reaction 2–7). Only doubly phosphorylated dimer is able to transfer one of the phosphate groups to protein HPr [9]. The mechanism of HPr phosphorylation (Fig. 2) is determined by reactions 8–10. The dynamics of the Enzyme I catalytic cycle is described by the following system of ordinary differential equations:

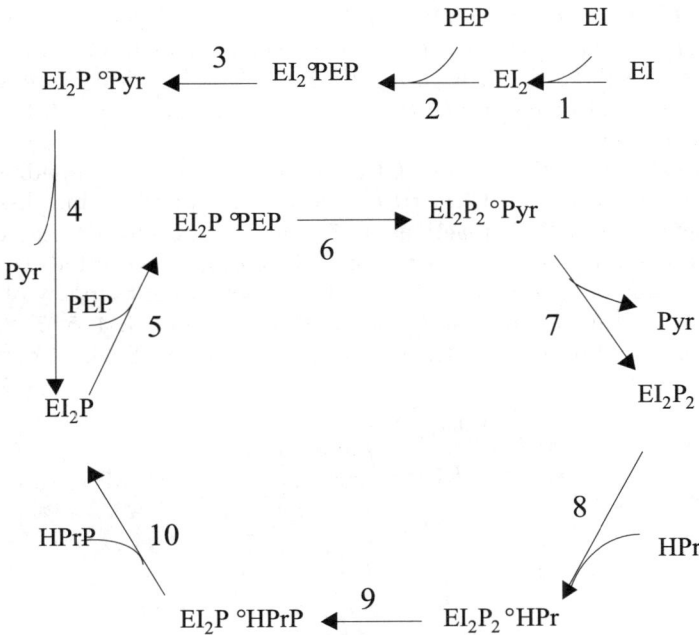

Fig. 2. Catalytic cycle of Enzyme I of PTS

$$dEI_2/dt = -v_2$$
$$dEI_2 \circ PEP/dt = v_2 - v_3$$
$$dEI_2P \circ Pyr/dt = v_3 - v_4$$
$$dEI_2P/dt = v_4 + v_{10} - v_5$$
$$dEI_2P \circ PEP/dt = v_5 - v_6 \qquad (3)$$
$$dEI_2P_2 \circ Pyr/dt = v_6 - v_7$$
$$dEI_2P_2/dt = v_7 - v_8$$
$$dEI_2P_2 \circ HPr/dt = v_8 - v_9$$
$$dEI_2P \circ HPr/dt = v_9 - v_{10}$$

where

$$v_2 = a_2 \cdot (EI_2 \cdot PEP/K_{d,0}^{PEP} - EI_2 \circ PEP)$$
$$v_3 = a_3 \cdot (EI_2 \circ PEP - EI_2P \circ Pyr/K_{t,0})$$
$$v_4 = a_4 \cdot (EI_2P \circ Pyr - EI_2P \cdot Pyr/K_{d,1}^{Pyr})$$
$$v_5 = a_5 \cdot (EI_2P \cdot PEP/K_{d,1}^{PEP} - EI_2P \circ PEP)$$
$$v_6 = k_1 \cdot EI_2P \circ PEP - k_{-1} \cdot EI_2P_2 \circ Pyr \qquad (4)$$
$$v_7 = a_7 \cdot (EI_2P_2 \circ Pyr - EI_2P_2 \cdot Pyr/K_{d,2}^{Pyr})$$
$$v_8 = a_8 \cdot (EI_2P_2 \cdot HPr/K_d^{HPr} - EI_2P_2 \circ HPr)$$
$$v_9 = k_2 \cdot EI_2P_2 \circ HPr - k_{-2} \cdot EI_2P \circ HPrP$$
$$v_{10} = a_{10} \cdot (EI_2P \circ HPrP - EI_2P \cdot HPrP/K_d^{HPrP})$$

Here, $EI_2P \circ Pyr$, $EI_2 \circ PEP$, EI_2, $EI_2P \circ PEP$, $EI_2P_2 \circ Pyr$, EI_2P, EI_2P_2, $EI_2P_2 \circ HPr$, $EI_2P \circ HPrP$ are the concentrations of Enzyme I's different states. $K_{d,0}^{PEP}$, $K_{d,1}^{Pyr}$, $K_{t,0}$, $K_{d,1}^{PEP}$, K_d^{HPrP}, $K_{d,2}^{Pyr}$, K_d^{HPr} are the corresponding dissociation constants; k_1, k_{-1}, k_2, k_{-2}, and a_i, $i = 2, 3, 4, 5, 7, 8, 10$, are rate constants.

We assume that the stages of the catalytic cycle corresponding to the association of substrates (PEP, HPr; reactions 2, 5, and 8) and dissociation of products (Pyr, HPrP; reactions 4, 7, and 10) are much faster than the processes of the intra-molecular transfer of phosphate (reactions 6, 9). This means that any of a_i, $i = 2, 3, 4, 5, 7, 8, 10$, is much larger than any of k_1, k_{-1}, k_2, k_{-2}. From this assumption it follows that stages 2, 3, 4, 5, 7, 8, and 10 are at quasi-equilibrium, i.e., the following relationships hold:

$$K_{d,0}^{PEP} = \frac{PEP \cdot EI_2}{EI_2 \circ PEP},$$
$$K_{t,0} = \frac{EI_2P \circ Pyr}{EI_2 \circ PEP},$$
$$K_{d,1}^{Pyr} = \frac{Pyr \cdot EI_2P}{EI_2P \circ Pyr},$$
$$K_{d,1}^{PEP} = \frac{PEP \cdot EI_2P}{EI_2P \circ PEP}, \qquad (5)$$
$$K_d^{HPrP} = \frac{HPrP \cdot EI_2P}{EI_2P \circ HPrP},$$
$$K_{d,2}^{Pyr} = \frac{Pyr \cdot EI_2P_2}{EI_2P_2 \circ Pyr},$$
$$K_d^{HPr} = \frac{HPr \cdot EI_2P_2}{EI_2P_2 \circ HPr}$$

and the system of differential equations (3) is reduced to the following two-component system:

$$dX/dt = v_9 - v_6$$
$$dY/dt = v_6 - v_9 \qquad (6)$$

where X and Y stand for the sums of Enzyme I states:

$$X = EI_2 + EI_2 \circ PEP + EI_2 \circ Pyr + EI_2P + EI_2P \circ PEP + \; + EI_2P \circ HPrP \qquad Y = EI_2P_2 \circ Pyr + EI_2P_2 + EI_2P_2 \circ HPr \qquad (7)$$

Using Eqs. (5) and (7) rate equations v_6 and v_9 can be rewritten as follows:

$$v_6 = \rho_1 \cdot PEP - \rho_{-1} \cdot Pyr$$
$$v_9 = \rho_2 \cdot HPr - \rho_{-2} \cdot HPrP$$

where:

$$\rho_1 = \cfrac{K_{t,0} \cdot \cfrac{PEP}{K_{d,1}^{PEP}} \cdot \cfrac{PEP}{K_{d,0}^{PEP}}}{\cfrac{Pyr}{K_{d,1}^{Pyr}} + \cfrac{PEP}{K_{d,o}^{PEP}} \cdot \cfrac{Pyr}{K_{d,1}^{Pyr}} + K_{t,0} \cdot \cfrac{PEP}{K_{d,0}^{PEP}} \cdot \cfrac{Pyr}{K_{d,1}^{Pyr}} + K_{t,0} \cfrac{PEP}{K_{d,0}^{PEP}} + K_{t,0}\cfrac{PEP}{K_{d,1}^{PEP}} \cdot \cfrac{PEP}{K_{d,0}^{PEP}} + K_{t,0}\cfrac{HPrP}{K_{d}^{HPrP}} \cdot \cfrac{PEP}{K_{d,0}^{PEP}}}$$

$$\rho_{-1} = \cfrac{\cfrac{Pyr}{K_{d,2}^{Pyr}}}{1 + \cfrac{Pyr}{K_{d,2}^{Pyr}} + \cfrac{HPr}{K_{d}^{HPr}}}, \quad \rho_2 = \cfrac{\cfrac{HPr}{K_{d}^{HPr}}}{1 + \cfrac{Pyr}{K_{d,2}^{Pyr}} + \cfrac{HPr}{K_{d}^{HPr}}}$$

$$\rho_{-2} = \cfrac{K_{t,0} \cdot \cfrac{HPrP}{K_{d}^{HPrP}} \cdot \cfrac{PEP}{K_{d,0}^{PEP}}}{\cfrac{Pyr}{K_{d,1}^{Pyr}} + \cfrac{PEP}{K_{d,o}^{PEP}} \cdot \cfrac{Pyr}{K_{d,1}^{Pyr}} + K_{t,0} \cdot \cfrac{PEP}{K_{d,0}^{PEP}} \cdot \cfrac{Pyr}{K_{d,1}^{Pyr}} + K_{t,0} \cfrac{PEP}{K_{d,0}^{PEP}} + K_{t,0}\cfrac{PEP}{K_{d,1}^{PEP}} \cdot \cfrac{PEP}{K_{d,0}^{PEP}} + K_{t,0}\cfrac{HPrP}{K_{d}^{HPrP}} \cdot \cfrac{PEP}{K_{d,0}^{PEP}}}$$

Equations (6) describe the dynamics of "reduced" two-component catalytic cycle of Enzyme I which is depicted in Fig. 3.

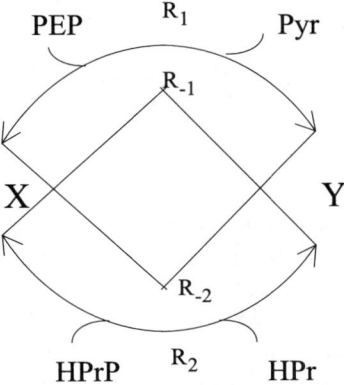

Fig. 3. "Reduced" catalytic cycle of Enzyme I of PTS. ρ_1, ρ_{-1}, ρ_2, and ρ_{-2} designate apparent rate constants

Applying this to the "reduced" catalytic cycle quasi-steady state approach we obtain the following equation (8):

$$v = \cfrac{EI_{tot} \cdot PEP \cfrac{K_{t,0} k_1 k_2}{K_{d,0}^{PEP} \cdot K_{d,1}^{PEP} \cdot K_d^{HPr}} \left[PEP \cdot HPr - \cfrac{Pyr HPr P K_{d,1}^{PEP} \cdot K_d^{HPr} k_{-1} k_{-2}}{K_{d,2}^{Pyr} \cdot K_d^{HPr} k_1 k_2} \right]}{\left\{ \begin{array}{l} K_{t,0} \cdot \cfrac{PEP}{K_{d,0}^{PEP}} \cdot \left(k_1 \cdot \cfrac{PEP}{K_{d,1}^{PEP}} + k_{-2} \cdot \cfrac{HPrP}{K_{d,1}^{HPrP}} \right) \cdot \left(1 + \cfrac{Pyr}{K_{d,2}^{Pyr}} + \cfrac{HPr}{K_d^{HPr}} \right) + \\[4mm] \left(k_{-1} \cfrac{Pyr}{K_{d,2}^{Pyr}} + \cfrac{HPr}{K_d^{HPr}} \cdot k_2 \right) \cdot \begin{pmatrix} \cfrac{Pyr}{K_{d,1}^{Pyr}} + \cfrac{PEP}{K_{d,0}^{PEP}} \cdot \cfrac{Pyr}{K_{d,1}^{Pyr}} + K_{t,0} \cdot \cfrac{PEP}{K_{d,0}^{PEP}} \cdot \cfrac{Pyr}{K_{d,1}^{Pyr}} + \\[3mm] K_{t,0} \cdot \cfrac{PEP}{K_{d,0}^{PEP}} + K_{t,0} \cdot \cfrac{PEP}{K_{d,1}^{PEP}} \cdot \cfrac{PEP}{K_{d,0}^{PEP}} + \\[3mm] K_{t,0} \cdot \cfrac{HPrP}{K_d^{HPrP}} \cdot \cfrac{PEP}{K_{d,0}^{PEP}} \end{pmatrix} \end{array} \right\}} \quad (8)$$

Equation (8) contains 12 parameters. To estimate these parameters we have used experimentally measured values of the total enzyme concentration [10], EI_{tot}, the equilibrium constant of Enzyme I [8], $K^E I_{eq}$, the rate constants of the overall process [11], k_b and k_f, and K_m's for PEP and HPr [12], as listed in Table 1.

The kinetic parameters of Eq. (8) can be expressed in terms of these measured parameters in the following manner. There is a standard procedure developed by the authors (in press) to obtain an interrelationship between parameters:

$$K_{d,1}^{PEP} = \alpha \cdot K_{d,0}^{PEP}, \quad K_{d,2}^{Pyr} = \beta \cdot K_{d,1}^{Pyr}, \quad k_1 = \gamma \cdot k_{-1} \cdot K_{t,0},$$

$$K_{d,0}^{PEP} = \frac{K_m^{PEP} + \varepsilon}{\alpha}, \quad k_{-1} = k_b \cdot K_{d,1}^{Pyr},$$

$$k_{-2} = \frac{\beta \cdot \gamma^2 \cdot k_f^2 \cdot K_d^{HPrP} \cdot K_m^{PEP}}{\alpha^2 \cdot k_b \cdot K_{eq}^{EI} \cdot K_m^{HPr}}, \quad k_2 = \frac{\gamma \cdot k_f \cdot K_m^{PEP} \cdot K_{d,1}^{Pyr}}{\varepsilon}, \quad (9)$$

$$K_{t,0} = \frac{k_f \cdot (K_m^{PEP} + \varepsilon)}{\alpha \cdot k_b \cdot K_{d,1}^{Pyr}}, \quad K_d^{HPr} = \frac{\alpha \cdot K_m^{HPr} \cdot K_{d,1}^{Pyr}}{\varepsilon}, \quad \varepsilon > 0$$

Equations (9) allow us to reduce the number of "free" parameters of Eq. (8) from 12 to just 6: α, β, ε, γ, $K_{d,1}^{Pyr}$, $K_{d,1}^{HPrP}$. The values of these "free" parameters are listed in Table 1.

Enzyme IIA^{glc} and Protein HPr

The doubly phosphorylated dimer of Enzyme I phosphorylates the protein HPr on histidine 15 [9]. Then, the phosphorylated HPr reacts with EIIA^{glc} resulting in the transfer of a phosphate group from HPrP to histidine 90 of EIIA^{glc} [9]. The phosphorylated Enzyme IIA^{glc}, EIIA^{glc}P, reacts with Enzyme IICB^{glc}. The reaction rates of binding HPrP with EIIA^{glc} (v_{HPrP_A}), the phosphate transfer ($v_{HPrP\_transfer}$), and the dissociation of HPr from EIIA^{glc}P (v_{HPr_AP}) are described by mass action:

$$v_{HPrP\_A} = k_{HPrP\_A} \cdot (HPrP \cdot EIIA^{glc} / K_d^{HPrP\_A} - HPrP\_A)$$
$$v_{HPrP\_transfer} = k_{HPrP\_transfer} \cdot (HPrP\_A - HPr\_AP / K_{eq}^{HPrP\_transfer})$$
$$v_{HPr\_AP} = k_{HPr\_AP} \cdot (HPr\_AP - HPr \cdot EIIA^{glc}P / K_d^{HPr\_AP})$$

$$(10)$$

Table 1. Kinetic parameters of kinetic model of glycolysis

Reaction number	Parameter	Parameter values estimated from fitting (f), measured experimentally (m), or choosen in arbitrary way (a).
1	EI_{tot}	0.03 [9], m
	k_b	800 [11], m
	k_f	3000 [11], m
	K_m^{PEP}	0.18 [12], m
	K_m^{HPr}	0.09 [12], m
	K_{eq}^{EI}	11 [8], m
	α	1.1 (a)
	β	1 (a)
	γ	2 (a)
	ε	0.001 (a)
	$K_{d,1}^{Pyr}$	5 (a)
	K_d^{HPrP}	0.1 (a)
2	k_{HPrP_A}	128700 [13], f
	$K_d^{HPrP\_A}$	0.000156 [13], f
3	$k_{HPrP\_transfer}$	75600 [13], f
	$K_{eq}^{HPrP\_transfer}$	0.84 [13], f
4	k_{HPr_AP}	2040 [13], f
	$K_d^{HPr\_AP}$	0.000289 [13], f
5–7	$EIICB_{tot}^{glc}$	1 (a)
	K_d^A	0.02 [14], m
	K_d^{AP}	0.002 [14], m
	K_d^{G6P}	0.001 [14], m
	K_d^{out}	0.002 [14], m
	K_t	1 [14] , m
	k_{Glc}	60000 [14], f
	k_{-Glc}	6 [14], f
	k_{10}	17000 [14], f
	k_{-10}	560 [14], f
	k_8	52000 [14], f
	k_{-8}	0.1 [14], f
	k_{14}	470 [14], f
	k_{11}	17000 [14], f
	k_{19}	80 [14], f
	k_{-19}	3 [14], f
8	V_{max}^{glk}	Varied
	K_{eq}^{glk}	10 (a)
	$K_{m,Glc}^{glk}$	0.15 [15], m
9	k_9	50 (a)
	$K_9^{glycolysis}$	10 (a)
10	k_{10}	50 (a)
	$K_{10}^{glycolysis}$	10 (a)
11	k_{11}	50 (a)
Conservation law	Total $EIIA^{glc}$	1 (a)
	Total HPr	0.3 [8], m
	Total $ADP+ATP$	4 [6], m
Buffered concentration	Glc_{out}	Varied
	Pyr	0.4 [8], m

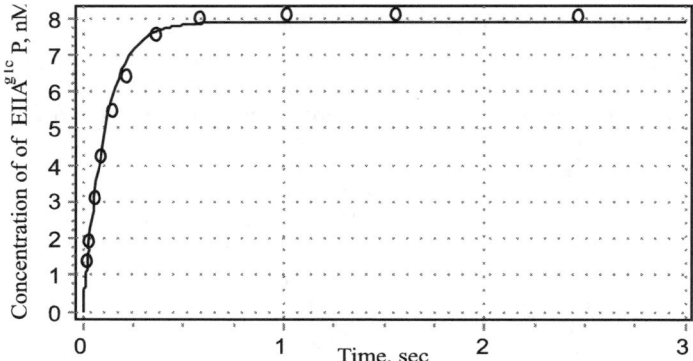

Fig. 4. Time dependence of EIIA^{glc} phosphorylation level measured experimentally in [13] (open circles) and resulting from kinetic model (line)

Values of the constants, k_{HPrP_A}, $k_{HPrP\_transfer}$, k_{HPr_AP}, $K_{eq}^{HPrP\_transfer}$, $K_d^{HPr\_AP}$, $K_d^{HPrP\_A}$, have been fitted (see Table 1) to describe the experimental data published in [13]. In this experiment, EIIA^{glc} phosphorylation has been measured upon addition of HPrP (Fig. 4, open circles). To describe the experiment and to estimate the kinetic parameters of Eq. (6), we have developed a kinetic model for transfer of the phosphate group from HPrP to EIIA^{glc}:

$$dHPrP/dt = -v_{HPrP\_A}$$
$$dEIIA^{glc}/dt = -v_{HPrP\_A}$$
$$dHPrP\_A/dt = v_{HPrP\_A} - v_{HPrP\_transfer}$$
$$dHPr\_AP/dt = v_{HPrP\_transfer} - v_{HPr\_AP}$$
$$dHPr/dt = v_{HPr\_AP}$$
$$dEIIA^{glc}P/dt = v_{HPr\_AP}$$
$$HPrP = 9.4\ \text{nM}$$
$$EIIA^{glc} = 86\ \text{nM}$$
$$HPrP\_A = HPr\_AP = HPr = EIIA^{glc}P = 0$$

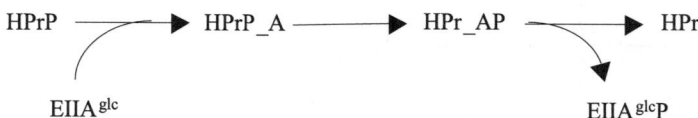

Fig. 5. Kinetic scheme of processes of EIIA^{glc} phosphorylation

The kinetic model corresponds to the kinetic scheme depicted in Fig. 5. The initial concentrations are taken from [13]. Fig. 4 shows the good quality of the correlation between the resulting simulation results (solid line) and the experimental data (open circles) [13].

Enzyme IICB^{glc}

Two domains, B and C, of Enzyme IICB^{glc} are incorporated into the inner membrane of *E. coli*. The trans-membrane domain C has a glucose binding site that can be exposed to either the periplasm or the cytoplasm. Domain B is exposed to the cytoplasm and carries cystein 421 that can be phosphorylated with EIIA^{glc}P. Enzyme IICB^{glc} catalyzes the transport and phosphorylation of glucose from the periplasm to the cytoplasm. The EIICB^{glc} catalytic cycle depicted in Fig. 6) consists of 13 states designated as x_i, $i = 1, 2, 3$; y_j, z_j, $j = 1, \ldots, 5$. Each state is determined by the sub-states of the glucose binding site belonging to domain C, sub-states of cysteine 421 situated on domain B, as well as sub-states of histidine 90 of EIIA^{glc} which can form a catalytically active complex with EIICB^{glc}. The glucose binding site can be exposed either to the periplasm (states y_5 and z_5) or to the cytoplasm (states x_i, $i = 1, 2, 3$; y_j, z_j, $j = 1, \ldots, 4$). Cystein 421 can be either phosphorylated (states y_j, z_j, $j = 4, 5$) or non-phosphorylated (states x_i, $i = 1, 2, 3$; y_j, z_j, $j = 1, 2, 3$). EIICB^{glc} can be free (states x_2, y_2, z_2) or in the form of a complex with EIIA^{glc} (states x_i, $i = 1, 3$; y_j, z_j, $j = 1, 3, \ldots, 5$), while histidine 90, in its turn, can be either phosphorylated (states x_3, y_3, z_3) or non-phosphorylated (states x_1, y_j, z_j, $j = 1, 4, 5$).

It was found [14] that free Enzyme IICB^{glc} is always in the state where the glucose binding site is exposed to cytoplasm (states x_2, y_2, z_2). When this occurs, EIICB^{glc} can bind cytoplasmic glucose, Glc$_{in}$ (reaction 2), or G6P (reaction 1) using the glucose binding site or EIICB^{glc} can form a complex with EIIA^{glc}P (reactions 3–5). Complexes of EIICB^{glc} with EIIA^{glc}P can also bind Glc$_{in}$ (reaction 7 in Fig. 6) or G6P (reaction 6) using the glucose binding site. Following the formation of the complex EIICB^{glc} with EIIA^{glc}P, a phosphate group is transferred from histidine 90 of EIIA^{glc} to cysteine 421 of EIICB^{glc} (reactions 8 and 9). As a result of this intra-molecular phosphate transfer, the complex becomes catalytically active – it is able to transport and phosphorylate glucose. If the glucose binding site is empty, then the phosphorylation of cysteine 421 leads to conformational changes in domain C resulting in the glucose binding site being exposed to the periplasm (reaction 11) where periplasmic glucose, Glc$_{out}$, binds (reaction 12). The EIICB^{glc} conformation changes how the glucose binding site (with glucose) is exposed to the cytoplasm (reaction 13). The EIICB^{glc}P and EIIA^{glc} complex can also be attained due to binding of the cytoplasmic glucose (reaction 10). Furthermore, a phosphate is transferred from cysteine 421 to the glucose (reaction 14) to form G6P and inactivate the EIICB^{glc}. Then the dissociation of G6P and the subsequent binding of Glc$_{in}$ (reactions 15 and 16) or the dissociation of EIIA^{glc}

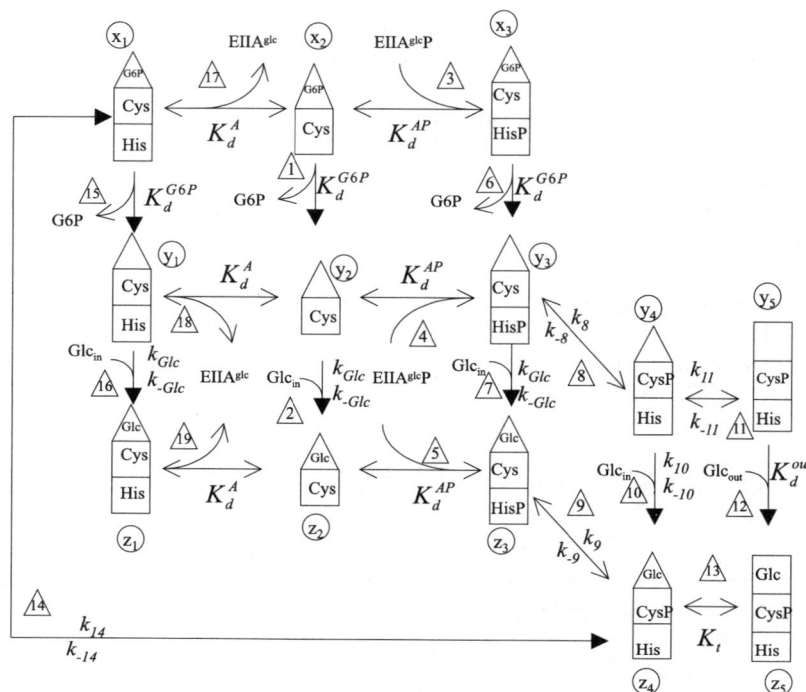

Fig. 6. Catalytic cycle of Enzyme IICB^{glc} of PTS. Each state of EIICB^{glc} is presented by two designations: structural and operational. Structural designation shows how states of EIICB^{glc} are determined by states of Cys 421, glucose and EIIA binding sites and His 90 of EIIA. Operational designations of EIICB^{glc} states are given in the main text. Each of the operational designations: $x[i], i = 1, 2, 3, y[j], z[j], j = 1, \ldots, 5$, is shown in open circles situated near to corresponding structural designations of the EIICB^{glc} states. Each reaction of the catalytic cycle is shown by numbers in open triangles. Dissociation or rate constants are situated near to each corresponding reaction.

from EIICB^{glc} (reaction 17–19) proceeds in random order resulting in the regeneration of free EIICB^{glc} and the completion of the catalytic cycle. Enzyme IICB^{glc} can function in three modes:

1. transport of periplasmic glucose to the cytoplasm with its concomitant phosphorylation to form G6P
   (transition via y2→y3→y4→y5→z5→z4→x1→y1→y2)
2. phosphorylation of cytoplasmic glucose to form G6P
   (transition via y2→y3→y4→z4→x1→y1→y2)
3. transport of periplasmic glucose to cytoplasm without its phosphorylation
   (transition via y4→y5→z5→z4→y4)

The contribution of each mode to the overall Enzyme IICB^{glc} reaction rate is determined by the concentrations of G6P, Glc$_{in}$, Glc$_{out}$, PEP, and Pyr.

We assume that the following processes of the catalytic cycle are at quasi-equilibrium:

- Glc$_{out}$ association (reaction 12),
- dissociation of G6P (reactions 1, 6, and 15),
- association/dissociation of EIICB^{glc} (reactions 3–5 and 17–19), and
- transition of the glucose binding site from the periplasm to the cytoplasm (reaction 13).

Applying the strategy described for the derivation of the Enzyme I rate equation, the following relationships have been produced:

$$
K_d^A = \frac{EIIA^{glc} \cdot x_2}{x_1} \quad K_d^A = \frac{EIIA^{glc} \cdot y_2}{y_1} \quad K_d^A = \frac{EIIA^{glc} \cdot z_2}{z_1}
$$

$$
K_d^{AP} = \frac{EIIA^{glc}P \cdot x_2}{x_3} \quad K_d^{AP} = \frac{EIIA^{glc}P \cdot y_2}{y_3} \quad K_d^{AP} = \frac{EIIA^{glc}P \cdot z_2}{z_3} \quad (11)
$$

$$
K_d^{G6P} = \frac{G6P \cdot y_1}{x_1} \quad K_d^{out} = \frac{Glc_{out} \cdot y_5}{z_5} \quad K_t = \frac{z_4}{z_5}
$$

In the scheme depicted in Fig. 6, K_d^A, K_d^A, K_d^A, K_d^{AP}, K_d^{AP}, K_d^{AP}, K_d^{G6P}, K_d^{out}, K_t stand for corresponding dissociation constants; k_{Glc}, k_{-Glc}, k_i, k_{-i}, $i = 8, \ldots, 11, 14$, are rate constants. Using Eqs. (11) the number of states of the catalytic cycle of Enzyme IICB^{glc} can be reduced to four (the "reduced" catalytic cycle is depicted in Fig. 7). Each state of the "reduced" catalytic cycle corresponds to the sum of states of the unreduced catalytic cycle:

$$
u_1 = x_1 + x_2 + x_3 + y_1 + y_2 + y_3
$$
$$
u_2 = y_4
$$
$$
w_1 = z_1 + z_2 + z_3
$$
$$
w_2 = y_5 + z_4 + z_5
$$

As shown in Fig. 7, four states of the "reduced" catalytic cycle are connected with six reactions whose rate constants are given by the following expressions:

$$
\rho_1 = \frac{k_8 \cdot K_d^{G6P} \cdot K_d^A}{\left(K_d^{G6P} + G6P\right) \cdot \left\{K_d^A \cdot \left(EAP + K_d^{AP}\right) + EA \cdot K_d^{AP}\right\}}
$$

$$
\rho_{-1} = k_{-8} \quad \rho_2 = k_{10}
$$

$$
\rho_{-2} = \frac{k_{-10}}{\left(1 + K_t\right) \cdot Glc_{out} + K_t \cdot K_d^{out}}
$$

$$
\rho_3 = k_{11}
$$

$$
\rho_{-3} = \frac{k_{-11} \cdot K_t \cdot K_d^{out}}{K_t \cdot K_d^{out} + \left(1 + K_t\right) \cdot Glc_{out}}
$$

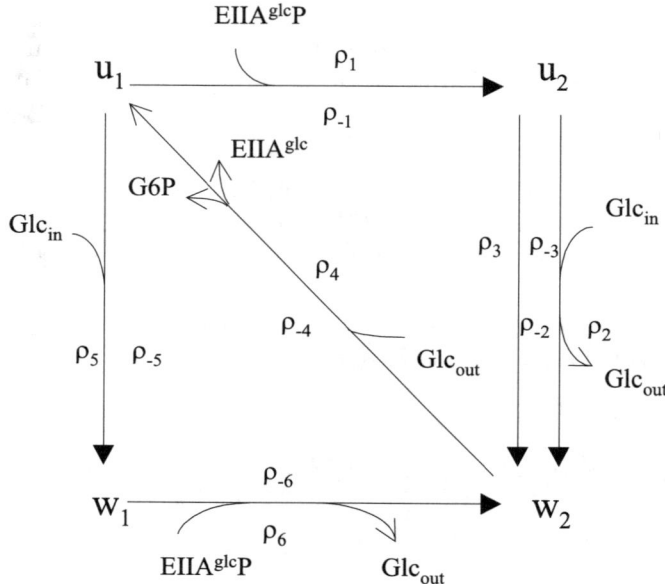

Fig. 7. "Reduced" catalytic cycle of Enzyme IICB^{glc} of PTS. ρ_1, ρ_{-1}, ρ_2, and ρ_{-2} designate apparent rate constants

$$\rho_4 = \frac{k_{14}}{Glc_{out}\left(1 + K_t\right) + K_t \cdot K_d^{out}}$$

$$\rho_{-4} = \frac{k_{-14} \cdot K_d^{AP}}{\left(K_d^{G6P} + G6P\right) \cdot \left\{K_d^A \cdot \left(EAP + K_d^{AP}\right) + EA \cdot K_d^{AP}\right\}}$$

$$\rho_5 = \frac{k_{Glc} \cdot K_d^{G6P}}{K_d^{G6P} + G6P} \quad \rho_{-5} = k_{-Glc}$$

$$\rho_6 = \frac{k_9 \cdot K_d^A}{K_d^A \cdot \left(K_d^{AP} + EAP\right) + EA \cdot K_d^{AP}}$$

$$\rho_{-6} = \frac{k_{-9}}{K_t \cdot K_d^{out} + \left(1 + K_t\right) \cdot Glc_{out}}$$

As mentioned, EIICB^{glc} can function in three different modes. In terms of states of the "reduced" catalytic cycle (see Fig. 7), these three modes can be related to the following transitions through states u_1, u_2, w_1, w_2:

1. $u_1 \rightarrow u_2 \rightarrow w_2 \rightarrow u_1$: transport of periplasmic glucose to cytoplasm with its concomitant phosphorylation to form G6P (v_{G6P}^{out})
2. $u_1 \rightarrow w_1 \rightarrow w_2 \rightarrow u_1$: phosphorylation of cytoplasmic glucose to form G6P (v_{G6P}^{in})
3. $u_2 \rightarrow w_2 \rightarrow u_2$: transport of periplasmic glucose to cytoplasm without its phosphorylation (v_{Glc}^{influx})

The rate equations derived from the reduced catalytic cycle are:

$$v_{Glc_{out}} = \frac{EIICB_{tot} \left\{ \begin{array}{l} (\rho_{-5} + \rho_6 \cdot EIIA^{glc}P) \cdot \rho_1 \cdot \rho_4 \cdot \\ \left[\begin{array}{l} \rho_3 \cdot \left(EIIA^{glc}P \cdot Glc_{out} - \frac{EA \cdot G6P}{K_{eq}^{EIICB}} \right) + \\ \rho_2 \cdot Glc_{out} \cdot \left(EIIA^{glc}P \cdot Glc_{in} - \frac{EIIA^{glc}G6P}{K_{eq}^{EIICB}} \right) \end{array} \right] + \\ \rho_1 \cdot \rho_3 \cdot \rho_{-5} \cdot \rho_{-6} \cdot EIIA^{glc}P \cdot (Glc_{out} - Glc_{in}) \end{array} \right\}}{(G6P)^2 \cdot \left\{ \begin{array}{l} K_d^A \cdot \left[EIIA^{glc}P \cdot \left(K_d^{G6P} + G6P \right) + \right. \\ \left. K_d^{AP} \cdot \left(K_d^{G6P} + G6P \right) \right] + \\ EIIA^{glc} \cdot K_d^{AP} \left(K_d^{G6P} + G6P \right) \end{array} \right\}}$$

$$v_{Glc_{in}} = \frac{EIICB_{tot} \left\{ \begin{array}{l} (\rho_{-1} + Glc_{in} \cdot \rho_2 + \rho_3) \cdot \rho_4 \cdot \\ Glc_{out} \cdot \left(Glc_{in} \cdot EIIA^{glc}P - \frac{EIIA^{glc} \cdot G6P}{K_{eq}^{EIICB}} \right) + \\ \rho_{-1} \cdot \rho_{-3} \cdot EIIA^{glc}P \cdot (Glc_{in} - Glc_{out}) \end{array} \right\}}{(G6P)^2 \cdot \left\{ \begin{array}{l} K_d^A \cdot \left[EIIA^{glc}P \cdot \left(K_d^{G6P} + G6P \right) + \right. \\ \left. K_d^{AP} \cdot \left(K_d^{G6P} + G6P \right) \right] + \\ EIIA^{glc} \cdot K_d^{AP} \left(K_d^{G6P} + G6P \right) \end{array} \right\}} \qquad (12)$$

$$v_{inf\,lux} = \frac{EIICB_{tot} \left\{ \begin{array}{l} (\rho_{-5} + \rho_6 \cdot EIIA^{glc}P) \cdot \rho_1 \cdot \rho_2 \cdot \rho_4 \cdot Glc_{out} \cdot \\ \left(\frac{EIIA^{glc} \cdot G6P}{K_{eq}^{EIICB}} - EIIA^{glc}P \cdot Glc_{in} \right) + \\ \left[(\rho_{-5} + \rho_6 \cdot EIIA^{glc}P) \cdot (\rho_1 \cdot EIIA^{glc}P + \right. \\ \left. \rho_{-4} \cdot EIIA^{glc} \cdot G6P) + \rho_5 \cdot \rho_6 \cdot Glc_{in} \cdot EIIA^{glc}P \right] \cdot \\ \rho_2 \cdot \rho_{-3} \cdot (Glc_{out} - Glc_{in}) \end{array} \right\}}{(G6P)^2 \cdot \left\{ \begin{array}{l} K_d^A \cdot \left[EIIA^{glc}P \cdot \left(K_d^{G6P} + G6P \right) + \right. \\ \left. K_d^{AP} \cdot \left(K_d^{G6P} + G6P \right) \right] + \\ EIIA^{glc} \cdot K_d^{AP} \left(K_d^{G6P} + G6P \right) \end{array} \right\}}$$

These equations describe the dependencies of the reaction rates of Enzyme IICB^{glc} functioning in different modes on substrate and product concentrations. Equations (12) contain 17 parameters. Some of them have been esti-

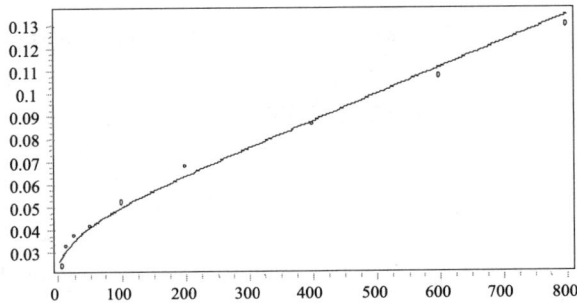

Fig. 8. Experimentally measured dependence of the reaction rates catalyzed by Enzyme IICB^{glc} on glucose concentration (open circles), and that generated with kinetic model (solid line)

mated in [14]. Others (listed in Table 1) were chosen to provide a good fit
of simulation results (solid line, Fig. 8) to the experimental data [14] (open
circles, Fig. 8).

Glycolysis

All the processes for model construction are shown in Fig. 9. For the purposes
of illustration, we have included only three processes for this model of glycolysis. The first process is phosphorylation of Glc_{in} catalyzed by glucokinase

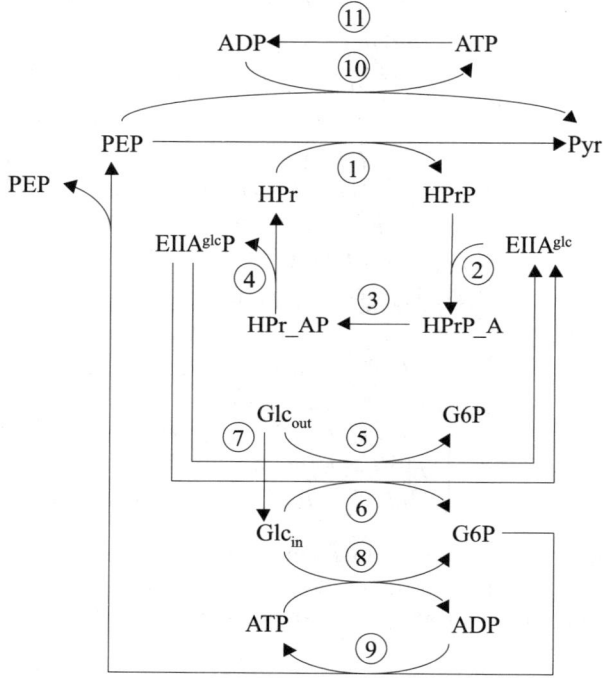

Fig. 9. Kinetic scheme of "simplified" glycolitic pathway

(reaction 8 in Fig. 9). Since K_m for ATP is five times greater than K_m for
Glc_{in} [15], the rate equation can be written as:

$$v_8 = V_{\max}^{glk} \cdot (Glc_{out} \cdot ATP - G6P \cdot ADP/K_{eq}^{glk})/(K_{m,Glc}^{glk} + Glc_{out}) \quad (13)$$

where $V_{\max}^{glk}$, K_{eq}^{glk}, and $K_{m,Glc}^{glk}$ are the kinetic constants of glucokinase (values
presented in Table 1).

The second process (reaction 9 in Fig. 9) mimics the glycolysis reactions
from G6P isomerization up to PEP formation. This reaction describes the production of two molecules of PEP and one ATP from one molecule of G6P and

one molecule of ADP. A reaction rate can be written from the corresponding mass action law:

$$v_9 = k_9 \cdot (G6P \cdot ADP - ATP \cdot PEP^2 / K_9^{glycolysis}) \qquad (14)$$

The third process is the reaction of PEP dephosphorylation catalyzed by pyruvate kinase (reaction 10). The reaction rate is derived from mass action:

$$v_{10} = k_{10} \cdot (PEP \cdot ADP - ATP \cdot Pyr / K_9^{glycolysis}) \qquad (15)$$

We also consider the process of ATP consumption (reaction 11) which is described according to the mass action law:

$$v_{11} = k_{11} \cdot ATP \qquad (16)$$

Values of the kinetic parameters in Eqs. (14)–(16) are listed in Table 1. The kinetic model of PTS and glycolysis shown in Fig. 9 is described by:

$$
\begin{aligned}
dHPrP/dt &= v_1 - v_2 \\
dEIIA^{glc}/dt &= v_5 + v_6 - v_2 \\
dHPrP\_A/dt &= v_2 - v_3 \\
dHPr\_AP/dt &= v_3 - v_4 \\
dHPr/dt &= v_4 - v_1 \\
dEIIA^{glc}P/dt &= v_4 - v_5 - v_6 \\
dGlc_{in}/dt &= v_7 - v_6 - v_8 \\
dG6P/dt &= v_6 + v_8 - v_9 \\
dPEP/dt &= 2v_9 - v_1 - v_{10} \\
dADP/dt &= v_8 + v_{11} - v_9 - v_{10} \\
dATP/dt &= v_9 + v_{10} - v_8 - v_{11}
\end{aligned}
\qquad (17)
$$

In Fig. 9, Glc$_{out}$ and Pyr are buffered, i.e., their concentrations do not change with time, and v_i, $i = 1, \ldots, 7$, are:

$$
\begin{aligned}
v_1 &= v_{EI}, \\
v_2 &= v_{HPrP\_A}, \\
v_3 &= v_{HPrP\_transfer}, \\
v_4 &= v_{HPr\_AP}, \\
v_5 &= v_{Glcout}, \\
v_6 &= v_{Glcin}, \\
v_7 &= v_{influx}.
\end{aligned}
$$

3.2 Determining Quasi-Steady States in Reduced Kinetic Models

We use the following method to determine the number of quasi-steady states in the kinetic model. Consider a system of differential equations corresponding to a kinetic model and assume that one of the intermediates X_i is constant. The assumption eliminates the specific differential equation for X_i. Yet, it does not change the functional dependencies of the other rate equations on X_i. There are still reactions producing and consuming X_i, since X_i is a concentration of an intermediate. Because X_i is an intermediate, and assumed to be in steady state (hence the $dx_i/dt = 0$ constancy assumption), we do not expect the dynamics to change much.

From Eq. (2), we designate the production and consumption fluxes for X_i as $v_{X_i}^{prod}$ and $v_{X_i}^{cons}$. Consider a *reduced* system where X_i is a constant so that (from Eq. 2) we obtain two functional relationships based upon X_i:

$$v_{X_i}^{prod}(X_i) = v_{X_i}^{cons}(X_i)$$

This effective quasi-steady state can only depend upon that set of the X_i values that represents the roots of this non-linear algebraic equation. Clearly there can be one root, many roots, or no bounded positive solutions. Where possible we strive for analytic solutions, but if that is not feasible a numerical root finder can always be used. The resulting quasi-steady states can be further characterized by their stability characteristics as being either stable or unstable. Such characterization is important to understand which artifacts are biologically meaningful and which are artifacts of the model. Usually, we perform stability analysis by numerically calculating eigenvalues and other relative parameters.

3.3 The Number of Quasi-Steady States Depends on the Expression Level of Glucokinase

We applied the previously described method for finding quasi-steady states to the kinetic model of PTS and glycolysis in *E. coli* (Fig. 9) with PEP as the intermediate held constant in the *reduced* system. We consider quasi-steady state dependencies of the rate of PEP consumption and production as a function of PEP:

$$v_{PEP}^{prod}(PEP) = 2v_9(PEP)$$

$$v_{PEP}^{cons}(PEP) = v_1(PEP) + v_{10}(PEP)$$

By varying PEP over a reasonable range from 0 to 0.2 mM we have found the quasi-steady states at different levels of glucokinase and periplasmic glucose, Glc_{out}. Figure 10A shows the dependencies of v_{PEP}^{prod} and v_{PEP}^{cons} on PEP and the three quasi-steady states that result for the original system (Fig. 9).

We found that two of these quasi-steady states are stable (indicated as "stable 0" and "stable 1" in Fig. 10) and one is unstable. Quasi-steady state "stable 1" is a physiological quasi-steady state, i.e., values of fluxes and concentrations at this quasi-steady state are comparable with those measured experimentally. The "stable 0" state is not physiologically meaningful since it is just the trivial null point of all zero fluxes and represents a "dead" cell. The unstable quasi-steady state cannot correspond to any meaningful biological state since any fluctuation, however small, will result in a transition to one of the other stable quasi-steady states. The expression level of glucokinase given by the V_{max}^{glk} value determines the number of quasi-steady states and their stability with a low glucose concentration ($\text{Glc}_{out} = 50\,\mu\text{M}$). When the glucokinase expression is low (V_{max}^{glk} is equal to $0.1\,\text{mM/min}$), then our kinetic model has three quasi-steady states (see Fig. 10A) only one of which is non-trivial and stable. Increasing the glucokinase expression level with low

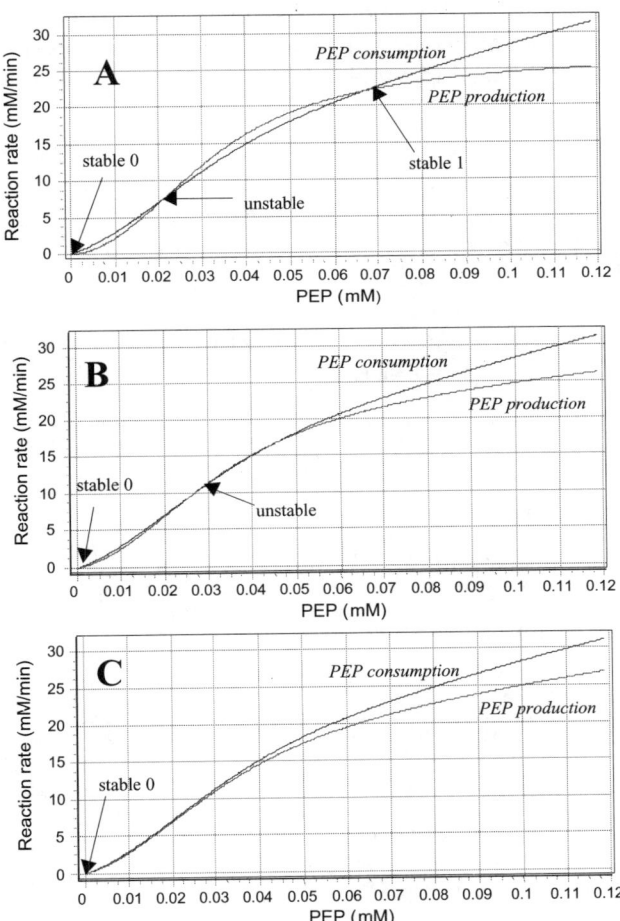

Fig. 10. Number of steady states of kinetic model of glycolysis at $\text{Glc}_{out}=0.05$ mM and different activities of glucokinase: (A) corresponds to $V_{max}=0.1$ mM/min, (B) corresponds to $V_{max}=120$ mM/min, (C) corresponds to $V_{max}=500$ mM/min

glucose results in a decrease of the distance between the physiological and unstable steady states and, finally, in a transition to the quasi-steady state corresponding to cell death. Indeed, when the glucokinase expression reaches a critical level of $V_{\max}^{glk}$ equal to $120\,\mathrm{mM/min}$ (this value depends on all other parameters of the model) the physiological quasi-steady state "stable 1" and unstable states merge and the resulting quasi-steady state will be unstable (Fig. 10B). Further increases in the glucokinase expression level to $V_{\max}^{glk}$ equal to $500\,\mathrm{mM/min}$ lead to the disappearance of this unstable quasi-steady state entirely (Fig. 10C).

Figure 11 shows how the level of glucokinase expression determines the number of quasi-steady states and their stability when glucose concentration is not less than $1\,\mathrm{mM}$. Indeed, when the glucokinase expression reaches a critical level of $V_{\max}^{glk}$ equal to $8\,\mathrm{mM/min}$ (this value depends on the values of

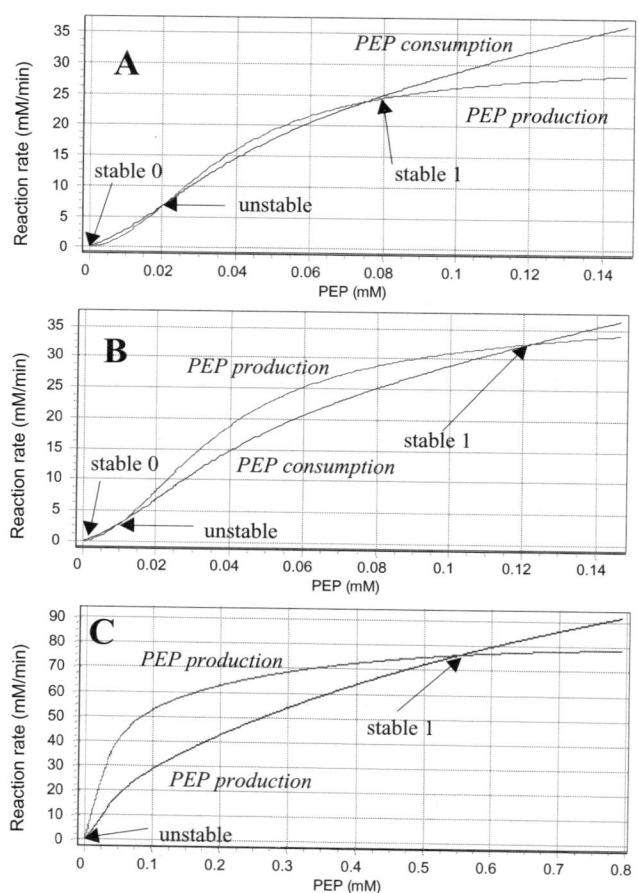

Fig. 11. Number of steady states of kinetic model of glycolysis at $\mathrm{Glc}_{out}=1$ mM and different activities of glucokinase: (A) corresponds to $V_{\mathrm{max}}=1$ mM/min, (B) corresponds to $V_{\mathrm{max}}=8$ mM/min, (C) corresponds to $V_{max}=100$ mM/min

all other parameters of the model) the non-physiological quasi-steady state and unstable quasi-steady state merge. The resulting quasi-steady state will be unstable (Fig. 11B). This analysis of the stability properties of the quasi-steady states as a function of glucokinase expression leads to a very simple biological interpretation: the glucokinase expression level can either lead to cell death or keep the cell alive depending on substrate availability. By using a bifurcation diagram, such as that shown in Fig. 12, an estimate can be made of the glucose concentrations and glucokinase expression levels that correspond to physiological quasi-steady states.

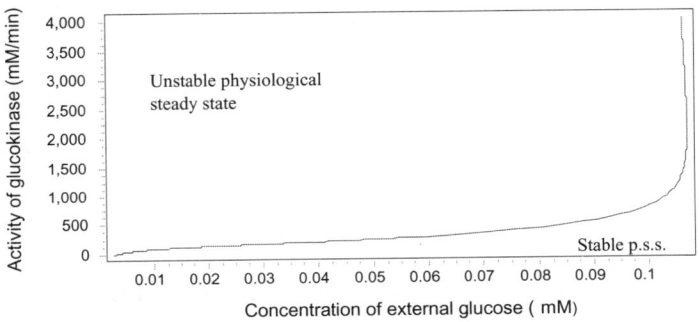

Fig. 12. Bifurcational diagram in two-dimensional parametric space, reflecting regions of stability of physiological steady state "stable 1"

4 Kinetic Model of the Aro pathway. Optimization of High-Throughput Screening Assays

The shikimate (Aro) pathway is the biosynthetic route to chorismate. It is an essential pathway in plants, fungi, and micro-organisms, yet it is absent in mammalian cells. This mammalian loss of the enzymes therefore provides attractive targets for potential new anti-microbial compounds [16]. One approach to discovering such compounds by high-throughput screening is to reconstruct the pathway in a homogeneous assay using purified enzymes to screen for inhibitors of all enzymes simultaneously. Using such a method, compounds that inhibit multiple target enzymes can be easily identified. By using a modelling approach we can adjust the enzyme concentrations to equally redistribute control among all enzymes of the system to optimize the resulting experimental enzyme assay screen. Numerical solutions of the models and analysis of their sensitivity were obtained using the software package DB-Solve [7], a mathematical simulation workbench that has been used for dynamic modelling of metabolic networks. Figure 13 presents the Aro system of four enzymes which catalyze steps 2 to 5 in the shikimate pathway:

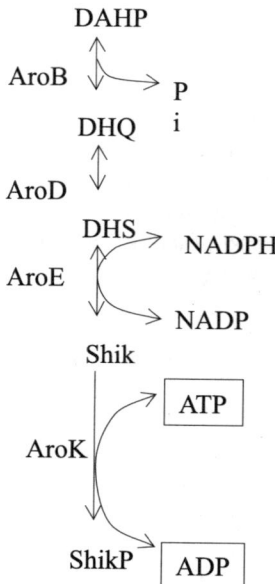

Fig. 13. Scheme of a four-reaction sequence of the shikimate pathway

1. 3-dehydroquinate synthase (3-DHQ synthase; AroB), a NAD$^+$- and Zn^{2+}-
   dependent enzyme that catalyzes the conversion of 3-deoxy-D-arabino-
   heptulosonic acid 7-phosphate (DAHP) to 3-dehydroquinate (DHQ),
2. 3-dehydroquinate dehydratase (3-DHQase; AroD), which catalyzes the de-
   hydration of DHQ to 3-dehydroshikimate (DHS),
3. shikimate dehydrogenase (AroE), which catalyzes the reversible reduction
   of DHS to shikimate using NADPH, and
4. shikimate kinase (AroK; EC 2.7.1.71) [7] which catalyzes the phosphorylation of the 3-hydroxyl group of shikimate using ATP.

The following system of differential equations describes the dynamics of the
pathway:

$$dDAHP/dt = -v_{AroB}$$
$$dDHQ/dt = v_{AroB} - v_{AroD}$$
$$dDHS/dt = v_{AroD} - v_{AroE}$$
$$dShik/dt = v_{AroE} - v_{AroK}$$
$$dDAHP/dt = v_{AroK}$$
$$dNADPH/dt = -v_{AroE}$$
$$dNADP/dt = v_{AroE}$$
$$dATP/dt = -v_{AroK}$$
$$dADP/dt = v_{AroK}$$

$$(18)$$

For AroB, only the NAD$^+$-bound form of the enzyme is active, and the rate
equation of active AroB is given by following equation:

$$v_{AroB} = \frac{k_{fB\circ NAD}.B^\circ NAD \cdot DAHP}{K_{mDAHP}\left(1 + \dfrac{DHQ}{K_{iDHQ}}\right) + DAHP}$$

The rate equation for AroD with competitive product inhibition is given by:

$$v_{AroD} = \frac{k_{fD} \cdot D \cdot DHQ}{K_{mDHQ} \cdot \left(1 + \dfrac{DHS}{K_{iDHS}}\right) + DHQ}$$

The AroE catalyzed reactions can be considered an ordered Bi-Bi mechanism [17]:

$$v_{AroE} = \frac{V_f \cdot V_r \left(A \cdot B - \frac{P \cdot Q}{K_{eq}}\right)}{V_r \cdot K_{iA} \cdot K_{mB} + V_r \cdot K_{mB} \cdot A + V_r \cdot K_{mA} \cdot B + \frac{V_f \cdot K_{mQ} \cdot P}{K_{eq}} + \frac{V_f \cdot K_{mP} \cdot Q}{K_{eq}} + {} \atop {} + V_r \cdot A \cdot B + \frac{V_f \cdot K_{mQ} \cdot A \cdot P}{K_{eq} \cdot K_{iA}} + \frac{V_f \cdot P \cdot Q}{K_{eq}} + \frac{V_r \cdot K_{mA} \cdot B \cdot Q}{K_{iQ}} + \frac{V_r \cdot A \cdot B \cdot P}{K_{iP}} + \frac{V_f \cdot B \cdot P \cdot Q}{K_{iB} \cdot K_{eq}}}$$

Here, A is NADPH, B is DHS, P is Shik, and Q is NADP. The equilibrium constant, K_{eq}, can be expressed by a Haldane relationship:

$$K_{eq} = (V_f \cdot K_{iQ} \cdot K_{mP})/(V_r \cdot K_{iA} \cdot K_{mB})$$

The rate equation we have used to describe the reaction catalyzed by AroK is an irreversible ordered Bi-Bi mechanism:

$$v_{AroK} = \frac{k_{fK} \cdot K \cdot Shik \cdot ATP}{\left(K_{mATP} \cdot \left(1 + \frac{ADP}{K_{iADP}}\right) + ATP\right) \cdot \left(K_{mShik} \cdot \left(1 + \frac{ShikP}{K_{iShikP}}\right) + Shik\right)}$$

The values of the parameters for the kinetic model were taken from [18] and are collected in Table 2. During the process of enzyme catalysis, there is a period of time over which the system will attain a quasi-steady state when

Table 2. Kinetic parameters of kinetic model of Aro pathway

Enzyme	Experimental estimated kinetic parameters	Kinetic parameters obtained by fitting to raw experimental data
AroB		$K_f = 5\,\text{s}^{-1}$, $K_{m,DAHP} = 3.38\,\mu\text{M}$ $K_{m,DHQ} = 2.2\,\mu\text{M}$
AroD	$K_f = 32\,\text{s}^{-1}$, $K_{m,DHQ} = 72\,\mu\text{M}$	$K_{m,DHS} = 692\,\mu\text{M}$
AroE	$K_f = 191\,\text{s}^{-1}$, $k_r = 46\,\text{s}^{-1}$ $K_{m,DHS} = 110\,\mu\text{M}$ $K_{m,NADPH} = 12.6\,\mu\text{M}$ $K_{m,Shik} = 95\,\mu\text{M}$, $K_{m,NADP} = 11\,\mu\text{M}$, $K_{i,NADPH} = 5\,\mu\text{M}$, $K_{i,NADP} = 5\,\mu\text{M}$	$K_{i,Shik} = 433\,\mu\text{M}$ $K_{i,DHS} = 44.8\,\mu\text{M}$
AroK	$K_f = 75\,\text{s}^{-1}$, $K_{m,Shik} = 28\,\mu\text{M}$ $K_{m,ATP} = 190\,\mu\text{M}$	$K_{m,ShikP} = 120\,\mu\text{M}$ $K_{m,ADP} = 90\,\mu\text{M}$

the concentration of the enzyme-bound intermediates do not change and the overall reaction rate remains essentially constant. This period of time occurs between the pre-quasi-steady state where the enzyme-bound intermediates build up to their quasi-steady state levels and when insufficient product has accumulated. Since the Aro pathway under investigation is a closed system and comprises four consecutive reactions it does not achieve steady state. However, under certain conditions, a quasi-steady state could occur, when the rates of change in concentration of each substrate and intermediate are equal and linear over a given period of time.

The results of numerical simulation (Fig. 14) of Eq. (18) show that transient concentration dependencies are approximately linear between 1000 and 3000 seconds. Below 1000 seconds, lag phases for all intermediates are analogous to a pre-quasi-steady state phase. Above 3000 seconds there is an accumulation of product that lowers the overall rate. Changes in the concentrations of all substrates, intermediates, and products are linear in the region around 2000 seconds. By setting the initial concentrations of DAHP, ShikP,

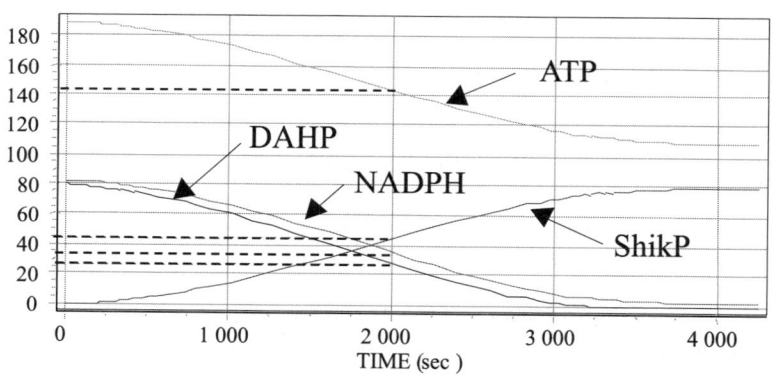

Fig. 14. Time dependencies of DAHP, ShikP, NADPH, and ATP calculated from the kinetic model. At time equal to 2000 s., the values of these metabolite concentrations are: DAHP=27 μM, NADPH=35 μM, ShikP=45 μM, ATP=144 μM

NADPH, NADP$^+$, ATP, and ADP (the "boundary intermediates") to their level at 2000 seconds we can set up a quasi-steady state situation. The true pre-quasi-steady state will be short and can be disregarded, because of the short time needed for the enzyme-bound intermediates to accumulate. The quasi-steady state rate around 2000 seconds is called the "quasi-stationary flux", or more simply just the "flux". To estimate the contribution of each individual enzyme controlling the flux under these substrate/intermediate concentrations, we investigated the sensitivity of the flux with respect to variations in the enzyme concentrations, and variations in the DAHP concentration. Concentrations of reactants/products were set to their quasi-steady state level as follows: [ShikP] at $45\,\mu$M, [NADPH] at $35\,\mu$M, [NADP] at $47\,\mu$M, and [NAD] at $10\,\mu$M. Each enzyme concentration was varied while the concentration of the other three enzymes remained fixed. Each plot in Fig. 15 shows the depen-

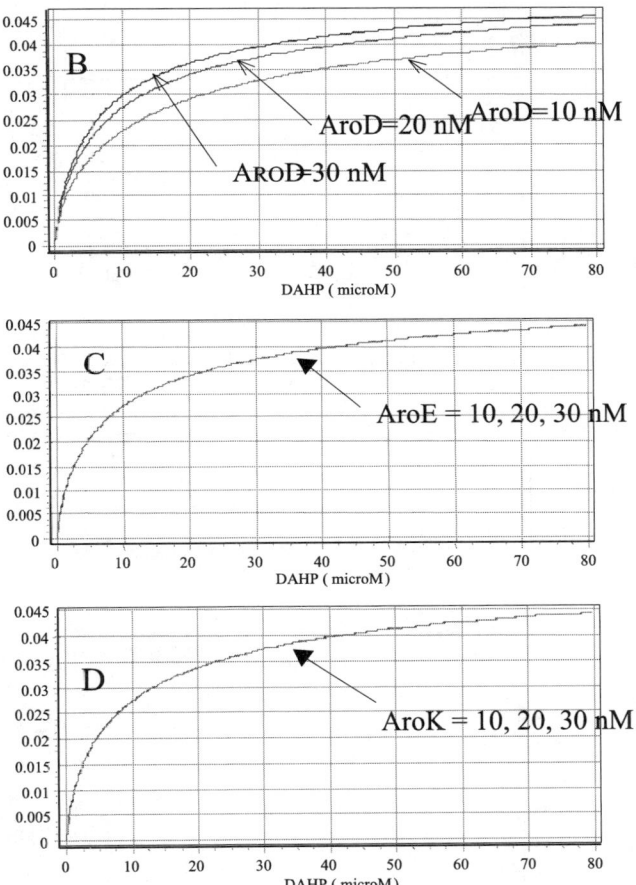

Fig. 15. Dependencies of quasi-steady state flux on initial concentration of DAHP: (A) AroD=AroE=AroK=20 nM, AroB is varied; (B) AroB=AroE=AroK=20 nM, AroD is varied; (C) AroD=AroB=AroK=20 nM, AroE is varied; (D) AroD=AroE=AroB=20 nM, AroK is varied

dency of the flux on the initial DAHP concentration. These curves are broadly hyperbolic, where the rate of flux is most sensitive at low DAHP concentrations, and gradually become less sensitive as the concentration of DAHP is increased. When concentrations of all enzymes are equal to 20 nM the flux is largely controlled by AroB ($\sim$70%) (Fig. 15A) and to a lesser extent by AroE ($\sim$30%) (Fig. 15B), whereas AroE and AroK exert no control (Fig. 15 C,D).

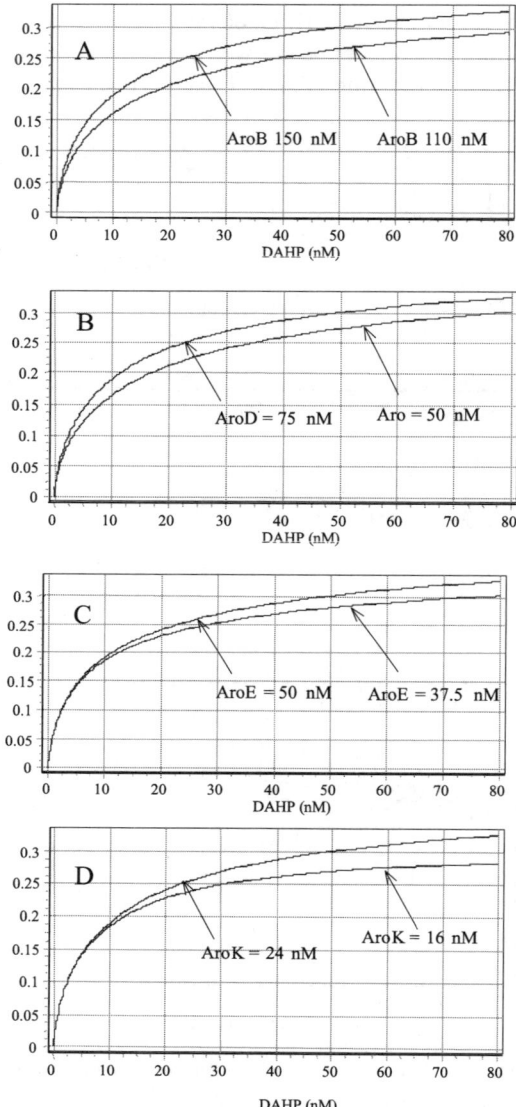

Fig. 16. Dependencies of quasi-steady state flux on initial concentration of DAHP:
(A) AroD=75 nM, AroE=50 nM, AroK=24 nM, AroB is varied; (B) AroE=50 nM,
AroK=24 nM, AroB=150 nM, AroD is varied; (C) AroD=75 nM, AroB=150 nM,
AroK=24 nM, AroE is varied; (D) AroD=75 nM, AroE=50 nM, AroB=150 nM,
AroK is varied

To shift control away from AroB and to redistribute control of the flux among all the enzymes we found that enzyme concentrations should be in following ratio $B : D : E : K = 2 : 1 : 0.75 : 0.32$. Indeed, as shown in Fig. 16, the flux becomes equally sensitive to variation in any enzyme concentration when the concentrations are: [B] 150 nM; [D] 75 nM; [E] 50 nM; [K] 24 nM. This ratio of enzyme concentrations defines the ideal conditions for inhibitor screening. The equal control of the flux among the enzymes means there is equal sensitivity to inhibition. Equal sensitivity will allow us to identify the best inhibitors of the pathway, the main and secondary targets, and the strength and type of inhibition.

5 Model of Mitochondrial Energy Metabolism. Prevention of Salicylate Hepatotoxic Effects

One of the most acute pharmacological problems is the investigation and mitigation of negative side-effects of drugs. The adverse effects of acetylsalicylic acid (aspirin) can be divided into following groups:

1. irritation of the gastric mucous membrane;
2. effects caused by the reduced biosynthesis of prostaglandins: ulcerogenic effects, kidney damage, increased blood pressure;
3. asthma and other allergic reactions; and
4. hemato- and hepatotoxicity [19], [20], [21].

Whereas there are some achievements [19] in correcting the side effects of the first three types of adverse effects, the more serious toxic effects are not so easily overcome. This is often a reason for the attrition of drugs during development. We would like to propose that kinetic modelling can be used to target some aspects of the hepatotoxicity problem.

It is known that utilization of salicylates in the liver can inhibit the beta-oxidation of fatty acids [22], decrease the pool of coenzyme A (CoA) [23], inhibit succinate dehydrogenase and alpha-ketoglutarate dehydrogenase [24], and increase the permeability of the inner mitochondrial membrane, thereby decreasing the proton motive force [25]. One reason for liver injury during aspirin therapy is disturbance of the hepatocyte's energy metabolism, particularly in the Krebs cycle. We have developed a kinetic model of the Krebs cycle to understand which aspirin modes-of-action are critical for hepatocyte insult and to estimate their contributions to changes in global regulatory properties of the mitochondrial energy metabolism. Using this model we have studied the influences of salicylates on the quasi-steady state flux in the Krebs cycle and examined possible ways to prevent such changes.

5.1 Kinetic Model of the Krebs Cycle

Traditionally, the Krebs cycle is described by a sequence of nine reactions resulting in the formation of oxaloacetate from citrate through cis-aconitate, isocitrate, ketoglutarate, succinyl-coenzyme A, succinate, fumarate, and malate. The cycle is closed because of the condensation of oxaloacetate with acetyl-coenzyme A and the formation of citrate. The cycle is:

1. Acetyl-coenzyme A is formed in the reactions of pyruvate, fatty acid, or amino acid oxidation. The slowest enzymes – citrate synthase and isocitrate dehydrogenase – determine the overall rate of the cycle.

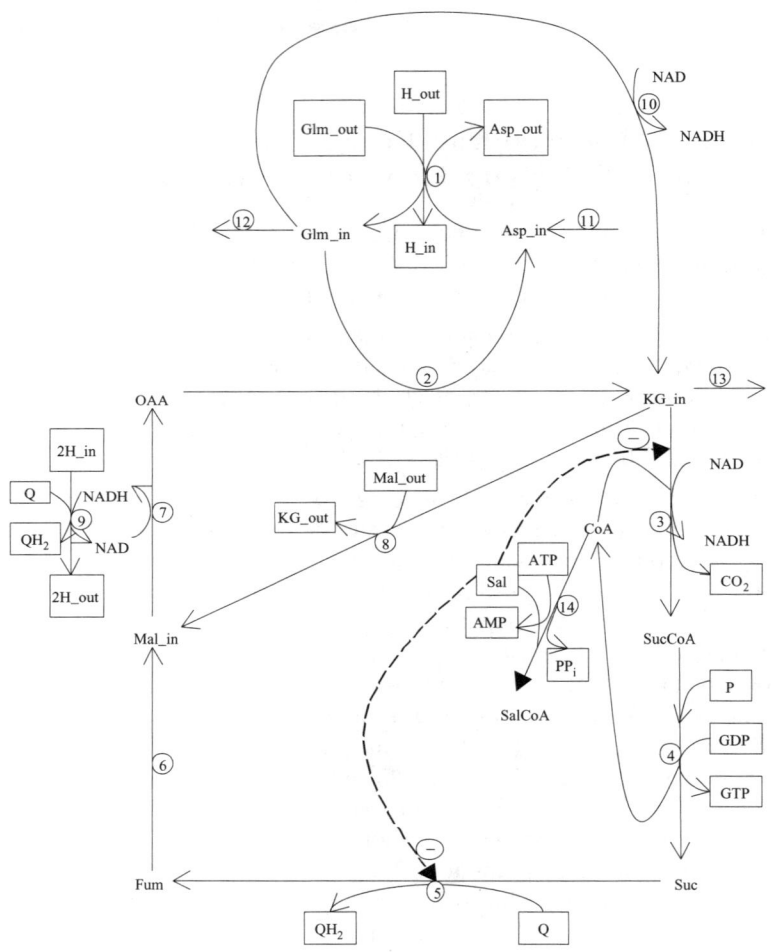

Fig. 17. Kinetic scheme of Krebs cycle and possible influences of salicylates (dashed lines). All processes are identified by circled numbers: 1. aspartate-glutamate carrier; 2. aspartate aminotransferase; 3. ketoglutarate dehydrogenase; 4. succinate thiokinase; 5. succinate dehydrogenase; 6. fumarase; 7. malate dehydrogenase; 8. dicarboxylate carrier; 9. complex I; 10. glutamate dehydrogenase; 11. aspartate influx; 12. glutamate influx; 13. ketoglutarate outflux; 14. salicylate-coenzyme A ligase

2. It was found that there is an alternative way to avoid such slow reactions [26]. That is, a trans-amination of glutamate and oxaloacetate with the subsequent formation of ketoglutarate (Fig. 17). In this case glutamate (Glu) is the carbon atom source.

3. Glu is transported to the mitochondrial matrix by the aspartate–glutamate carrier, AGC (reaction 1 in Fig. 17), which exchanges external glutamate, Glu_{out}, with an internal aspartate, Asp_{in}, carrying a proton from the inter-membrane space of the mitochondria to the matrix.

4. Glutamate donates its amino group to oxaloacetate, OAA, to form α-ketoglutarate, KG. This reaction (reaction 2 in Fig. 17) is catalyzed by aspartate aminotransferase, AspAT.

5. The next reaction, an oxidative decarboxylation of KG to succinyl-coenzyme A, SucCoA (reaction 3), is catalyzed by α-ketoglutarate dehydrogenase, KGDH.

6. KG may also be exchanged with external malate, Mal_{out} (reaction 8), catalyzed by dicarboxylate carrier.

7. SucCoA is broken up to form succinate, Suc (reaction 4). This reaction is associated with phosphorylation of GDP and catalyzed by succinate thiokinase, STK.

8. Suc is oxidized to fumarate, Fum, with succinate dehydrogenase, SDH (reaction 5).

9. Fum is hydrated to Mal in reaction 6, catalyzed by fumarase, FUM.

10. Oxidation of Mal to OAA (reaction 7) catalyzed by malate dehydrogenase, MDH, closes the Krebs cycle.

11. The oxidation of reduced pyridine nucleotides (reaction 9) formed in dehydrogenase reactions is catalyzed by NADH-coenzyme Q-reductase (complex I) – the first enzyme of the respiratory chain.

The model also includes a glutamate dehydrogenase, GDH (reaction 10), and reactions connecting the Krebs cycle with amino acid metabolism. These are effluxes from Glu_{in} and KG_{in} and influx to Asp_{in} (reactions 11, 12, 13), which determine the anabolic function of the cycle. In mitochondria, the Glu and KG are involved in catabolism of glutamine, proline, and arginine. Asp is formed from asparagine in the asparaginase reaction.

Glu_{in}, Asp_{in}, OAA, KG_{in}, SucCoA, CoA, Suc, Fum, Mal_{in}, NAD, and NADH are the model variables. Concentrations of Glu_{out}, Asp_{out}, H_{out}, H_{in}, Mal_{out}, KG_{out}, CO_2, P, ATP, ADP, Ca^{2+}, GTP, GDP, Q, and QH_2 are model parameters. There are two conservation laws in the system:

$$NAD + NADH = const \text{ (conservation of pyridine nucleotides)}$$
$$CoA + SucCoA = const \text{ (conservation of CoA)} \tag{19}$$

The model is described by the following system of differential equations:

$$\frac{dGlm\_in}{dt} = V_1 - V_2 - V_{10} - V_{12} \qquad \frac{dAsp\_in}{dt} = V_2 - V_1 + V_{11}$$

$$\frac{dOAA}{dt} = V_7 - V_2 \qquad \frac{dKG\_in}{dt} = V_2 - V_3 - V_8 + V_{10} - V_{13}$$

$$\frac{dSucCoA}{dt} = V_3 - V_4 \qquad \frac{dSuc}{dt} = V_4 - V_5 \qquad \frac{dFum}{dt} = V_5 - V_6 \qquad (20)$$

$$\frac{dMal\_in}{dt} = V_6 - V_7 + V_8 \qquad \frac{dCoA}{dt} = V_4 - V_3$$

$$\frac{dNAD}{dt} = -V_3 - V_7 + V_9 - V_{10} \qquad \frac{dNADH}{dt} = V_3 + V_7 - V_9 + V_{10}$$

5.2 Description of the Krebs Cycle Kinetics

To derive the enzymatic rate laws we drew upon kinetic data from the literature where, in most cases, there was information about probable mechanisms and their kinetic characteristics such as Michaelis, inhibition, or equilibrium constants, and specific activities. Values of these literature parameters are presented in Table 3.

Aspartate–glutamate carrier (AGC). The inner mitochondrial membrane is not permeable to glutamate. Its protonated form can be transferred to the matrix only by a membrane carrier in exchange for an aspartate. The rate equation for AGC was derived by assuming that it functions through a random Ter-Ter mechanism [27] (according to the Cleland classification [28]):

$$V_1 = \frac{AGC \cdot \left(k_1 k_2 \frac{Glu_{out}}{K_d^{Glu_{out}}} \frac{Asp_{in}}{K_d^{Asp_{in}}} \frac{H_{out}}{K_d^{H_{out}}} - k_{-1} k_{-2} \frac{Glu_{in}}{K_d^{Glu_{in}}} \frac{Asp_{out}}{K_d^{Asp_{out}}} \frac{H_{in}}{K_d^{H_{in}}}\right)}{\left(k_1 \frac{Glu_{out}}{K_d^{Glu_{out}}} \frac{Asp_{in}}{K_d^{Asp_{in}}} \frac{H_{out}}{K_d^{H_{out}}} + k_{-2}\right)\left(1 + \frac{H_{in}}{K_d^{H_{in}}}\right)\left(1 + \frac{Glu_{in}}{K_d^{Glu_{in}}}\right)\left(1 + \frac{Asp_{out}}{K_d^{Asp_{out}}}\right) + \left(k_1 \frac{Glu_{in}}{K_d^{Glu_{in}}} \frac{Asp_{out}}{K_d^{Asp_{out}}} \frac{H_{in}}{K_d^{H_{in}}} + k_2\right)\left(1 + \frac{H_{out}}{K_d^{H_{out}}}\right)\left(1 + \frac{Glu_{out}}{K_d^{Glu_{out}}}\right)\left(1 + \frac{Asp_{in}}{K_d^{Asp_{in}}}\right)} \qquad (21)$$

where

$$K_{eq}^1 = K_{eq0}^1 * e^{\delta_2 \frac{\Delta\psi}{RT/F}}, \quad K_d^{H_{in}} = K_{d0}^{H_{in}} * e^{\delta_1 \frac{\Delta\psi}{RT/F}}, \quad K_d^{H_{out}} = K_{d0}^{H_{out}} * e^{-\delta_3 \frac{\Delta\psi}{RT/F}},$$

$$k_1 = k_1^0 * e^{(1-\alpha)\delta_2 \frac{\Delta\psi}{RT/F}}, \quad k_{-1} = k_{-1}^0 * e^{-\alpha\delta_2 \frac{\Delta\psi}{RT/F}}$$

and $\Delta\psi > 0$ is the trans-membrane potential; T is the absolute temperature; R is the universal gas constant; F is Faraday's constant; δ_i is the part of the potential consumed by the i-th stage; α is the part of the potential that influences reverse reaction. Some parameter values can be taken from the literature: K_{eq} [29], K_{m,Glu_out}, K_{m,Glu_in}, K_{m,Asp_out}, and K_{m,Asp_in} [27]. The remaining parameters, the concentration of AGC, K_{m,H_out}, K_{m,H_in},

Table 3. Kinetic parameters of kinetic model of Krebs cycle

Enzyme	Kinetic parameters known from literature	Kinetic parameters estimated via fitting to experimental data
AGC	K_{m,Glu_out}=0.25 mM [27] K_{m,Glu_in}=3 mM [27] K_{m,Asp_out}=0.12 mM [27] K_{m,Asp_in}=2.8 mM [27] K_{eq}=1 [27]	AGC=0.2 mM [30] K_{m,H_out}=0.0004 mM K_{m,H_in}=0.0001 mM K_{m,Glu_out}=15 mM $k_{1,0}=10^5$ min^{-1} $k_{-1,0}=10^4$ min^{-1} $k_2=10^5$ min^{-1} $k_{-2}=10^4$ min^{-1}
AspAT	k_1=42000 min^{-1}, k_{-1}=36000 min^{-1} [31] k_f=22400 min^{-1}, k_r=17217 min^{-1} [31] K_{m,Glu_in}=38 mM [31] K_{m,Asp_in}=4.5 mM [31] $K_{m,OAA}$=0.035 mM [31] K_{m,KG_in}=1.3 mM [31]	AspAT=0.02 mM [30]
KGDH	$K_{m,KG}$=1.5 mM [32] $K_{m,CoA}$=0.02 mM [32] $K_{m,NAD}$=0.075 mM [32] $K_{i,ATP}$=0.1 mM [32] $K_{i,ADP}$=0.3 mM [32] $K_{i,Ca}$=0.006 mM [32] $K_{m,SucCoA}$=0.095 mM [32] $K_{m,NADH}$=0.024 mM [32]	V_f=24 mM/min [30]
STK	$K_{m,Suc}$=0.4–0.8 mM [33] $K_{m,SucCoA}$=0.01–0.06 mM [33] $K_{m,GDP}$=0.002–0.008 mM [33] $K_{m,GTP}$=0.005–0.01 mM [33] $K_{m,P}$=0.2–0.7 mM [33] $K_{m,CoA}$=0.005–0.02 mM [33]	$K_{m,Suc}$=0.4 mM $K_{m,CoA}$=0.005 mM $K_{m,SucCoA}$=0.04 mM $K_{m,GDP}$=7.4e-3 mM $K_{m,GTP}$=0.01 mM $K_{m,P}$=0.7 mM STK=0.02 mM [30] k_1=1.28e8 min^{-1} k_{-1}=1.49e3 min^{-1} k_2=16.29 min^{-1} k_{-2}=1 min^{-1}
SDH	k_i=1200 min^{-1}, k_{-i}=0.02 min^{-1} [36] k_f=10000 min^{-1}, k_r=102 min^{-1} [36] SDH=0.05 mM [36] $K_{m,Suc}$=0.4 mM [35] $K_{m,Q}$=0.002 mM [37] $K_{m,QH2}$=0.0045 mM [37] K_{m,Mal_in}=2.2 mM [38]	$K_{m,Fum}$=0.1 mM k_2=10000 min^{-1} k_{-2}=102 min^{-1} SDH=0.1 mM [30]
FUM	$K_{m,Fum}$=0.0026, K_{m,Mal_in}=0.009 [38] k_f=138000 min^{-1} [38] k_r=102000 min^{-1} [38]	FUM=0.009 mM [30]
MDH	k_f=5.4e5 min^{-1}, k_r=8.6e3 min^{-1} [39] MDH=9.03e-4 mM [39] $K_{m,OAA}$=0.0795 mM [39] $K_{m,Mal}$=0.386 mM [39] $K_{m,NAD}$=0.0599 mM [39] $K_{m,NADH}$=0.26 mM [39] $K_{i,OAA}$=0.0055 mM [39] $K_{i,Mal}$=0.36 mM [39] $K_{i,NAD}$=1.1 mM [39] $K_{i,NADH}$=0.0136 [39], K_{eq}=8000 [39]	
KMC	k_f=325 min^{-1}, k_r=309 min^{-1} [40] K_{m,Mal_out}= 1.36 mM [40] K_{m,Mal_in}=0.71 mM [40] K_{m,KG_in}=0.17 mM [40] K_{m,KG_out}=0.31 mM [40]	k_1=1000 min^{-1} k_{-1}=918.4 min^{-1} k_2=481.5 min^{-1} k_{-2}=465.7 min^{-1} KMC=0.2 mM [30]
Complex I	E_0=0.027 mM [41] $K_{m,NADH}$=0.0087 mM [41] $K_{m,Q}$=11 mM [41]	$K_{m,NAD}$=1 mM [30] $K_{m,QH2}$=100 mM $K_{m,H\_out}=10^{-4}$ mM $K_{m,H\_in}=10^{-6}$ mM $k_{1,-1}=10^5$ min^{-1} $k_{2,-2}=10^5$ min^{-1} $k_{3,-3}=10^5$ min^{-1}
GDH	k_f=25 min^{-1} [45], k_r=8889 min^{-1} [44] K_{m,Glu_in}=1 mM, K_{m,KG_in}=1 mM [46] $K_{m,NAD}$=0.05 mM [46] $K_{m,NADH}$=0.01 mM [46] $K_{m,NH4}$=3.2 mM [43]	GDH=0.024 mM [30]
SCL	$k_{cat,14}$=1325 min^{-1} [48]	

$k_{1,0}$, $k_{-1,0}$, k_2, and k_{-2} are unknown and can be estimated from fitting the model to experimental data [30] (see below).

Aspartate aminotransferase (AspAT) catalyzes the trans-amination of Glu and OAA with the formation of KG and Asp. AspAT kinetics is best described using a ping pong Bi-Bi mechanism [31] with the resulting rate equation:

$$V_2 = AspAT \cdot \frac{k_f^2 \frac{OAA}{K_{OAA}} \frac{Glu_{in}}{K_{Glu}} - k_r^2 \frac{Asp_{in}}{K_{Asp}} \frac{KG_{in}}{K_{KG}}}{\left(k_f \frac{Glu_{in}}{K_{Glu}} + k_r \frac{Asp_{in}}{K_{Asp}}\right)\left(1 + \frac{k_r}{k_{-1}} \frac{KG_{in}}{K_{KG}} + \frac{k_f}{\frac{k_1 * k_f}{k_1 - k_f}} \frac{OAA}{K_{OAA}}\right) + \left(k_r \frac{KG_{in}}{K_{KG}} + k_f \frac{OAA}{K_{OAA}}\right)\left(1 + \frac{k_f}{k_1} \frac{Glu_{in}}{K_{Glu}} + \frac{k_r}{\frac{k_{-1} * k_r}{k_{-1} - k_r}} \frac{Asp_{in}}{K_{Asp}}\right)} \tag{22}$$

Parameters which were found in the literature [31]: K_{m,Asp_in}, K_{m,KG_in}, $K_{m,OAA}$, k_r, K_{m,Glu_in}, k_f, k_1, and k_{-1} (see Table 3).

α-ketoglutarate dehydrogenase (KGDH) catalyzes the irreversible reaction of oxidative decarboxylation of KG with the formation of SucCoA and the reduction of NAD. The enzyme is described according to a ping pong Ter-Ter mechanism [32]. The influence of activators (ADP and Ca^{2+}) and inhibitors (ATP, NADH, SucCoA) were taken into account. It is assumed that the enzyme exists in both active and inactive forms. ADP or Ca^{2+} transforms the enzyme to its active form, whereas ATP transforms the enzyme to its inactive form. The composite rate equation can be written:

$$V_3 = \frac{V_f \frac{KG}{K_m^{KG}} \frac{CoA}{K_m^{CoA}} \frac{NAD}{K_m^{NAD}}}{\frac{CoA}{K_m^{CoA}} \frac{NAD}{K_m^{NAD}}\left(\frac{KG}{K_m^{KG}} + \frac{1 + \frac{ATP}{K_i^{ATP}}}{1 + \frac{ADP}{K_i^{ADP}} + \frac{Ca}{K_i^{Ca}}}\right) + \frac{KG}{K_m^{KG}}\left(\frac{CoA}{K_m^{CoA}} + \frac{NAD}{K_m^{NAD}}\right)\left(1 + \frac{NADH}{K_m^{NADH}} + \frac{SucCoA}{K_m^{SucCoA}}\right)} \tag{23}$$

Parameters known from the literature were $K_{m,NAD}$, $K_{m,NADH}$, $K_{m,SucCoA}$, $K_{m,CoA}$, $K_{m,KG}$, $K_{i,ATP}$, $K_{i,ADP}$, and $K_{i,Ca}$ [32]. V_f is an unknown parameter which can be estimated from fitting our model to experimental data [30] (see below).

Succinate thiokinase (STK). The rate is described according to a random Bi-Uni-Uni-Bi mechanism [33] with a rate equation:

$$V = \frac{STK \cdot \left(k_1 k_2 \frac{SucCoA}{K_d^{SucCoA}} \frac{GDP}{K_d^{GDP}} \frac{P}{K_d^P} - k_{-1} k_{-2} \frac{Suc}{K_d^{Suc}} \frac{GTP}{K_d^{GTP}} \frac{CoA}{K_d^{CoA}}\right)}{\left(k_1 \frac{SucCoA}{K_d^{SucCoA}} \frac{GDP}{K_d^{GDP}} + k_{-2} \frac{GTP}{K_d^{GTP}} \frac{CoA}{K_d^{CoA}}\right)\left(1 + \frac{Suc}{K_d^{Suc}} + \frac{P}{K_d^P}\right) + \left(1 + \frac{GDP}{K_d^{GDP}} + \frac{GTP}{K_d^{GTP}}\right)\left(1 + \frac{CoA}{K_d^{CoA}} + \frac{SucCoA}{K_d^{SucCoA}}\right)\left(k_2 \frac{P}{K_d^P} + k_{-1} \frac{Suc}{K_d^{Suc}}\right)} \tag{24}$$

Parameters known from the literature were $K_{m,Suc}$, $K_{m,CoA}$, $K_{m,SucCoA}$, $K_{m,P}$, $K_{m,GDP}$, and $K_{m,GTP}$ [33] (see Table 3). Values of k_1, k_{-1}, and k_2

were estimated from fitting the rate equation to the experimental data published in [33]. Figure 18 shows two series of experimental points and theoretical curves obtained from Eq. (24).

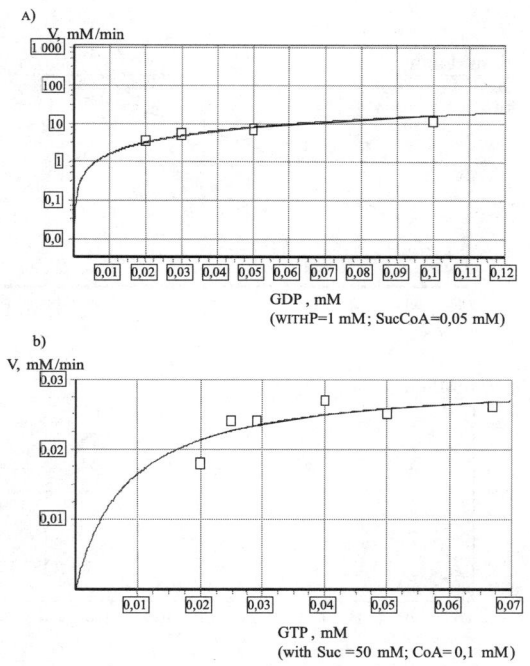

Fig. 18. Dependencies of forward (a) and reverse (b) initial rate of succinate thiokinase on GDP (a) and ATP (b) concentrations. Squares designate experimental data taken from [33], solid line represents modelling results

Succinate dehydrogenase (SDH). The rate was described according to a random Bi-Bi mechanism with the formation of two dead-end complexes taking into account that OAA and Mal inhibit the enzyme in a complex manner:

$$V_5 = SDH \cdot \frac{\left(k_f \frac{Suc}{K_d^{Suc}} \frac{Q}{K_d^Q} - k_r \frac{Fum}{K_d^{Fum}} \frac{QH_2}{K_d^{QH_2}} \right) \left(k_{-i} + k_{-2} \frac{\frac{QH_2}{K_d^{QH_2}}}{1 + \frac{Q}{K_d^Q} + \frac{QH_2}{K_d^{QH_2}}} \right)}{\left(k_2 \frac{Mal_{in}}{K_d^{Mal_{in}}} \frac{Q}{K_d^Q} + k_i OAA \left(1 + \frac{Q}{K_d^Q} + \frac{QH_2}{K_d^{QH_2}} \right) \right) + \atop + \left(1 + \frac{Suc}{K_d^{Suc}} + \frac{Fum}{K_d^{Fum}} + \frac{Mal_{in}}{K_d^{Mal_{in}}} \right) \times \atop \times \left(k_{-i} \left(1 + \frac{Q}{K_d^Q} + \frac{QH_2}{K_d^{QH_2}} \right) + k_{-2} \frac{QH_2}{K_d^{QH_2}} \right)}$$ (25)

Parameters known from the literature [34–37] are k_i, k_{-i}, SDH, $K_{m,Suc}$, $K_{m,Q}$, $K_{m,QH2}$, and K_{m,Mal_in}. Parameters SDH, $K_{m,Fum}$, k_f, k_r, k_2, and k_{-2} are unknown and can be estimated from fitting our model to experimental data [30] (see below).

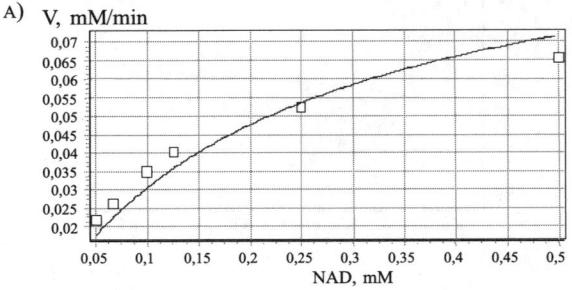

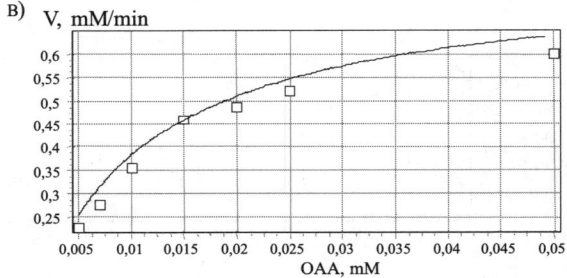

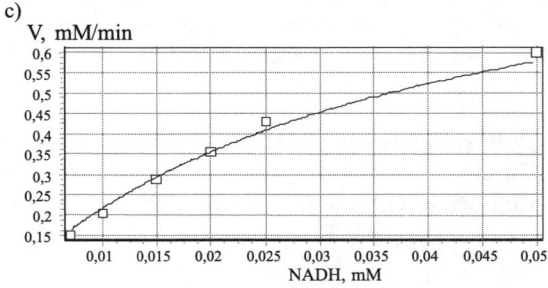

Fig. 19. Dependencies of initial rate of malate dehydrogenase on NAD (a), OAA (b), NADH (c). Solid line represents modelling results and squares designate experimental data [39] measured under following conditions: (a) Mal=1 mM; OAA=NADH=0; (b) NADH=0.025 mM; NAD=Mal=0; (c) OAA=0.02 mM; Mal=1.5 mM, NAD=0.

Fumarase (FUM). The transformation of Fum to Mal was described as a Uni-Uni mechanism [38]:

$$V_6 = FUM \cdot \frac{k_f \cdot \dfrac{Fum}{K_m^{Fum}} - k_r \cdot \dfrac{Mal_{in}}{K_m^{Mal}}}{1 + \dfrac{Fum}{K_m^{Fum}} + \dfrac{Mal_{in}}{K_m^{Mal}}} \tag{26}$$

All the parameters were taken from the literature [38] (see Table 3).

Malate dehydrogenase (MDH). The rate was described according to an ordered Bi-Bi mechanism with an NAD complex isomerization [39]:

$$V_7 = MDH \cdot \frac{k_f \cdot k_r \left(\frac{Mal_{in} \cdot NAD}{K_{eq}} - NADH \cdot OAA \right)}{\begin{array}{c} K_{i,NADH} \cdot K_{OAA} \cdot k_r + K_{OAA} \cdot k_r \cdot NADH + \\ + K_{NADH} \cdot k_r \cdot OAA + k_r \cdot NADH \cdot OAA + \\ + \frac{K_{NAD} \cdot k_f \cdot Mal_{in}}{K_{eq}} + \frac{K_{Mal} \cdot k_f \cdot NAD}{K_{eq}} + \frac{k_f \cdot Mal_{in} \cdot NAD}{K_{eq}} + \\ + \frac{K_{NAD} \cdot k_f \cdot NADH \cdot Mal_{in}}{K_{i,NADH} \cdot K_{eq}} + \frac{K_{NADH} \cdot k_r \cdot OAA \cdot NAD}{K_{iq}} + \\ + \frac{k_r \cdot NADH \cdot OAA \cdot Mal_{in}}{K_{i,Mal}} + \frac{k_f \cdot OAA \cdot Mal_{in} \cdot NAD}{K_{i,OAA} \cdot K_{eq}} \end{array}} \tag{27}$$

Parameters known from the literature [39] were k_f, k_r, MDH, $K_{m,OAA}$, $K_{m,Mal}$, $K_{m,NAD}$, $K_{m,NADH}$, $K_{i,OAA}$, $K_{i,Mal}$, $K_{i,NAD}$, $K_{i,NADH}$, and K_{eq} (see Table 3). Figure 19 shows three series of experimental points and model curves obtained from Eq. (27).

Dicarboxylate carrier (KMC) catalyzes the exchange of external Mal for intramotochondrial KG with a rate law described according to a random Bi-Bi mechanism [40]:

$$V = KMC \frac{V_f^2 \frac{k_2}{k_1} \frac{KG_{in}}{K_m^{KG_{in}}} \frac{Mal_{out}}{K_m^{Mal_{out}}} - V_r^2 \frac{k_{-2}}{k_{-1}} \frac{KG_{out}}{K_m^{KG_{out}}} \frac{Mal_{in}}{K_m^{Mal_{in}}}}{\begin{array}{c} \left(\frac{V_f^2}{k_1} \frac{KG_{in}}{K_m^{KG_{in}}} \frac{Mal_{out}}{K_m^{Mal_{out}}} + k_{-2} \right) \left(1 + \frac{V_r}{k_{-1}} \frac{KG_{out}}{K_m^{KG_{out}}} \right) \times \\ \times \left(1 + \frac{V_r}{k_{-1}} \frac{Mal_{in}}{K_m^{Mal_{in}}} \right) + \left(k_2 + \frac{V_r^2}{k_{-1}} \frac{KG_{out}}{K_m^{KG_{out}}} \frac{Mal_{in}}{K_m^{Mal_{in}}} \right) \times \\ \times \left(1 + \frac{V_f}{k_1} \frac{KG_{in}}{K_m^{KG_{in}}} \right) \left(1 + \frac{V_f}{k_1} \frac{Mal_{out}}{K_m^{Mal_{out}}} \right) \end{array}} \tag{28}$$

where k_f, k_r, K_{m,Mal_out}, K_{m,Mal_in}, K_{m,KG_in}, and K_{m,KG_out} are known from the literature [40]. KMC, k_1, k_1, k_2, and k_{-2} are unknown parameters which can be estimated from fitting to experimental data [30] (see below).

Complex I has a rate best described by a ping pong Ter-Ter mechanism [41].

$$V = E_0 \frac{\begin{array}{c}(k_{-1}\frac{NAD}{K_d^{NAD}} + k_2 \frac{H_{in}^4}{K_d^{4H_{in}}}) \times \\ \times (k_1 k_2 k_3 \frac{NADH}{K_d^{NADH}} \frac{H_{in}^4}{K_d^{4H_{in}}} \frac{Q}{K_d^Q} - k_{-1}k_{-2}k_{-3}\frac{NAD}{K_d^{NAD}} \frac{H_{out}^4}{K_d^{4H_{out}}} \frac{QH_2}{K_d^{QH_2}})\end{array}}{\begin{array}{c}\left(k_{-1}k_3 \frac{NAD}{K_d^{NAD}} \frac{Q}{K_d^Q} + k_2 k_3 \frac{H_{in}^4}{K_d^{4H_{in}}} \frac{Q}{K_d^Q} + k_{-1}k_{-2}\frac{NAD}{K_d^{NAD}} \frac{H_{out}^4}{K_d^{4H_{out}}}\right) \times \\ \times \left(k_1 \frac{NADH}{K_d^{NADH}}\left(1 + \frac{NAD}{K_d^{NAD}} + \frac{H_{in}^4}{K_d^{4H_{in}}}\right) + \right. \\ + \left(k_{-1}\frac{NAD}{K_d^{NAD}} + k_2 \frac{H_{in}^4}{K_d^{4H_{in}}}\right)\left(1 + \frac{NADH}{K_d^{NADH}} + \frac{QH_2}{K_d^{QH_2}}\right)\left.\right) \\ + \left(k_1 k_2 \frac{NADH}{K_d^{NADH}} \frac{H_{in}^4}{K_d^{4H_{in}}} \frac{Q}{K_d^Q} + k_{-1}k_{-3}\frac{NAD}{K_d^{NAD}} \frac{QH_2}{K_d^{QH_2}} + \right. \\ + k_2 k_{-3} \frac{QH_2}{K_d^{QH_2}} \frac{H_{in}^4}{K_d^{4H_{in}}}\right)\left(k_{-2}\frac{H_{out}^4}{K_d^{4H_{out}}}(1 + \frac{NAD}{K_d^{NAD}} + \frac{H_{in}^4}{K_d^{4H_{in}}}) + \right. \\ + \left(k_{-1}\frac{NAD}{K_d^{NAD}} + k_2 \frac{H_{in}^4}{K_d^{4H_{in}}}\right)\left(1 + \frac{Q}{K_d^Q} + \frac{H_{in}^4}{K_d^{4H_{in}}}\right)\left.\right)\end{array}} \quad (29)$$

The dependence of Complex I on the potential $\Delta\psi$ is given by classical thermodynamic dependencies:

$$K_{eq}^2 = K_{eq0}^2 \cdot e^{-\delta_2 \frac{\Delta\psi}{RT/F}}, \quad K_d^{H_{in}} = K_{d0}^{H_{in}} \cdot e^{\delta_1 \frac{\Delta\psi}{RT/F}},$$

$$K_d^{H_{out}} = K_{d0}^{H_{out}} \cdot e^{-\delta_3 \frac{\Delta\psi}{RT/F}}, \quad (30)$$

$$k_2 = k_2^0 \cdot e^{-\alpha\delta_2 \frac{\Delta\psi}{RT/F}}, \quad k_{-2} = k_{-2}^0 \cdot e^{(1-\alpha)\delta_2 \frac{\Delta\psi}{RT/F}}$$

The enzyme concentration E_0, and parameters $K_{m,NADH}$ and $K_{m,Q}$ are known from the literature [41] (see Table 3). $K_{m,NAD}$, $K_{m,QH2}$, K_{m,H_out}, K_{m,H_in}, k_1, k_{-1}, k_2, k_{-2}, k_3, and k_{-3} are unknown parameters which can be estimated by fitting to experimental data [30] (see below).

Glutamate dehydrogenase (GDH) catalyzes the oxidative deamination of Glu according to a random Bi-Ter mechanism with the formation of two dead-end complexes [42–45]:

$$V = GDH\frac{k_f \frac{Glu}{K_m^{Glu}} \frac{NAD}{K_m^{NAD}} - k_r \frac{KG}{K_m^{KG}} \frac{NADH}{K_m^{NADH}} \frac{NH_4}{K_m^{NH_4}}}{\begin{array}{c}1 + \frac{NAD}{K_m^{NAD}} + \frac{Glu}{K_m^{Glu}} + \frac{Glu}{K_m^{Glu}} \frac{NAD}{K_m^{NAD}} + \frac{KG}{K_m^{KG}} \frac{NADH}{K_m^{NADH}} \frac{NH_4}{K_m^{NH_4}} + \\ + \frac{KG}{K_m^{KG}} \frac{NH_4}{K_m^{NH_4}} + \frac{KG}{K_m^{KG}} \frac{NADH}{K_m^{NADH}} + \frac{NADH}{K_m^{NADH}} \frac{NH_4}{K_m^{NH_4}} + \frac{KG}{K_m^{KG}} + \\ + \frac{NADH}{K_m^{NADH}} + \frac{NH_4}{K_m^{NH_4}} + \frac{KG}{K_m^{KG}} \frac{NAD}{K_m^{NAD}} + \frac{Glu}{K_m^{Glu}} \frac{NADH}{K_m^{NADH}}\end{array}} \quad (31)$$

GDH, k_f, k_r, K_{m,Glu_in}, $K_{m,NAD}$, K_{m,KG_in}, $K_{m,NADH}$, and $K_{m,NH4}$ (see Table 3) are known from the literature [42, 43, 45].

Biosynthetic effluxes and influxes (reactions 11–13 in Fig. 17) are described according to the mass action laws. Amino acid concentrations (S_1, S_2, S_3) are parameters of the model:

$$V_{11} = k_{14} \left(S_1 - \frac{1}{K_{eq}^{14}} Asp_{in} \right)$$

$$V_{12} = k_{12} \left(Glu_{in} - \frac{1}{K_{eq}^{12}} S_2 \right)$$

$$V_{13} = k_{13} \left(KG_{in} - \frac{1}{K_{eq}^{13}} S_3 \right)$$

We assume the values of the parameters of these equations are equal to: $S_1=1$, $S_2=1$, $S_3=1$, $k_{11}=20$ 1/min, $k_{12}=20$ 1/min, $k_{13}=20$ 1/min, $K_{eq,11}=1$, $K_{eq,12}=1$, and $K_{eq,13}=1$.

Estimation of Unknown Parameters Using *In Vivo* Experimental Data. All unknown parameters whose values cannot be estimated from *in vitro* data published in the literature can be determined by fitting the whole model of the Krebs cycle to experimental data where the glutamate consumption rate by mitochondrial suspensions has been measured [30]. These rates reflected the overall Krebs cycle functioning rate, so we were able to find the concentrations of mitochondrial enzymes and other unknown parameters. Model simulations resulted in the best description of *in vivo* experimental data (Fig. 20).

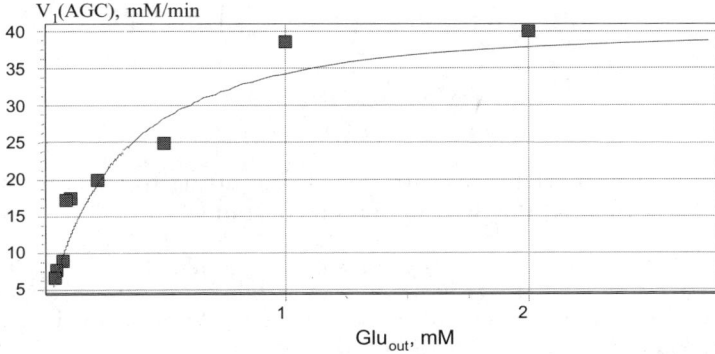

Fig. 20. Dependence of the steady state rate of Glu consumption on its concentration resulted from modelling (solid line) and measured in mitochondrial suspension (squares)

5.3 Results

Salicylates can cause liver failure by poisoning, by prolonged application in therapeutic doses, or by side-effects of other medications. This type of hepatotoxicity is an important challenge for medical research. Salicylates can inhibit the Krebs cycle in different ways (dashed arrows and reaction 14 in Fig. 17). To understand which of these effects are critical for hepatocyte viability and their contribution to the inhibition of the Krebs cycle, we have studied each mechanism separately.

1. It is known that aspirin is a competitive inhibitor of KGDH [24]. It prevents KG_{in} binding to the enzyme. To take this into account we changed the rate equation of KGDH by introducing non-linear competitive inhibition by KGDH:

$$V = \cfrac{V_f \frac{KG}{K_m^{KG}} \frac{CoA}{K_m^{CoA}} \frac{NAD}{K_m^{NAD}}}{\frac{CoA}{K_m^{CoA}} \frac{NAD}{K_m^{NAD}} \left(\frac{KG}{K_m^{KG}} + \frac{1 + \frac{ATP}{K_i^{ATP}} + \frac{Sal}{K_i^{Sal}}}{1 + \frac{ADP}{K_i^{ADP}} + \frac{Ca}{K_i^{Ca}}} \right) + \frac{KG}{K_m^{KG}} \left(\frac{CoA}{K_m^{CoA}} + \frac{NAD}{K_m^{NAD}} \right) \left(1 + \frac{NADH}{K_m^{NADH}} + \frac{SucCoA}{K_m^{SucCoA}} \right)} \tag{32}$$

Using experimental data [24] we estimated the inhibition constant by numerical simulations as $K_i^{Sal} = 0.05$ mM.

2. Because salicylates inhibit the succinate dehydrogenase [24] we modified the kinetic parameters of Eq. (25) as:

$$K_m^{Suc} = K_{m,0}^{Suc} \left(1 + \frac{Sal}{K_{im}^{Sal}} \right)$$

$$K_m^Q = K_{m,0}^Q \left(1 + \frac{Sal}{K_{im}^{Sal}} \right)$$

$$V_f^{SDH} = \frac{V_{f,0}^{SDH}}{\left(1 + \frac{Sal}{K_{iv}^{Sal}} \right)}$$

where Sal is the salicylate concentration. Using fitting to experimental data [24] we have estimated the SDH inhibition constants:

$$K_{im}^{Sal} = 2.5 \text{ mM}, \quad K_{iv}^{Sal} = 8 \text{ mM}$$

3. Salicylates decrease the pool of coenzyme A. Reaction 14 (Fig. 17), catalyzed by Sal-coenzymeA ligase (SCL), describes the formation of the SalCoA complex from Sal and CoA. This reaction consumes energy as ATP. Using the known values of the SCL molecular weight [46] and specific activity [47] we calculated its turnover number (k_{cat}). The rate of SCL was described by mass action:

$$V_{14} = k_{cat,14}(Sal \cdot CoA \cdot ATP - \frac{1}{K_{eq,14}} SalCoA \cdot AMP \cdot PP_i)$$

where $k_{cat,14}$=1325 1/min [24] and $K_{eq,14}$=300.

4. The uncoupling effect of Sal [25]. Since salicylates are weak acids they are able to uncouple mitochondrial oxidative phosphorylation thereby decreasing the electric potential across the inner mitochondrial membrane [48] and

slowing the rate of ATP production. This uncoupling effect manifests itself as concerted changes in the following parameters:

Without salicylates:	With salicylates:
$\Delta\psi=0.127$ V	$\Delta\psi=0$
$H_{in}=10^{-5}$ mM	$H_{in} = H_{out}=10^{-4}$ mM
$H_{out}=10^{-4}$mM	
$ATP=5$ mM	$ATP=0.1$ mM
$ADP=5$ mM	$ADP=9.9$ mM
$GTP=1$ mM	$GTP=0.1$ mM
$GDP=1$ mM	$GDP=1.9$ mM
$Q=19$ mM	$Q=1$ mM
$QH_2=1$ mM	$QH_2=19$ mM

Now that the salicylate effects are in the model, their inhibitory effects on the quasi-steady state fluxes of the Krebs cycle can be studied. Figure 21 shows the dependencies of these state fluxes with respect to glutamate consumption (reaction 1, Fig.17) and NADH oxidation (reaction 9, Fig.17) on the concentration of external malate. It is easy to see that the only uncoupling effect of salicylates results in a substantial decrease in flux through the Krebs cycle, whereas an inhibition of KGDH, SDH, or CoA sequestration does not influence the mitochondrial energy metabolism.

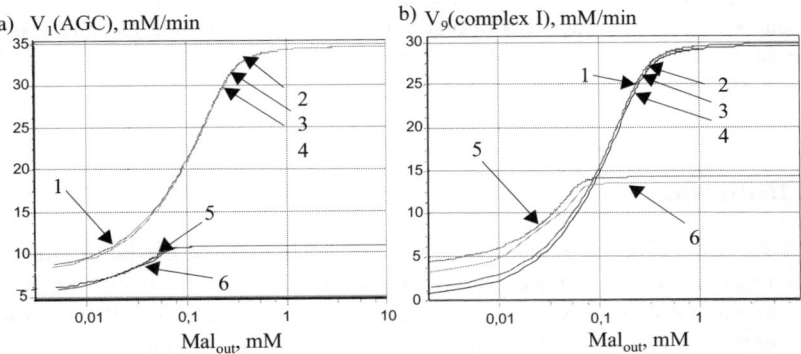

Fig. 21. Dependencies of steady state rates of Glu consumption (a) and NADH oxidation (b) on external malate concentration. Curve 1 results from the model that does not take into account any salicylate influences on the Krebs cycle. Curves 2, 3. 4, and 5 correspond to the case when separate effects of salicylates on the Krebs cycle are considered in the model: (2) CoA sequestration in reaction 14; (3) KGDH inhibition; (4) SDH inhibition; (5) encoupling effect of salicylates. Curve 6 results from the model taking into account all possible effects of salicylates

The next goal was to find out a way to prevent the inhibition of the Krebs cycle by salicylates. Figure 22 shows that an increase in the concentration of glutamine (curve c) and asparagine (curve d) results in a restoration of the Krebs cycle flux to a normal level (curve a). This result can be verified experimentally by examining the influence of increasing concentrations of glutamine and asparagine on hepatocyte viability.

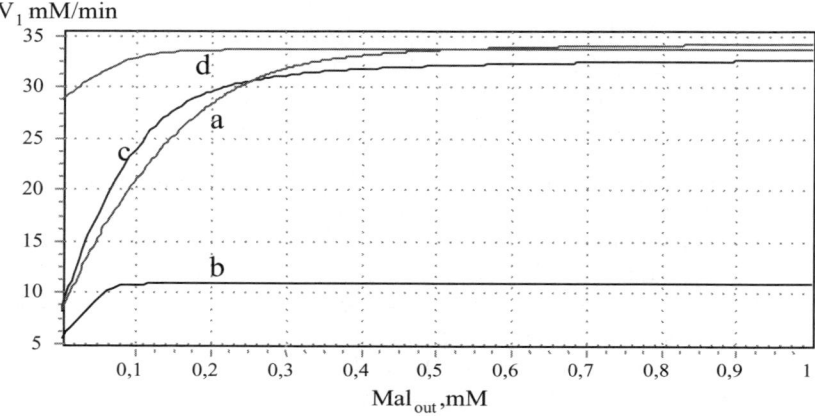

Fig. 22. Dependence of steady state rate of Glu consumption on external Mal concentrations. These were calculated with a model that did not take into account salicylates influences on Krebs cycle (a) and one that did (b). Other curves were calculated from the model taking into account both all possible ways of salicylate influences and a 4-fold increased level of glutamine (c), and an 8-fold increased level of asparagine (d)

6 Discussion

In this chapter, we have tried to show that mathematical modelling in biology in general and metabolic kinetic models in particular have been moving from pure academic exercises to a powerful predictive technology. We are only at the beginning of a new era, where such tools can be applied to a variety of challenging practical problems in medical and pharmaceutical research. As systems biology and other technologies deliver more and more data, and more is known about biological pathways, such techniques will be critical. The complexity of biological systems, even with our current understanding, is no longer intuitive and only mathematical approaches will allow the fine dissection of behaviour. Indeed, it is both the hypothesis testing and hypothesis generation capabilities of mathematical modelling that give it such wide practicability. In this chapter, we have tried to provide some examples of both.

References

1. Perna NT, Plunkett G III, Burland V, Mau B, Glasner JD, Rose DJ, Mayhew GF, Evans PS, et al. Genome sequence of enterohaemorrhagic Escherichia coli O157:H7. Nature 409(6819) (2001) 529-33.
2. Cheung VG, Nowak N, Jang W, Kirsch IR, Zhao S, Chen XN, Furey TS, Kim UJ, Kuo WL, et al. Integration of cytogenetic landmarks into the draft sequence of the human genome. Nature 409(6822) (2001) 953-8.
3. Barbasiewicz A, Liu L, Lang BF, Burger G. Building a genome database using an object-oriented approach. In Silico Biol.2(3) (2002) 213-7.
4. Selkov E, Basmanova S, Gaasterland T, Goryanin I, Gretchkin Y, Maltsev N, Nenashev V, Overbeek R, Panyushkina E, Pronevitch L, Selkov E, Yunus I. (1996) The metabolic pathway collection from EMP: the enzymes and metabolic pathways database. Nucleic Acids Res. 24 (1996) 26-28.
5. Schomburg I, Chang A, Schomburg D. BRENDA, Enzyme Data and Metabolic Information. Nucleic Acids Res. 30 (2002) 47-49.
6. Neidhardt FC. *Escherichia coli and Salmonella typhimurium: Cellular and Molecular Biology.* Vol.1 (1987) 3-6, American Society for Microbiology, Washington DC.
7. Goryanin I, Hodgman TC, Selkov E. Bioinformatics 15 (1999) 749-58.
8. Postma PW, Lengeler JW, Jacobson JR. *Escherichia coli and Salmonella typhimurium: Cellular and Molecular Biology.* Vol.1 (1996) 1149-74, American Society for Microbiology, Washington DC.
9. Postma PW, Lengeler JW, Jacobson GR. Phosphoenolpyruvate: carbohydrate phosphotransferase systems of bacteria. Microbiol Rev 57(3) (1993) 543-94.
10. Rohwer JM, Meadow ND, Roseman S, Westerhoff HV, Postma PW. Understanding glucose transport by the bacterial phosphoenolpyruvate: glycose phosphotransferase system on the basis of kinetic measurements in vitro. J Biol Chem 275(45) (2000) 34909-21.
11. Hoving H, Lolkema JS, Robillard GT. Escherichia coli phosphoenolpyruvate-dependent phosphotransferase system: equilibrium kinetics and mechanism of enzyme in phosphorylation. Biochemistry 20(1) (1981) 87-93.
12. Waygood EB, Steeves T. Enzyme I of the phosphoenolpyruvate: sugar phosphotransferase system of Escherichia coli. Purification to homogeneity and some properties. Can.J.Biochem 58(1) (1980) 40-8.
13. Meadow ND, Roseman S. Rate and equilibrium constants for phosphoryltransfer between active site histidines of Escherichia coli HPr and the signal transducing protein IIIGlc. J.Biol.Chem 271(52) (1996) 33440-5.
14. Lolkema JS, ten Hoeve-Duurkens RH, Robillard GT. Quasi-steady state kinetics of mannitol phosphorylation catalyzed by enzyme IImtl of the Escherichia coli phosphoenolpyruvate-dependent phosphotransferase system. J.Biol.Chem 268(24) (1993) 17844-9.
15. Meyer D, Schneider-Fresenius C, Horlacher R, Peist R, Boos W. Molecular characterization of glucokinase from Escherichia coli K-12. J.Bacteriol. 179(4) (1997) 1298-306.
16. Davies GM, Barrett-Bee KJ, Jude DA, Lehan M, Nichols WW, Pinder ...
17. Pittard AJ. Biosynthesis of the Aromatic Amino Acids. In Neihardt FC.(Ed.): *Escherchia coli and Salmonella: Cellular and Molecular Biology* 458-460.

18. Noble M, Goryanin I, Demin O, Kolupaev A, Sinha Yu, Tew D, Earnshaw D, Tobin F, DeWolf WE. An *In Silico* Model of the Shikimate Pathway Exemplifies a Novel Approach for Optimizing High-Throughput Screening of Enzyme Pathways. In press.

19. Prescott LF. Effects of non-narcotic analgesics on the liver. Drugs 32 (1986) 129-47.

20. Benson GD. Hepatotoxicity following the therapeutic use of antipyretic analgesics. Am.J.Med. 75(5A) (1983) 85-93.

21. Temple AR. Acute and chronic effects of aspirin toxicity and their treatment. Arch.Intern.Med. 141 (1981) 364-9.

22. Fromenty B, Pessayre D. Inhibition of mitochondrial beta-oxidation as a mechanism of hepatotoxicity. Pharmacol.Therapeut. 67(1) (1995) 101-54.

23. Vessey DA, Hu J, Kelly M. Interaction of salicylate and ibuprofen with the carboxylic acid: CoA ligases from bovine liver mitochondria. J.Biochem.Toxicol. 11(2) (1996) 73-8.

24. Kaplan EH, Kennedy J, Davis J. Effects of salicylate and other benzoates on oxidative enzymes of the tricarboxylic acid cycle in rat tissue homogenates. Arch.Biochem. 51 (1954) 47-61.

25. Miyahara JT, Karler R. Effect of salicylate on oxidative phosphorylation and respiration of mitochondrial fragments. Biochem.J. 97 (1965) 194.

26. Kondrashova MN. Structuro-kinetic organization of the tricarboxylic acid cycle in the active functioning of mitochondria. Biofizika 34(3) (1989) 450-8.

27. Dierks T, Kramer R. Asymmetric orientation of the reconstituted aspartate/glutamate carrier from mitochondria. Biochim.Biophys.Acta 937 (1988) 112-26.

28. Cleland, WW. The kinetics of enzyme-catalyzed reactions with two or more substrates or products. Biochim.Biophys.Acta 67 (1963) 104-37.

29. Strayer L. Biochemistry 2 (1985).

30. Lemasters JJ, Hackenbrock CR, Thurman RG, Westerhoff HV. (Eds.) *Integration of mitochondrial function* (1988) New York and London, 497-509.

31. Cascante M, Cortes A. Kinetic studies of chicken and turkey liver mitochondrial aspartate aminotransferase. // Biochem. J., 1988, 250, p. 805-812.

32. McCormack JG, Denton RM. The effect of calcium ions and adenine nucleotides on the activity of pig heart a-oxoglutarate dehydrogenase complex. Biochem.J. 180 (xxxx) 533-44.

33. Cha S, Parks RE. Succinic Thiokinase II. Kinetic studies: initial velocity, product inhibition, and effect of arsenate. JBC 239(6) (1964) 1968.

34. Rasschaert J., Malaisse W.J. Hexose metabolism in pancreatic islets: succinate dehydrogenase activity in islet homogenates. Cell Biochem.Funct. 11(3) (1993) 155-8.

35. Vinogradov AD. Succinate-ubiquinone reductase site of the respiratory chain. Biokhimiia 51(12) (1986) 1944-73.

36. Comparison of Catalytic Activity and Inhibitors of Quinone Reactions of Succinate Dehydrogenase (Succinate-Ubiquinone Oxidoreductase) and Fumarate Reductase (Menaquinol-Fumarate Oxidoreductase) from Escherichia coli. Arch.Biophys.Biochem. 369(2) (1999) 223-32.

37. Belikova YO, Kotlyar AB, Vinogradov AD. Oxidation of malate by the mitochondrial succinate-ubiquinone reductase. Biochim.Biophys. Acta 936(1) (1988) 1-9.

38. Alberty RA. Fumarase. The Enzymes 5(B) (1961) 531-44.
39. Heyde E, Ainsworth S. Kinetic Studies on the Mechanism of the Malate Dehydrogenase Reaction. J.Biol.Chem. 243 (1968) 2413-23.
40. Indiveri C, Dierks T, Kramer R, Palmieri F. Reaction mechanism of the reconstituted oxoglutarate carrier from bovine heart mitochondria. Eur.J.Biochem. 198 (1991) 339-47.
41. Fato R, Estornell E, di Bernardo S, Palotti F, Castelli GP, Lenaz G. Quasi-steady-state Kinetics of the Reduction of Coenzyme Q Analogs by Complex I (NADH:Ubiquinone Oxidoreductase) in Bovine Heart Mitochondria and Submitochondrial Particles. Biochemistry 35 (1996) 2705-16.
42. Silverstein E, Sulebele G. Equilibrium Kinetic Study of the Catalytic Mechanism of Bovine Liver Glutamate Dehydrogenase. Biochemistry 12(11) (1973) 2164-71.
43. Di Franco A. Reaction mechanism of L-glutamate dehydrogenase. Transient complexes in the oxidative deamination of L-glutamate catalyzed by NAD(P)-dependent L-glutamate dehydrogenase. Eur.J.Biochem. 45(2) (1974) 407-24.
44. Jonker A, Geerts WJ, Charles R, Lamers WH, Van Noorden CJ. The dynamics of local kinetic parameters of glutamate dehydrogenase in rat liver. Histochem. Cell Biol. 106(4) (1996) 437-43.
45. Shatilov VR. Mechanism of the glutamate dehydrogenase reaction. Biokhimiia 48(7) (1983) 1059-66.
46. Vessey DA, Hu J. Isolation from bovine liver mitochondria and characterization of three distinct carboxylic acid: CoA ligases with activity toward xenobiotics. J.Biochem.Toxicol. 10(6) (1995) 329-37.
47. Killenberg PG, Davidson ED, Webster LT. Evidence for a medium-chain fatty acid: Coenzyme A ligase (Adenosine Monophosphate) that activates salicylate. Mol.Pharmacol. 7 (1971) 260-8.
48. Andreyev AY, Bondareva TO, Dedukhova VI, Mokhova EN, Skulachev VP, Tsofina LM, Volkov NI, Vygodina TV. The ATP/ADP-antiporter is involved in the uncoupling effect of fatty acids on mitochondria. Eur.J. Biochem. 182(3) (1989) 585-92.

Modelling Gene Assembly in Ciliates

Andrzej Ehrenfeucht[1], Tero Harju[2], Ion Petre[3], David M. Prescott[4],
and Grzegorz Rozenberg[5,1]

[1] Department of Computer Science, University of Colorado, Boulder, USA
   andrzej@cs.colorado.edu
[2] Department of Mathematics, University of Turku, FIN-20014 Turku, Finland
   harju@utu.fi
[3] Department of Computer Science, Åbo Akademi University, Turku, Finland
   ipetre@abo.fi
[4] Department of Molecular, Cellular and Developmental Biology,
   University of Colorado, Boulder, CO 80309-0347, USA
   prescotd@spot.colorado.edu
[5] Leiden Institute for Advanced Computer Science, Leiden University,
   Niels Bohrweg 1, 2333 CA Leiden, the Netherlands
   rozenber@liacs.nl

Summary. The gene assembly process that transforms the micronuclear genome to
the macronuclear genome in stichotrichs ciliates is a complex DNA processing which
is very interesting from the computational point of view. We shall give a survey of
the notions and results concerning three molecular operations, known as *ld*, *hi*, and
dlad, that have been postulated to accomplish the gene assembly process. These
operations are modelled on three abstraction levels: MDS descriptors, legal strings,
and overlap graphs. It turns out that these three abstractions are equivalent as far
as the operational modelling of gene assembly is concerned.

1 Introduction

Ciliates (ciliated protozoans) have been around for about two billion years.
These single-cell organisms are presently divided into at least 8000 species,
but it is quite likely that the number of (still unknown) different species of
ciliates is a multiple of this number. Ciliates are present in almost all aquatic
environments on the earth. For our computational and modelling purposes,
the most important feature of ciliates is their *nuclear dualism*: they possess
two types of nucleus that are functionally different. The RNA transcripts
to operate the cell are provided by the *macronucleus*, but ciliates also have
a *micronucleus*, which is activated only during the mating between ciliates
belonging to the same species.

The modelling discussed in this chapter concerns one group of ciliates
called *stichotrichs*, in which the micronucleus and macronucleus are orga-

nized in remarkably different ways. A micronucleus contains about 100 chromosomes, each of which contains a single very long DNA molecule, hundreds of thousands base pairs (bp) long. Chromosomes in the macronucleus are much shorter ranging in size from 200 to around 15 000 bp. Also, each chromosome is present in 1000 or more copies. The genes in the micronucleus are not functional and they are separated from each other by long stretches of spacer DNA. Each micronuclear gene consists of segments (called *macronuclear destined segments* or *MDSs*) of the macronuclear version of this gene separated by noncoding segments of DNA (called *internal eliminated segments* or *IESs*).

During sexual reproduction the micronuclear genome is translated into the macronuclear genome. This transformation is a very intricate DNA processing during which over 100 000 IESs are excised, and the MDSs are ligated into functional genes. This process of gene assembly (converting micronuclear genes into their expressible macronuclear form) is intriguing from the computational point of view. For example, it turns out that the process of transformation of a micronuclear gene into its macronuclear form uses one of the central data structures of computer science – linked lists.

The computational aspects of gene assembly in ciliates were first studied by Landweber and Kari [8, 9] – the focus of this study was the computational power (as studied in theoretical computer science) of the proposed intermolecular model of operations that accomplish gene assembly. A different, intramolecular model of operations for the assembly was initiated by Ehrenfeucht, Prescott, and Rozenberg in [7, 17]. This model is based on three molecular operations *ld*, *hi*, and *dlad* that were formalized on three different levels: *MDS descriptors*, *legal strings*, and *overlap graphs*. It was proved by Ehrenfeucht, Petre, Prescott, and Rozenberg [5] that every micronuclear gene can be assembled using these operations.

The MDS descriptors give a rather faithful and easily recognizable description of the micronuclear genes. On the other hand, the legal strings and the overlap graphs are more abstract, and therefore the formal operations in these models are considerably simpler compared to the operations that act on MDS descriptors.

In this chapter we shall review only the basic abstraction problems concerning possible formalisms of gene assembly. We refer the interested reader to [2] for the problems concerning the assembly pattern of subsets of our operations, to [3] for general graph theoretic methods for gene assembly, and to [6] for invariant properties of the assembly process. For background on DNA molecules, we refer the reader to [10] and [11].

2 Preliminaries on Strings

Let Σ be an *alphabet*, that is, a set of symbols called *letters*. Also, let $\overline{\Sigma} = \{\overline{a} \mid a \in \Sigma\}$ be a copy of Σ that is disjoint from Σ. A *string* $v = a_1 a_2 \ldots a_n$ over Σ is any sequence of letters, $a_i \in \Sigma$ for $i = 1, 2, \ldots, n$. The set of all strings over Σ is denoted by Σ^*; it contains also the *empty string* Λ that has no letters. We let

$$\Sigma^{\maltese} = (\Sigma \cup \overline{\Sigma})^*$$

be the set of all strings over $\Sigma \cup \overline{\Sigma}$. Each element $v \in \Sigma^{\maltese}$ is called a *signed string* over Σ. The signing of letters, $a \mapsto \overline{a}$, can be extended by first setting $\overline{\overline{a}} = a$, for each $a \in \Sigma$, and then letting $\overline{u} = \overline{a}_n \overline{a}_{n-1} \ldots \overline{a}_1$ for each $u = a_1 a_2 \ldots a_n \in \Sigma^{\maltese}$, where $a_i \in \Sigma \cup \overline{\Sigma}$ for each i. The signed string $\overline{u}$ is the *inversion* of u. We say that a letter $a \in \Sigma \cup \overline{\Sigma}$ *occurs* in v if either a or $\overline{a}$ is a substring of v. Let the *domain* of v be the set

$$\mathrm{dom}(v) = \{a \in \Sigma \mid a \text{ occurs in } v\}.$$

A function $\tau \colon \Sigma^{\maltese} \to \Gamma^{\maltese}$, where Σ and Γ are alphabets, is a *morphism* if $\tau(uv) = \tau(u)\tau(v)$ for all $u, v \in \Sigma^{\maltese}$, and τ is a *substitution* if, moreover, $\overline{\tau(u)} = \tau(\overline{u})$ for all $u \in \Sigma^{\maltese}$. Note that the images of the letters $a \in \Sigma$ determine the substitution τ. Two strings $u \in \Sigma^{\maltese}$ and $v \in \Gamma^{\maltese}$ are *isomorphic* if there exists an injective substitution $\tau \colon \Sigma^{\maltese} \to \Gamma^{\maltese}$ with $\tau(\Sigma) \subseteq \Gamma$ such that $\tau(u) = v$.

Let $\xi \colon \Sigma^{\maltese} \to \Sigma^*$ be the morphism that removes the bars from the letters: for all $a \in \Sigma$,

$$\xi(a) = a = \xi(\overline{a}).$$

A signed string $v \in \Sigma^{\maltese}$ is a *signing* of a string $u \in \Sigma^*$ if $\xi(v) = u$.

For a string $v = a_1 a_2 \ldots a_n \in \Sigma^*$, a string u is a *permutation* of v if there exists a permutation $(i_1 \, i_2 \ldots i_n)$ of $\{1, 2, \ldots, n\}$ such that $u = a_{i_1} a_{i_2} \ldots a_{i_n}$. Moreover, a signing of a permutation of v is said to be a *signed permutation of v*. A string $v \in \Sigma^*$ is a *double occurrence string* if every letter $a \in \mathrm{dom}(v)$ occurs exactly twice in v. A signing of a nonempty double occurrence string is called a *legal string*. If a legal string $u \in \Sigma^{\maltese}$ contains one occurrence of $a \in \Sigma$ and one of $\overline{a}$, then a is said to be *positive* in u; otherwise, a is *negative* in u.

Example 1. (1) The signed string $u = 433\overline{3}\overline{5}45 \in \{3, 4, 5\}^{\maltese}$ is isomorphic to $v = 233\overline{3}\overline{4}24$. The isomorphism τ is defined by $\tau(4) = 2$, $\tau(3) = 3$, and $\tau(5) = 4$. Also, u is a signing of the string 433545, since $\xi(u) = 433545$.

(2) In the legal string $u = 2\,\overline{4}\,3\,2\,\overline{5}\,3\,\overline{4}\,5$ letters 2 and 5 are positive while 3 and 4 are negative. The string $w = 2\,4\,3\,\overline{2}\,\overline{5}\,3\,5$ is not legal, since it has only one occurrence of 4. $\square$

3 The Gene Assembly Operations in Ciliates

Each gene in a micronuclear chromosome in ciliates consists of a sequence of MDSs that are separated by IESs, and some of the MDSs may even be inverted (see Prescott [11, 12]). In the gene assembly process each micronuclear gene is translated into a macronuclear gene by excising all IESs and by splicing the MDSs in their orthodox order $M_1, M_2, \ldots, M_\kappa$ (see [11, 12, 16]).

Example 2. The actin I gene in *Sterkiella nova* has the following MDS/IES micronuclear structure: $M_3 I_1 M_4 I_2 M_6 I_3 M_5 I_4 M_7 I_5 M_9 I_6 \overline{M}_2 I_7 M_1 I_8 M_8$, which is represented in Fig. 1. There each MDS is represented by a rectangle and each IES is represented by a simple line. Note that the MDS M_2 is inverted in the micronuclear MDS/IES structure of this gene. The macronuclear version of this gene is given in Fig. 2. There the IESs have been excised and the MDSs have been ligated in the orthodox order $M_1 M_2 \ldots M_9$ (as a matter of fact these MDSs are "glued on common ends" – this is made more precise in the following paragraph).   □

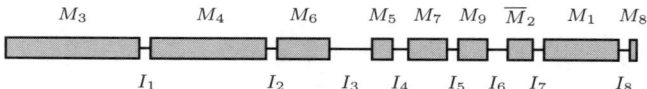

Fig. 1. The micronuclear version of the actin I gene of *Sterkiella nova*

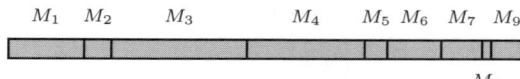

Fig. 2. The macronuclear version of the actin I gene of *Sterkiella nova*

Each MDS M_i of a micronuclear gene can be represented in the form

$$M_i = (\pi_i, \mu_i, \pi_{i+1}), \quad \text{where } \pi_1 = \frac{b}{b}, \ \pi_i = \frac{p_i}{p_i} \text{ (for } 1 < i < \kappa\text{), and } \pi_{\kappa+1} = \frac{e}{e}.$$

Here π_i, μ_i, and π_{i+1} represent double stranded molecules; π_i and π_{i+1} are the (*incoming* and *outgoing*) *pointers* and μ_i is the *body* of the MDS M_i. The *markers* b and e (and their inversions $\overline{b}$ and $\overline{e}$) give the locations where the macronuclear DNA gene is to be excised from the micronuclear molecule (after it has been assembled). In the macronuclear gene the MDSs $M_1, M_2, \ldots, M_\kappa$ are spliced together by attaching M_j and M_{j+1} on the pointer π_{j+1} for each j.

We now describe the molecular operations *ld*, *hi*, and *dlad* introduced in [7, 17].

1. The operation *(loop, direct repeat)-excision*, or *ld*, for short, is applied to a molecule with a *direct repeat pattern* $(- \pi - \pi -)$ of a pointer (see Fig. 3): either the two occurrences of π are separated by one IES or they are at the two opposite ends of the part of the molecule containing MDSs. The molecule is folded into a loop in such a way that the occurrences of π are aligned and then the operation proceeds as shown in Fig. 4. The *ld* operation yields two molecules: a linear and a circular one where the circular molecule consists of either the whole gene or one IES only.

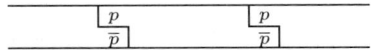

Fig. 3. Direct repeat pattern for pointer $\pi = \frac{p}{\overline{p}}$

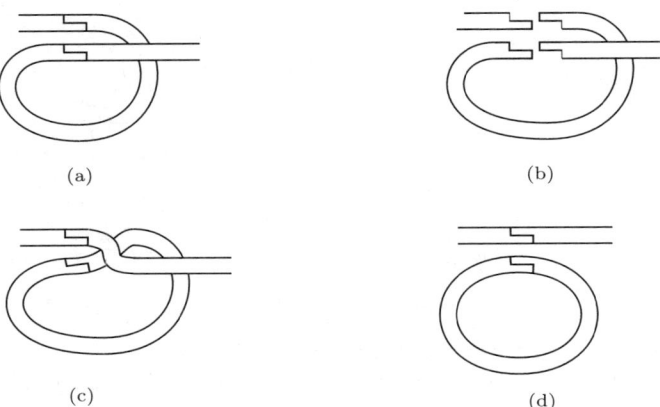

(a)   (b)

(c)   (d)

Fig. 4. The *ld* operation

2. The operation *(hairpin, inverted repeat)-excision/reinsertion*, or *hi*, for short, is applied to a molecule with an *inverted repeat pattern* $(- \pi - \overline{\pi} -)$ of a pointer (see Fig. 5). The molecule is folded in such a way that the two occurrences of π are aligned and the operation proceeds as in Fig. 6. This operation yields only one molecule.

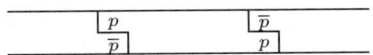

Fig. 5. Inverted repeat pattern for pointer $\pi = \frac{p}{\overline{p}}$

3. The operation *(double loop, alternating direct repeat)-excision/reinsertion*, or *dlad*, for short, is applied to a molecule with an *alternating direct repeat pattern* $(- \pi - \pi' - \pi - \pi' -)$ for two pointers π and π' (see Fig. 7). The molecule is folded into a double loop in such a way that π is aligned in one loop and π' is aligned in the other loop. The operation then proceeds as in Fig. 8. Also *dlad* yields one molecule.

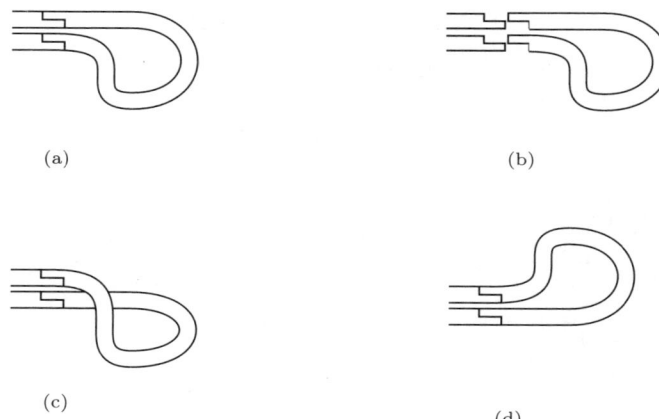

Fig. 6. The operation *hi*

Fig. 7. Alternating direct repeat pattern for pointers $\pi = \frac{p}{\overline{p}}$ and $\pi' = \frac{q}{\overline{q}}$

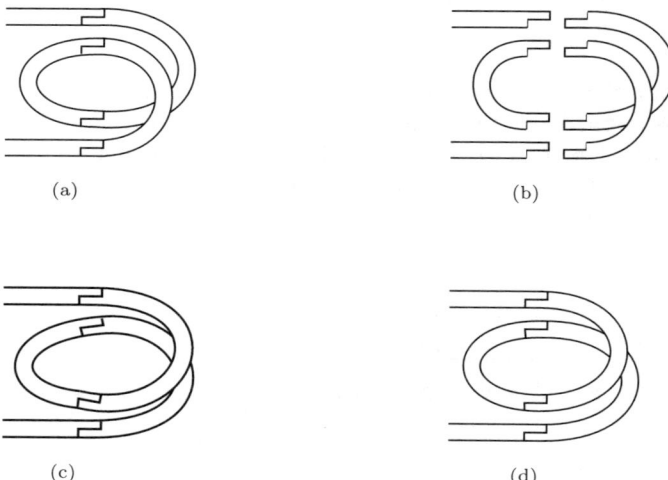

Fig. 8. The operation *dlad*

4 Modelling by MDS Descriptors

The process of gene assembly in ciliates can be considered as a process of assembling MDSs so that they are sorted in an orthodox order. Hence the information about the structure of a micronuclear gene, or an intermediate precursor of a macronuclear gene, is given by the sequence of MDSs of this gene.

Example 3. The representation of the actin I gene in *Sterkiella nova* from Example 2 can be simplified to $\alpha = M_3 M_4 M_6 M_5 M_7 M_9 \overline{M}_2 M_1 M_8$ by omitting all IESs.   □

We shall use the alphabets

$$\Theta_\kappa = \{\, M_{i,j} \mid 1 \le i \le j \le \kappa \,\}$$

to denote the MDSs for all $\kappa \ge 1$. The signed strings in $\Theta^{\maltese}$ are called *MDS arrangements*. Elements $M_{i,i}$ then denote the micronuclear MDSs; they will be called *elementary MDSs*, and they are often rewritten as M_i. Letters $M_{i,j}$ with $i \ne j$ (i.e., $j > i$) denote *composite MDSs* that are formed by splicing together the MDSs $M_i, M_{i+1}, \ldots, M_j$. An MDS arrangement α of the form

$$\alpha = M_{1,i_2-1} M_{i_2,i_3-1} \ldots M_{i_n,\kappa} \tag{1}$$

is said to be *orthodox*. Note that an orthodox arrangement does not contain any inverted MDSs. A signed permutation of an orthodox arrangement is a *realistic arrangement* of *size* κ. Moreover, a signed permutation of $M_1 M_2 \ldots M_\kappa$, is a *micronuclear arrangement*.

Example 4. The arrangement $M_{1,4} M_{5,5} M_{6,8}$ is orthodox (of size 8). The MDSs $M_{1,4}$ and $M_{6,8}$ are composite, and the MDS $M_{5,5}$ is elementary. Also, the arrangement $\alpha = M_{3,5} \overline{M}_{9,11} \overline{M}_{1,2} M_{12} \overline{M}_{6,8}$ is realistic, since it is a signed permutation of the orthodox arrangement $M_{1,2} M_{3,5} M_{6,8} M_{9,11} M_{12}$.   □

An assembled gene has no pointers left, since it has no IESs, but the micronuclear form of a gene has pointers. Thus the gene assembly process can be analysed by representing the MDS arrangements by the sequence of pointers present in this gene. We can thus simplify the formal framework by denoting each MDS by the ordered pair of its pointers and markers only, i.e., $M_{i,j} = (\pi_i, \mu, \pi_j)$ is represented as (p_i, p_j), and its inversion $\overline{M}_{i,j} = (\overline{\pi}_j, \overline{\mu}, \overline{\pi}_i)$ is represented as $(\overline{p}_j, \overline{p}_i)$ for all $i \le j$.

Let $\Psi = \{b, e, \overline{b}, \overline{e}\}$ denote the set of the *markers* (b stands for "beginning", and e for "end"). For each $\kappa \ge 2$, let

$$\Delta_\kappa = \{2, \ldots, \kappa\} \quad \text{and} \quad \overline{\Delta}_\kappa = \{\overline{2}, \ldots, \overline{\kappa}\},$$

and

$$\Pi_\kappa = \Delta_\kappa \cup \overline{\Delta}_\kappa \quad \text{and} \quad \Pi_{ex,\kappa} = \Pi_\kappa \cup \Psi.$$

Also, let $\Delta = \{2, 3, \ldots\}$ and $\Pi = \Delta \cup \overline{\Delta}$. The letters in Π are called *pointers*. Each pointer $p \in \Pi$ determines its *pointer set* $\mathbf{p} = \{p, \overline{p}\}$. In the sequel, unless explicitly stated otherwise, we assume that the index κ is clear from the context. In order to present the encodings of the MDSs by pairs of pointers (and markers), we introduce the alphabet

$$\Gamma_\kappa = \{\,(b,e),(\overline{e},\overline{b})\,\} \;\cup\; \{\,(i,j),(\overline{j},\overline{i})\;\mid\;2\le i<j\le\kappa\,\}$$
$$\cup\;\{\,(b,i),(\overline{i},\overline{b}),(i,e),(\overline{e},\overline{i})\;\mid\;2\le i\le\kappa\,\}.$$

A string over Γ_κ is called an *MDS descriptor*.

We define a morphism $\psi_\kappa \colon (\Theta_\kappa)^{\maltese} \to \Gamma_\kappa^*$ as follows:

$$\psi_\kappa(M_{1,\kappa}) = (b,e) \qquad \text{and} \qquad \psi_\kappa(\overline{M}_{1,\kappa}) = (\overline{e},\overline{b}),$$
$$\psi_\kappa(M_{1,i}) = (b,i+1) \quad \text{and} \quad \psi_\kappa(\overline{M}_{1,i}) = (\overline{i+1},\overline{b}),$$
$$\psi_\kappa(M_{i,\kappa}) = (i,e) \qquad \text{and} \qquad \psi_\kappa(\overline{M}_{i,\kappa}) = (\overline{e},\overline{i}),$$
$$\psi_\kappa(M_{i,j}) = (i,j+1) \quad \text{and} \quad \psi_\kappa(\overline{M}_{i,j}) = (\overline{j+1},\overline{i}),$$

where $1 < i \le j < \kappa$.

Example 5. By Example 4 the arrangement $\alpha = M_{3,5}\overline{M}_{9,11}\overline{M}_{1,2}M_{12}\overline{M}_{6,8}$ is realistic. By applying the function ψ_κ to α, we obtain the corresponding MDS descriptor $\psi_{12}(\alpha) = (3,6)(\overline{12},\overline{9})(\overline{3},\overline{b})(12,e)(\overline{9},\overline{6})$. □

In most instances only the order of the pointers is essential in the MDS descriptors. Because of this, we now define isomorphism of MDS descriptors as follows. Let $\delta = (x_1,x_2)\ldots(x_{2n-1},x_{2n})$ and $\delta' = (y_1,y_2)\ldots(y_{2n-1},y_{2n})$ be two MDS descriptors. They are *isomorphic*, if

(1) $(x_i \in \Delta \iff y_i \in \Delta)$ and $(x_i \in \Psi \implies y_i = x_i)$,
(2) $\xi(x_i) < \xi(x_j) \iff \xi(y_i) < \xi(y_j)$ for $x_i, x_j \notin \Psi$.

Example 6. The MDS descriptors $(4,5)(\overline{8},\overline{6})(b,4)$ and $(2,3)(\overline{5},\overline{4})(b,2)$ satisfy the above requirements, and thus they are isomorphic.

An MDS descriptor δ is said to be *realistic* if δ is isomorphic with an MDS descriptor $\psi_\kappa(\alpha)$ for some κ and a micronuclear arrangement α. The following lemma is then clear.

Lemma 1. *Let δ be an MDS descriptor. Then it is realistic if and only if $\delta = \psi_\kappa(\alpha)$ for some κ and a realistic MDS arrangement α of size κ.*

Let $\delta = (x_1,x_2)(x_3,x_4)\ldots(x_{2n-1},x_{2n})$ be a realistic MDS descriptor. We say that a pointer p *occurs* in δ if p occurs in the string $x_1x_2x_3x_4\ldots x_{2n-1}x_{2n}$. For each pointer $p \in \Delta$, there are either no occurrences or exactly two occurrences from the pointer set $\mathbf{p}$ in δ. If there are two occurrences, let these be $x_i, x_j \in \mathbf{p} = \{p,\overline{p}\}$ for $1 \le i < j \le 2n$. The *p-interval* of δ is then defined to be the set

$$\delta_{(p)} = \{x_i, x_{i+1}, \ldots, x_j\}.$$

If $x_i = x_j$, then p is *negative* in δ; otherwise (i.e., if $x_i = \overline{x}_j$) p is *positive* in δ.

We can now formalize the molecular gene assembly operations by MDS descriptors. Corresponding to the three molecular operations *ld*, *hi*, and *dlad*, we have three operations ld, hi, and dlad on MDS descriptors. In the illustrations for these operations, rectangles denote MDSs (and their pointers are indicated), the zigzag line denotes a segment of a molecule that may contain both MDSs and IESs, and a simple straight line represents an IES.

1. For each $p \in \Pi_\kappa$, the ld-*rule* for p is defined as follows:

$$\mathsf{ld}_p(\delta_1(q,p)(p,r)\delta_2) = \delta_1(q,r)\delta_2, \tag{11}$$

$$\mathsf{ld}_p((p,q)\delta_1(r,p)) = (r,q)\delta_1, \tag{12}$$

where $q,r \in \Pi_{ex,\kappa}$ and $\delta_1, \delta_2 \in (\Gamma_\kappa)^*$.

The case (11) is a *simple* ld-*rule*. It applies to adjacent occurrences of p. The case (12) is called a *boundary* hi-*rule*, and it applies to two occurrences of p at the ends of the descriptor. Both of these cases are illustrated in Fig. 9.

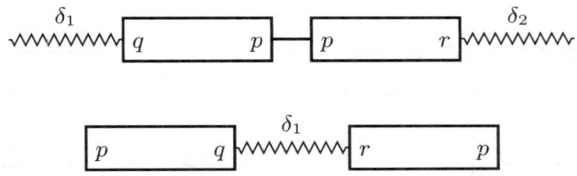

Fig. 9. The MDS/IES structure to which ld_p is applicable

2. For each $p \in \Pi_\kappa$, the hi-*rule* for p is defined as follows:

$$\mathsf{hi}_p(\delta_1(p,q)\delta_2(\overline{p},\overline{r})\delta_3) = \delta_1\overline{\delta_2}(\overline{q},\overline{r})\delta_3, \tag{h1}$$

$$\mathsf{hi}_p(\delta_1(q,p)\delta_2(\overline{r},\overline{p})\delta_3) = \delta_1(q,r)\overline{\delta_2}\delta_3, \tag{h2}$$

where $q,r \in \Pi_{ex,\kappa}$ and $\delta_i \in (\Gamma_\kappa)^*$ for each $i = 1,2,3$.

In these cases, illustrated in Fig. 10, an occurrence of p is the incoming pointer, and the occurrence of $\overline{p}$ is the outgoing pointer.

3. For each $p,q \in \Pi_\kappa$, $p \neq q$, the dlad-*rule* for p and q is defined as follows:

$$\mathsf{dlad}_{p,q}(\delta_1(p,r_1)\delta_2(q,r_2)\delta_3(r_3,p)\delta_4(r_4,q)\delta_5) = \delta_1\delta_4(r_4,r_2)\delta_3(r_3,r_1)\delta_2\delta_5 \tag{d1}$$

$$\mathsf{dlad}_{p,q}(\delta_1(p,r_1)\delta_2(r_2,q)\delta_3(r_3,p)\delta_4(q,r_4)\delta_5) = \delta_1\delta_4\delta_3(r_3,r_1)\delta_2(r_2,r_4)\delta_5 \tag{d2}$$

$$\mathsf{dlad}_{p,q}(\delta_1(r_1,p)\delta_2(q,r_2)\delta_3(p,r_3)\delta_4(r_4,q)\delta_5) = \delta_1(r_1,r_3)\delta_4(r_4,r_2)\delta_3\delta_2\delta_5 \tag{d4}$$

$$\mathsf{dlad}_{p,q}(\delta_1(r_1,p)\delta_2(r_2,q)\delta_3(p,r_3)\delta_4(q,r_4)\delta_5) = \delta_1(r_1,r_3)\delta_4\delta_3\delta_2(r_2,r_4)\delta_5 \tag{d4}$$

$$\mathsf{dlad}_{p,q}(\delta_1(p,r_1)\delta_2(q,p)\delta_4(r_4,q)\delta_5) = \delta_1\delta_4(r_4,r_1)\delta_2\delta_5 \tag{d5}$$

$$\mathsf{dlad}_{p,q}(\delta_1(p,q)\delta_3(r_3,p)\delta_4(q,r_4)\delta_5) = \delta_1\delta_4\delta_3(r_3,r_4)\delta_5 \tag{d6}$$

$$\mathsf{dlad}_{p,q}(\delta_1(r_1,p)\delta_2(q,r_2)\delta_3(p,q)\delta_5) = \delta_1(r_1,r_2)\delta_3\delta_2\delta_5 \tag{d7}$$

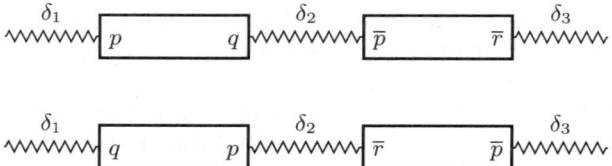

Fig. 10. The MDS/IES structure to which hi_p is applicable

where $r_i \in \Pi_{ex,\kappa}$ and $\delta_i \in (\Gamma_\kappa)^*$ for each i.

In each of the above instances, pointers p and q overlap. The cases (d1)–(d4) are illustrated in Fig. 11 and the cases (d5)–(d7) in Fig. 12.

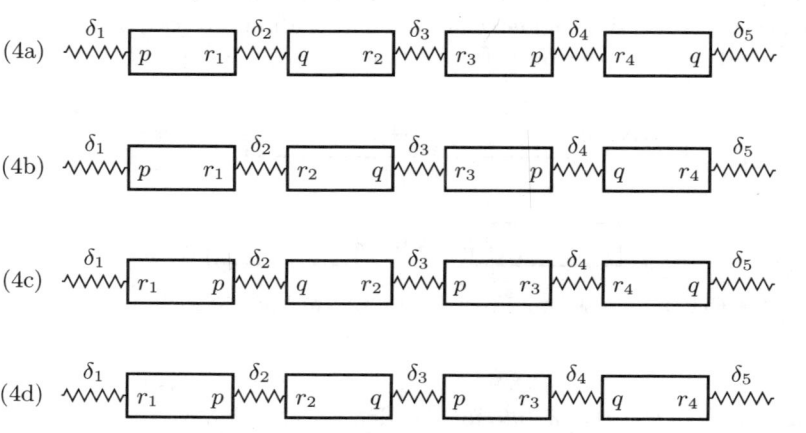

Fig. 11. The MDS/IES structures (d1)–(d4) to which $dlad_{p,q}$ is applicable

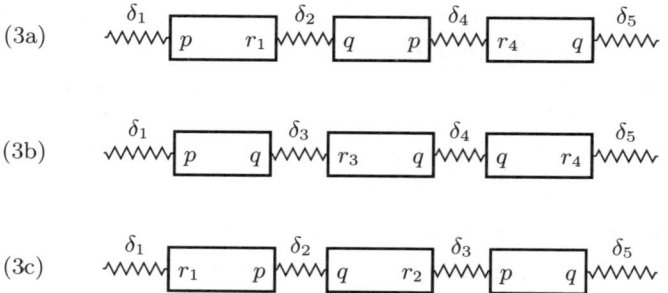

Fig. 12. The MDS/IES structures (d5)–(d7) to which $dlad_{p,q}$ is applicable

The following lemma is clear by the form of the operations and the definition of isomorphism of MDS descriptors.

Lemma 2. *Let δ be a realistic MDS descriptor, p and q be pointers in δ, and $\varphi \in \{\mathsf{Id}_p, \mathsf{hi}_p, \mathsf{dlad}_{p,q}\}$. If φ is applicable to δ, then also $\varphi(\delta)$ is realistic.*

Let $\varphi = \varphi_k \ldots \varphi_1$ be a composition of operations $\varphi_i \in \{\mathsf{Id}_p, \mathsf{hi}_p, \mathsf{dlad}_{p,q} \mid p, q \in \Pi\}$. We say that φ is *applicable* to an MDS descriptor δ if φ_1 is applicable to δ, and φ_i is applicable to $\varphi_{i-1} \ldots \varphi_1(\delta)$. In this case, we also say that φ is a *reduction* of δ. Moreover, φ is *successful* for δ if either $\varphi(\delta) = (b, e)$ or $\varphi(\delta) = (\overline{e}, \overline{b})$.

Example 7. The operations $\mathsf{Id}_{\overline{3}}$, hi_4, $\mathsf{dlad}_{5,\overline{2}}$ are applicable to the MDS descriptor $\delta = (4,5)(\overline{2}, \overline{b})(5, e)(\overline{4}, \overline{3})(\overline{3}, \overline{2})$. Indeed, $\mathsf{Id}_{\overline{3}}(\delta) = (4,5)(\overline{2}, \overline{b})(5, e)(\overline{4}, \overline{2})$, $\mathsf{hi}_4(\delta) = (\overline{e}, \overline{5})(b, 2)(\overline{5}, \overline{3})(\overline{3}, \overline{2})$, and $\mathsf{dlad}_{5,\overline{2}}(\delta) = (4, e)(\overline{4}, \overline{3})(\overline{3}, \overline{b})$. Also,

$$\mathsf{hi}_4(\mathsf{dlad}_{5,\overline{2}}(\delta)) = (\overline{e}, \overline{3})(\overline{3}, \overline{b}) \quad \text{and} \quad \mathsf{Id}_{\overline{3}}(\mathsf{hi}_4\, \mathsf{dlad}_{5,\overline{2}}(\delta)) = (\overline{e}, \overline{b}),$$

and hence the composition $\mathsf{Id}_{\overline{3}}\, \mathsf{hi}_4\, \mathsf{dlad}_{5,\overline{2}}$ is successful for δ.   □

A single application of the three operations Id, hi, and dlad shortens the MDS descriptor. The following universality theorem was first proved in [4].

Theorem 1. *Each realistic MDS descriptor δ has a successful reduction.*

We now come back to the boundary Id-rule. If we have a boundary occurrence of pointer p in an MDS descriptor $\delta = (p, q)\delta_1(r, p)$ as in (12), then only the boundary rule is applicable to p. Moreover, if a rule φ is applied to any other pointer or pointers (in δ_1), then $\varphi(\delta)$ still has pointer p as its boundary. Therefore, as far as assembly of the macronuclear gene is concerned, we may assume that the Id-rule for the boundary pointer p is applied as the last one in the assembly:

$$\mathsf{Id}_p((p, m)(m', p)) = (m', m), \tag{12'}$$

where $m, m' \in \Psi$. Consequently, in the sequel we shall assume that *the boundary Id_p-rule for pointer p will have the form (12')*.

5 Modelling by Legal Strings

The formal model of gene assembly of ciliates can be considerably simplified by considering legal strings instead of MDS descriptors. This simplification was introduced and studied in [1] and [5]. We transform the MDS descriptors to legal strings by removing the parentheses from the descriptors and deleting the markers. We obtain in this way legal strings as the formalism for describing the sequences of pointers present in the micronuclear and the intermediate molecules.

In order to formalize the above transformation, first let $\varsigma_\kappa \colon (\Delta_{ex,\kappa})^{\maltese} \to (\Delta_\kappa)^{\maltese}$ be a morphism defined by $\varsigma_\kappa(x) = \Lambda$ if $x \in \Psi$ and $\varsigma_\kappa(x) = x$, otherwise. Then let $\mu_\kappa(x, y) = \varsigma_\kappa(x)\varsigma_\kappa(y)$ for all $x, y \in \Pi_{ex,\kappa}$. In the sequel, for each realistic MDS descriptor $\delta \in (\Gamma_\kappa)^*$, we let

$$w_\delta = \mu_\kappa(\delta).$$

We say that a legal string u is *realistic* if there exists a realistic MDS descriptor δ such that $u = w_\delta$.

Example 8. (1) The legal string of the following realistic MDS descriptor $\delta = (5, e)(b, 2)(2, 3)(\overline{5}, \overline{4})(3, 4)$ is equal to $w_\delta = 5\,2\,2\,3\,\overline{5}\,\overline{4}\,3\,4$.

(2) Let $\delta = (4, 5)(\overline{2}, \overline{b})(5, e)(\overline{4}, \overline{2})$ be an MDS descriptor. Then we have that $\mu_5(4, 5) = 4\,5$, $\mu_5(\overline{2}, \overline{b}) = \overline{2}$, $\mu_5(5, e) = 5$, and $\mu_5(\overline{4}, \overline{2}) = \overline{4}\,\overline{2}$. Therefore $w_\delta = 4\,5\,\overline{2}\,5\,\overline{4}\,\overline{2}$.   □

Example 9. Consider the actin I gene of *Sterkiella nova* (see Examples 2 and 3), which is described by the MDS arrangement $M_3M_4M_6M_5M_7M_9\overline{M}_2M_1M_8$, and hence by the MDS descriptor

$$\delta = (3, 4)(4, 5)(6, 7)(5, 6)(7, 8)(9, e)(\overline{3}, \overline{2})(b, 2)(8, 9).$$

The legal string obtained from δ is then given by $w_\delta = 3\,4\,4\,5\,6\,7\,5\,6\,7\,8\,9\,\overline{3}\,\overline{2}\,2\,8\,9$.   □

The following example shows that there are legal strings that are not realistic.

Example 10. (1) The string $u = 2\,3\,4\,3\,2\,4$ is legal and $\mathrm{dom}(u) = [2, \kappa]$ for $\kappa = 4$. However, as is easy to see, it is not realistic.

(2) The legal string $u = 22$ is realistic as it is given by the realistic MDS descriptor $\psi_2(M_1M_2) = (b, 2)(2, e)$.   □

We now introduce the *string pointer reduction system* as a formal system modelling the transformation of pointers during the gene assembly process. This system consists of three sets of reduction rules operating on legal strings, corresponding to the micronuclear operations ld, hi, and dlad.

- The *string negative rule* snr_p for a pointer $p \in \Pi$ is applicable to a legal string of the form $u = u_1ppu_2$, where $u_1, u_2 \in \Delta^{\maltese}$. The result of this operation is

$$\mathsf{snr}_p(u_1ppu_2) = u_1u_2. \tag{2}$$

- The *string positive rule* spr_p for a pointer $p \in \Pi$ is applicable to a legal string of the form $u = u_1pu_2\overline{p}u_3$, where $u_1, u_2, u_3 \in \Delta^{\maltese}$. The result of this operation is

$$\mathsf{spr}_p(u_1pu_2\overline{p}u_3) = u_1\overline{u}_2u_3. \tag{3}$$

- The *string double rule* $\mathsf{sdr}_{p,q}$ for pointers $p, q \in \Pi$ with $\mathbf{p} \neq \mathbf{q}$ is applicable to a legal string of the form $u = u_1 p u_2 q u_3 p u_4 q u_5$, where $u_i \in \Delta^{\circledast}$ for each i. The result of this operation is

$$\mathsf{sdr}_{p,q}(u_1 p u_2 q u_3 p u_4 q u_5) = u_1 u_4 u_3 u_2 u_5. \tag{4}$$

We also use the following notations for the applications of the above operations (assuming that the operations are indeed applicable):

$$u \xrightarrow{\ \mathsf{snr}_p\ } \mathsf{snr}_p(u), \quad u \xrightarrow{\ \mathsf{spr}_p\ } \mathsf{spr}_p(u), \quad u \xrightarrow{\ \mathsf{sdr}_{p,q}\ } \mathsf{sdr}_{p,q}(u).$$

For a legal string u, a composition $\varphi = \varphi_n \ldots \varphi_1$ of operations from the set $\{\mathsf{snr}_p, \mathsf{spr}_p, \mathsf{sdr}_{p,q} \mid p, q \in \Pi\}$ is *applicable to u* if φ_1 is applicable to u, and φ_i is applicable to $\varphi_{i-1} \ldots \varphi_1(u)$ for all $1 < i \leq n$. We also say that a composition φ is a *string reduction* of u if φ is applicable to u, and that φ is *successful* for u if $\varphi(u) = \Lambda$, the empty string.

Note that we have only three types of rules for the legal strings. This is to be contrasted with the case of the MDS descriptors for which we had altogether 11 (two for ld and for hi, and seven for dlad).

Example 11. The string $u = 5\,2\,2\,3\,\overline{5}\,\overline{4}\,3\,4$ is legal. The rule snr_2 is applicable to u, and $\mathsf{snr}_2(u) = 5\,3\,\overline{5}\,\overline{4}\,3\,4$. Also, the rules spr_5 and $\mathsf{spr}_{\overline{4}}$ are applicable to u. For these rules, we have $\mathsf{spr}_5(w) = \overline{3}\,\overline{2}\,\overline{2}\,\overline{4}\,3\,4$ and $\mathsf{spr}_{\overline{4}}(w) = 5\,2\,2\,3\,\overline{5}\,3$. Notice that the rule spr_4 is *not applicable* to u, since the string $4\,\overline{4}$ is not a scattered substring of u. For $\varphi = \mathsf{snr}_5\,\mathsf{spr}_3\,\mathsf{spr}_{\overline{4}}\,\mathsf{snr}_2$, we have $\varphi(u) = \mathsf{snr}_5\,\mathsf{spr}_3\,\mathsf{spr}_{\overline{4}}(5\,3\,\overline{5}\,\overline{4}\,3\,4) = \mathsf{snr}_5\,\mathsf{spr}_3(5\,3\,\overline{5}\,3) = \mathsf{snr}_5(5\,5) = \Lambda$, or

$$5\,2\,2\,3\,\overline{5}\,\overline{4}\,3\,4 \xrightarrow{\ \mathsf{snr}_2\ } 5\,3\,\overline{5}\,\overline{4}\,3\,4 \xrightarrow{\ \mathsf{spr}_{\overline{4}}\ } 5\,3\,\overline{5}\,3 \xrightarrow{\ \mathsf{spr}_3\ } 5\,5 \xrightarrow{\ \mathsf{snr}_5\ } \Lambda,$$

and hence φ is successful for u. $\square$

It was shown in [1] and [5] that there is a one-to-one correspondence between the operations ld_p, hi_p, and $\mathsf{dlad}_{p,q}$ on the realistic MDS descriptors and the operations snr_p, spr_p, and $\mathsf{sdr}_{p,q}$ on legal strings. The following theorem gives intertranslatability of the operations on the MDS descriptors and the operations on legal strings.

Theorem 2. *Let δ be a realistic MDS descriptor.*

1. *Let $p \in \Pi$ be a pointer. The rule ld_p is applicable to δ if and only if snr_p is applicable to w_δ. In this case,*

$$\mathsf{snr}_p(w_\delta) = w_{\mathsf{ld}_p(\delta)}.$$

2. *Let $p \in \Pi$ be a pointer. The rule hi_p is applicable to δ if and only if spr_p is applicable to w_δ. In this case,*

$$\mathsf{spr}_p(w_\delta) = w_{\mathsf{hi}_p(\delta)}.$$

3. *Let $p, q \in \Pi$ be pointers such that $\mathbf{p} \neq \mathbf{q}$. The rule $\mathsf{dlad}_{p,q}$ is applicable to δ if and only if $\mathsf{sdr}_{p,q}$ is applicable to w_δ. In this case,*

$$\mathsf{sdr}_{p,q}(w_\delta) = w_{\mathsf{dlad}_{p,q}(\delta)}.$$

Theorem 1 and Theorem 2 together give universality of the string pointer reduction system:

Theorem 3. *Each legal string v has a successful string reduction.*

We also have the following corollary of Lemma 2 using the above results.

Corollary 1. *Let φ be a string reduction for a realistic legal string v. Then also the image $\varphi(v)$ is realistic.*

6 Modelling by Overlap Graphs

We shall now present a model for the gene assembly in ciliates by using (overlap) graphs. The transition from the string pointer reduction system to the graph pointer reduction system, to be considered in this section, is another step towards a more abstract model.

Let $u = a_1 a_2 \ldots a_n \in \Delta^{\maltese}$ be a legal string of pointers. If $p \in \mathrm{dom}(u)$, then there are indices $1 \leq i < j \leq n$ such that $\xi(a_i) = p = \xi(a_j)$. We call the substring $u_{(p)} = a_i a_{i+1} \ldots a_j$ the p-*interval* of u. Pointers p and q *overlap in* u if the p-interval and the q-interval overlap; that is, if $u_{(p)} = a_{i_1} \ldots a_{j_1}$ and $u_{(q)} = a_{i_2} \ldots a_{j_2}$, then either $i_1 < i_2 < j_1 < j_2$ or $i_2 < i_1 < j_2 < j_1$. We denote by $O_u(p)$ (resp. $O_u^+(p)$, $O_u^-(p)$) the set of the letters (resp. positive, negative letters) overlapping with p in u. For technical reasons we include p in $O_u(p)$: if p is positive in u, then $p \in O_u^+(p)$, and if p is negative in u, then $p \in O_u^-(p)$.

Example 12. Let $u = 2\,4\,3\,5\,3\,\overline{2}\,\overline{6}\,\overline{5}\,4\,6$. Then the 2-interval of u is the substring $u_{(2)} = 2\,4\,3\,5\,3\,\overline{2}$. Since $u_{(2)}$ contains one occurrence of 4, the letters 2 and 4 overlap in u. Also, 2 overlaps with 5, but does not overlap with 3 and 6.   □

For a finite set V, let $E(V) = \{\{x, y\} \mid x, y \in V, x \neq y\}$ be the set of all sets of two elements in V. A *signed graph* $\gamma = (V, E, \sigma)$ consists of a set V of *vertices*, a set $E \subseteq E(V)$ of *edges*, together with a labelling $\sigma \colon V \to \{-, +\}$ of the vertices, called the *signing* of γ. Vertices labelled by $+$ are *positive*, and those labelled by $-$ are *negative*. We sometimes write $x^{\sigma(x)}$ for the signed vertex x instead of $\sigma(x) = +$ or $\sigma(x) = -$.

Let $S \subseteq V$ be a subset of the vertices of a signed graph $\gamma = (V, E, \sigma)$. The graph $\gamma' = (V, E', \sigma')$ is obtained from γ by *complementing* S, if for all pairs $\{x, y\}$ of V with $x \neq y$, we have $\{x, y\} \in E'$ if and only if

$$\{x, y\} \in E \ \text{ and } \ x \notin S \text{ or } y \notin S \ \text{ or }$$
$$\{x, y\} \notin E \ \text{ and } \ x \in S \text{ and } y \in S,$$

and $\sigma'(p) = -\sigma(p)$ for all $p \in S$, and $\sigma'(p) = \sigma(p)$ if $p \notin S$. Moreover, if for a vertex $p \in V$ we complement the neighbourhood $N_\gamma(p)$ of p, then we have the (signed) *local complement* $\mathsf{loc}_p(\gamma)$ at p (in γ).

Example 13. Let the signed graph γ be as in Fig. 13(a). Then the graph which is obtained from γ by complementing the set S of shaded vertices is given in Fig. 13(b). □

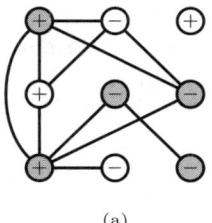
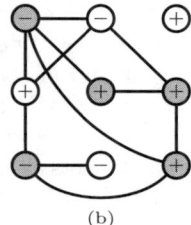

(a)   (b)

Fig. 13. (a) Signed graph γ, and (b) the graph where the set of shaded vertices is complemented

Recall that $\mathbf{p} = \{p, \bar{p}\}$ for each pointer p. Moreover, for each legal string $v \in \Delta^{\maltese}$, let $\mathbf{P}_v = \{\mathbf{p} \mid p \in \mathrm{dom}(v)\}$. We define the *overlap graph* of v as the signed graph $\gamma_v = (\mathbf{P}_v, E, \sigma)$ with the vertex set $\mathbf{P}_v$ such that $\{\mathbf{p}, \mathbf{q}\} \in E$ if and only if p and q overlap in v, and

$$\sigma(\mathbf{p}) = \begin{cases} + , & \text{if } p \text{ is positive in } v, \\ - , & \text{if } p \text{ is negative in } v. \end{cases}$$

Example 14. Let u be the legal string in Example 12. The overlap graph γ_u is then given in Fig. 14, where the signs are given as superscripts of the vertices. □

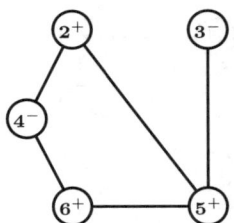

Fig. 14. The overlap graph of $u = 2\,4\,3\,5\,3\,\overline{2}\,\overline{6}\,\overline{5}\,4\,6$

We shall first define the *graph pointer reduction system* and then relate it to the string pointer reduction system on legal strings. Finally, using the results of the previous section we prove that the graph pointer reduction system is universal.

- The *graph negative rule* for a vertex p is applicable to γ if p is isolated and negative in γ. The result $\mathbf{gnr}_p(\gamma)$ is the signed graph $\mathbf{gnr}_p(\gamma)$ obtained from γ by removing the vertex p.
- The *graph positive rule* for a vertex p is applicable to γ if p is positive in γ. The result $\mathbf{gpr}_p(\gamma)$ is the signed graph $\mathbf{gpr}_p(\gamma)$ obtained from the local complement $\mathsf{loc}_p(\gamma)$ by removing the vertex p.
- The *graph double rule* for different vertices p and q is applicable to γ if p and q are adjacent and negative in γ. The result $\mathbf{gdr}_{p,q}(\gamma)$ is the signed graph, where $\mathbf{gdr}_{p,q}(\gamma) = (V \setminus \{p,q\}, E', \sigma')$ is obtained as follows: σ' equals σ restricted to $V \setminus \{p,q\}$, and E' is obtained from E by complementing the edges between the sets $N_\gamma(p)$ and $N_\gamma(q)$. This means that the status of a pair $\{x,y\}$ (for $x, y \in V \setminus \{p,q\}$) as an edge will change if and only if

$$x \in N_\gamma(p) \setminus N_\gamma(q) \quad \text{and} \quad y \in N_\gamma(q),$$
$$x \in N_\gamma(p) \cap N_\gamma(q) \quad \text{and} \quad y \in N_\gamma(q) + N_\gamma(p),$$
$$x \in N_\gamma(q) \setminus N_\gamma(p) \quad \text{and} \quad y \in N_\gamma(p),$$

where $+$ denotes the symmetric difference of the two neighbourhoods (i.e., $N_\gamma(q) \setminus N_\gamma(p) \cup N_\gamma(p) \setminus N_\gamma(q)$).

Example 15. Consider the legal string $w = 3\,\bar{5}\,2\,6\,5\,4\,7\,3\,6\,7\,2\,\bar{4}$. Then $\gamma = \gamma_w$ is given in Fig. 15(a), where the signs of the vertices are represented by superscripts. Recall that the vertices of γ_w are the pointer sets $\mathbf{p} = \{p, \bar{p}\}$ for $p \in [2, 7]$. The operation $\mathbf{gpr}_4$ is applicable to γ, since the vertex $\mathbf{4}$ is positive in the graph. The neighbourhood of $\mathbf{4}$ is $N_\gamma(\mathbf{4}) = \{\mathbf{2, 3, 6}\}$. The graph $\mathbf{gpr}_4(\gamma)$ is given in Fig. 15(b). Also, $\mathbf{gdr}_{2,3}$ is applicable to γ, since these vertices are negative and adjacent. Now $N_\gamma(\mathbf{2}) = \{\mathbf{3, 4, 5, 6}\}$ and $N_\gamma(\mathbf{3}) = \{\mathbf{2, 4, 6, 7}\}$. Therefore $N_\gamma(\mathbf{2}) \cap N_\gamma(\mathbf{3}) = \{\mathbf{4}\}$, $N_\gamma(\mathbf{2}) \setminus N_\gamma(\mathbf{3}) = \{\mathbf{5}\}$, and $N_\gamma(\mathbf{3}) \setminus N_\gamma(\mathbf{2}) = \{\mathbf{6, 7}\}$. Then the graph $\mathbf{gdr}_{2,3}(\gamma)$ is given in Fig. 15(c). $\square$

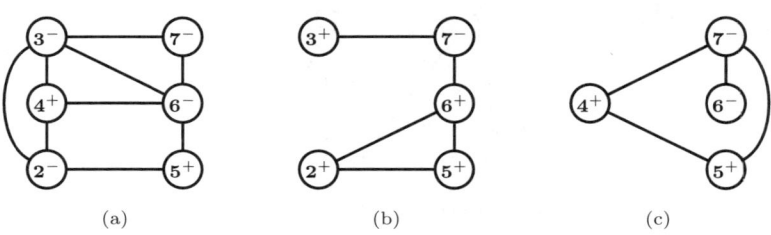

(a)   (b)   (c)

Fig. 15. (a) γ, (b) $\mathbf{gpr}_4(\gamma)$, and (c) $\mathbf{gdr}_{2,3}(\gamma)$

Let γ be a signed graph on the set V of vertices. A composition $\varphi = \varphi_n \ldots \varphi_1$ of operations from $\{\mathsf{gnr}_p, \mathsf{gpr}_p, \mathsf{gdr}_{p,q} \mid p, q \in V\}$ is *applicable to* γ, if φ_1 is applicable to γ, and φ_i is applicable to $\varphi_{i-1} \ldots \varphi_1(\gamma)$ for all $1 < i \leq n$. We say that a composition φ is a *graph reduction* for γ, if φ is applicable to γ, and that φ is *successful* for γ, if $\varphi(\gamma)$ is the empty graph with no vertices.

Example 16. The overlap graph $\gamma = \gamma_w$ given in Fig. 15(a) is reduced to the empty graph by the composition $\mathsf{gpr}_5 \, \mathsf{gpr}_6 \, \mathsf{gpr}_7 \, \mathsf{gpr}_4 \, \mathsf{gdr}_{2,3}$. Indeed, $\gamma_1 = \mathsf{gdr}_{2,3}(\gamma)$ is given in Fig. 15(c), and $\gamma_2 = \mathsf{gpr}_4(\gamma_1)$ has a negative vertex $\mathbf{5}$, positive vertices $\mathbf{6}, \mathbf{7}$ and a single edge $\{\mathbf{5}, \mathbf{6}\}$. Next $\gamma_3 = \mathsf{gpr}_7(\gamma_2)$ is γ_2 with the isolated positive vertex $\mathbf{7}$ removed. Then $\mathsf{gpr}_6(\gamma_3)$ is the graph with a single (positive) vertex $\mathbf{5}$, which is, finally removed by gnr_5. $\square$

The above operations are universal for signed graphs:

Theorem 4. *Let γ be any signed graph. Then there exists a successful graph reduction φ for γ.*

We now study the relationship between the string pointer reduction system and the graph pointer reduction system. Recall that the vertices of the overlap graphs are pointer sets $\mathbf{p}$.

The following theorem was proved in [1] and [5], and it gives the translation from the legal strings to overlap graphs.

Theorem 5. *Let w be a legal string of pointers.*

1. *Let p be a pointer. If snr_p is applicable to w, then $\mathsf{gnr}_{\mathbf{p}}$ is applicable to γ_w. In this case,*
$$\gamma_{\mathsf{snr}_p(w)} = \mathsf{gnr}_{\mathbf{p}}(\gamma_w).$$

2. *Let p be a pointer. If spr_p is applicable to w, then $\mathsf{gpr}_{\mathbf{p}}$ is applicable to γ_w. In this case,*
$$\gamma_{\mathsf{spr}_p(w)} = \mathsf{gpr}_{\mathbf{p}}(\gamma_w).$$

3. *Let $p, q \in \Pi$ be pointers such that $\mathbf{p} \neq \mathbf{q}$. If $\mathsf{sdr}_{p,q}$ is applicable to w, then $\mathsf{gdr}_{\mathbf{p},\mathbf{q}}$ is applicable to γ_w. In this case,*
$$\gamma_{\mathsf{sdr}_{p,q}(w)} = \mathsf{gdr}_{\mathbf{p},\mathbf{q}}(\gamma_w).$$

Therefore, for each string reduction $\varphi = \varphi_n \ldots \varphi_1$ of a legal string w there exists a corresponding graph reduction $\varphi' = \varphi'_n \ldots \varphi'_1$ for the overlap graph γ_w, where the correspondence $\varphi_i \mapsto \varphi'_i$ is obtained by the following cases:

$$\mathsf{snr}_p \mapsto \mathsf{gnr}_{\mathbf{p}}, \quad \mathsf{spr}_p \mapsto \mathsf{gpr}_{\mathbf{p}}, \quad \mathsf{sdr}_{p,q} \mapsto \mathsf{gdr}_{\mathbf{p},\mathbf{q}} \ .$$

Consequently, if φ is successful for w, then φ' is successful for γ_w.

The converse correspondence of the graph reduction rules to the string reduction rules is not as straightforward. Indeed, the mapping $w \mapsto \gamma_w$ from the legal strings to the overlap graphs is not injective; that is, the same overlap graph represents several legal strings.

In the following we denote by $\mathbf{p}(w)$ the *first occurrence* of p or $\bar{p}$ in w.

Example 17. For the legal string $w = 2\,3\,\overline{4}\,\overline{5}\,3\,5\,\overline{4}\,2$ we have $\mathbf{2}(w) = 2$, $\mathbf{3}(w) = 3$, $\mathbf{4}(w) = \overline{4}$, and $\mathbf{5}(w) = \overline{5}$. □

We begin with the positive rules spr_p and $\mathsf{gpr_p}$, and leave the more problematic negative rules as the last case.

Theorem 6. *Let w be a legal string.*

1. *Let p be a pointer. If $\mathsf{gpr_p}$ is applicable to γ_w, then $\mathsf{spr}_{\mathbf{p}(w)}$ is applicable to w. In this case,*
$$\gamma_{\mathsf{spr}_{\mathbf{p}(w)}(w)} = \mathsf{gpr_p}(\gamma_w).$$

2. *Let p and q be pointers such that $\mathbf{p} \neq \mathbf{q}$. If $\mathsf{gdr_{p,q}}$ is applicable to γ_w, then $\mathsf{sdr}_{\mathbf{p}(w),\mathbf{q}(w)}$ is applicable to w. In this case,*
$$\gamma_{\mathsf{sdr}_{\mathbf{p}(w),\mathbf{q}(w)}(w)} = \mathsf{gdr_{p,q}}(\gamma_w).$$

There is no similar result for the negative rules snr_p and $\mathsf{gnr_p}$. Indeed, consider the legal string $w = pqqp$. Now the vertex $\mathbf{p}$ of the overlap graph γ_w is negative and it is isolated in γ_w. Hence $\mathsf{gnr_p}$ is applicable to γ_w. However, snr_p is not applicable to the legal string w.

We shall now show that one can *translate* successful graph reductions into successful string reductions. For this we need the notion of permutation of a composition.

We say that a composition φ of string operations (resp. graph operations) is *canonical* if it has the form $\varphi = \rho\omega$, where $\rho = \mathsf{snr}_{p_n}\,\mathsf{snr}_{p_{n-1}}\ldots\mathsf{snr}_{p_1}$ consists of string negative rules only, and the composition ω has no string negative rules (resp. the graph negative rules $\mathsf{gnr_p}$).

We then have

Lemma 3. *Let w be a legal string. Then there exists a canonical string reduction successful for w.*

The corresponding result for graphs is given in the following lemma.

Lemma 4. *Let γ be a signed graph, and let $\varphi = \varphi_n \ldots \varphi_1$ be a successful graph reduction for γ. Then there exists a permutation $(i_1 \ldots i_n)$ of $[1,n]$ such that $\varphi' = \varphi_{i_n} \ldots \varphi_{i_1}$ is a canonical graph reduction successful for γ.*

Two compositions $\rho = \rho_n \ldots \rho_1$ and $\rho' = \rho'_n \ldots \rho'_1$ of string operations (resp. graph operations) are said to be *permutations* of each other if there is a permutation $(i_1 \ldots i_n)$ of $[1,n]$ such that $\rho'_j = \rho_{i_j}$.

Lemma 5. *Let $\varphi = \rho\omega$ be a canonical graph reduction, where ρ consists of the graph negative rules in φ. If φ is successful for γ, then so is $\varphi' = \rho'\omega$ for each permutation ρ' of ρ.*

Note that Lemma 5 does not have a corresponding result for strings. Indeed, for $w = qppq$, the string reduction $\mathsf{snr}_q \, \mathsf{snr}_p$ is successful for w, while $\mathsf{snr}_p \, \mathsf{snr}_q$ is not.

The following theorem gives the translation from the overlap graphs to legal string in the case of the double rules.

Theorem 7. *Let w be a legal string, and let $p_1, \ldots, p_n$ be pointers. If $\varphi = \mathsf{gnr}_{p_n} \cdots \mathsf{gnr}_{p_1}$ is successful for γ_w, then there exists a permutation $(i_1 \ldots i_n)$ of $[1, n]$ such that $\varphi' = \mathsf{snr}_{p_{i_n}(w)} \cdots \mathsf{snr}_{p_{i_1}(w)}$ is successful for w and $\varphi'' = \mathsf{gnr}_{p_{i_n}} \cdots \mathsf{gnr}_{p_{i_1}}$ is successful for γ_w.*

As a conclusion, let w be a legal string for which there exists a successful graph composition φ for γ_w. Then there exists a permutation $\varphi'' = \varphi_n \ldots \varphi_1$ of φ which can be translated to a successful string reduction $\varphi' = \varphi'_n \ldots \varphi'_1$ for w by the following translations:

$$\mathsf{gnr}_{\mathbf{p}} \mapsto \mathsf{snr}_{\mathbf{p}(w)}, \quad \mathsf{gpr}_{\mathbf{p}} \mapsto \mathsf{spr}_{\mathbf{p}(w)}, \quad \mathsf{gdr}_{\mathbf{p,q}} \mapsto \mathsf{sdr}_{\mathbf{p}(w),\mathbf{q}(w)} \ .$$

Acknowledgements. G. Rozenberg gratefully acknowledges partial support by NSF grant 0121422.

References

1. Ehrenfeucht, A., Harju, T., Petre, I., Prescott, D. M., Rozenberg, G., Formal systems for gene assembly in ciliates. *Theor. Comput. Sci.* **292**, 199–219 (2003).
2. Ehrenfeucht, A., Harju, T., Petre, I., Rozenberg, G., Characterizing the micronuclear gene patterns in ciliates. *Theory Comput. Syst.* **35**, 501–519 (2002).
3. Ehrenfeucht, A., Harju, T., Rozenberg, G., Gene assembly through cyclic graph decomposition. *Theor. Comput. Syst.* **281**, 325–349 (2002).
4. Ehrenfeucht, A., Petre, I., Prescott, D. M., Rozenberg, G., Universal and simple operations for gene assembly in ciliates. In *Words, Sequences, Languages: Where computer science, biology and linguistics come across*, ed by Mitrana, V., Martin-Vide, C. (Kluwer Academic Publishers, Dortrecht/Boston 2001), 329–342.
5. Ehrenfeucht, A., Petre, I., Prescott, D. M., Rozenberg, G., String and graph reduction systems for gene assembly in ciliates. *Math. Struct. Comput. Sci.* **12**, 113–134 (2001).
6. Ehrenfeucht, A., Petre, I., Prescott, D. M., Rozenberg, G., Circularity and other invariants of gene assembly in ciliates. In *Words, semigroups, and transductions*, ed by Ito, M., Păun, Gh., Yu, S. (World Scientific, Singapore 2001) pp 81–97.
7. Ehrenfeucht, A., Prescott, D. M., Rozenberg, G., Computational aspects of gene (un)scrambling in ciliates. In *Evolution as Computation*, ed by Landweber, L., Winfree, E. (Springer, Berlin Heidelberg New York 2001) pp 45–86.
8. Landweber, L. F., Kari, L., The evolution of cellular computing: nature's solution to a computational problem. In *Proceedings of the 4th DIMACS meeting on DNA based computers*, ed by Kari, L., Rubin, H., Wood, D., Philadelphia, 1998, pp 3–15.

9. Landweber, L. F., Kari, L., Universal molecular computation in ciliates. In *Evolution as Computation*, ed by Landweber L., Winfree E. (Springer, Berlin Heidelberg New York 2002).

10. Păun, Gh., Rozenberg, G., Salomaa, A., *DNA Computing* (Springer, Berlin Heidelberg New York 1998).

11. Prescott, D. M., Cutting, splicing, reordering, and elimination of DNA sequences in hypotrichous ciliates. *BioEssays* **14**, 317–324 (1992).

12. Prescott, D. M., The unusual organization and processing of genomic DNA in hypotrichous ciliates. *Trends Genet.* **8**, 439–445 (1992).

13. Prescott, D. M., The DNA of ciliated protozoa. *Microbiol. Rev.* **58**(2), 233–267 (1994).

14. Prescott, D. M., The evolutionary scrambling and developmental unscabling of germlike genes in hypotrichous ciliates. *Nucleic Acids Res.* **27**, 1243–1250 (1999).

15. Prescott, D. M., Genome gymnastics: unique modes of DNA evolution and processing in ciliates. *Nat. Rev. Genet.* **1**(3), 191–198 (2000).

16. Prescott, D. M., DuBois, M., Internal eliminated segments (IESs) of Oxytrichidae. *J. Eukariot. Microbiol.* **43**, 432–441 (1996).

17. Prescott, D. M., Ehrenfeucht, A., Rozenberg, G., Molecular operations for DNA processing in hypotrichous ciliates. *Eur. J. Protistol.* **37**, 241–260 (2001).

18. Prescott, D. M., Rozenberg, G., How ciliates manipulate their own DNA – A splendid example of natural computing. *Nat. Comput.* **1**, 165–183 (2002).

Towards Molecular Programming – a Personal Report on DNA8 and Molecular Computing

Masami Hagiya

JST CREST and Department of Computer Science, Graduate School of
Information Science and Technology, University of Tokyo, Japan
hagiya@is.s.u-tokyo.ac.jp

Summary. According to the author's view, research in the field of DNA and
molecular computing is expanding into broader applications in nanotechnology and
biotechnology, and the principles and methods for designing molecular systems with
information-processing capability for such applications are considered important.
We call research into designing such molecular systems *molecular programming*.
This chapter first summarizes a recent international conference in this field, *DNA8,
the Eighth International Meeting on DNA Based Computers*, to support the above,
rather personal view on the field. Next, it reviews existing models of DNA and
molecular computation and analyses the results of these models, and then briefly
describes some methods for molecular programming, including sequence design. It
finally touches on molecular machines made of DNA, one of the current focuses of
molecular programming.

1 Introduction

Several years have passed since the direction of research in the field of DNA
and molecular computing began to expand from purely mathematical computation into much broader applications in nanotechnology and biotechnology [11, 12]. This means that principles and methods developed in the field
are now being applied to non-mathematical problems, such as constructing
molecular machines and analysing the human genome. A few years ago, while
developing his DNA computer for solving 3-SAT problems [47], Suyama in-
sisted that if a DNA computer could solve 3-SAT problems accurately, then
the computer could be used to analyse the human genome and produce reliable
results that could be used for medical diagnosis [34, 40].

Another possible direction in research is towards applying evolutionary
computation to molecular evolution. It is expected that applying methods
established in evolutionary computation, such as the genetic algorithm or
programming, to molecular evolution will be fruitful [31]. Since evolution can
be regarded as a kind of computation, this direction is considered to be yet
another application of molecular computing in a broad sense.

These new directions in the field were also topics at a recent international conference on DNA and molecular computing, i.e., *DNA 8, the Eighth International Meeting on DNA Based Computers* [14], which was held in Sapporo, Japan, on June 10–13, 2002. Here, I briefly touch on the conference and summarize current research trends from a rather personal view.

Research in this field is gradually being directed towards developing principles and methods for designing molecules and molecular systems that solve not only purely computational problems, but also the broader problems mentioned above. To solve such problems, a computing perspective is still important, because some kind of information processing is always involved. For example, molecular machines should have information-processing capability at the molecular level. Analyzing human genomes is nothing but processing information on genes and their products. Any molecular system that is capable of information processing is considered a subject of this field.

In order to construct such molecular systems, we need to have a deep understanding of what molecules can compute. This is exactly the question that researchers in the field of DNA and molecular computing have been hoping to answer, and many models of DNA and molecular computation have been proposed. In this chapter, I very briefly summarize the models used for molecular computation and the results of analyses of the computational power of the models, including computability and complexity.

The increasing power of these models is allowing the field to expand as they are applied to broader applications. This requires research on design and construction. I use the term *molecular programming* to include research into designing molecules and molecular systems with information-processing capabilities. The word *programming* suggests that designing molecules and molecular systems is like programming electronic computers. A typical example of molecular programming is designing DNA nucleotide sequences. Sequence design is one of the most important aspects of DNA computing, because it greatly influences the accuracy and efficiency of DNA computation. By hybridization, DNA molecules interact with each other and form complex structures or even machines encoded in their sequences. Moreover, the sequences of molecules with enzymatic functions, such as RNA and proteins, control their own reactions. Designing the sequences of such molecules is simply programming their behaviour.

Needless to say, programming molecules and molecular systems involves many complex computational problems that are themselves good research subjects in computer science. In this chapter, I briefly describe some methods of molecular programming, including sequence design. Finally, I touch on molecular machines made of DNA, which are currently the main target of molecular programming.

2 DNA8

DNA8, the Eighth International Meeting on DNA Based Computers, was held on June 10–13, 2002, at Hokkaido University, in Sapporo, Japan [14]. There were about 100 participants, including around 50 from overseas. I chaired the programming committee, and Azuma Ohuchi, of Hokkaido University, chaired the organizing committee. This conference series was organized under the aegis of the steering committee chaired by Grzegorz Rozenberg of the University of Leiden.

The sessions at DNA8 are listed below, in order for the reader to grasp a rough idea of the conference.

- Molecular Evolution
- Computing by Self-Assembly
- DNA Computing and Nanotechnology
- Applications to Graph Problems
- Applications to Biotechnology and Engineering
- Nucleic Acid Sequence Design
- Theory
- Autonomous Molecular Computation

The sessions *Computing by Self-Assembly, DNA Computing and Nanotechnology*, and *Autonomous Molecular Computation* were all concerned with DNA nanotechnology, an emerging subfield of nanotechnology that is based on DNA and related molecules. In particular, the session *Autonomous Molecular Computation* was about molecules that move, i.e., molecular machines.

There were two particularly interesting talks in the session on *Applications to Biotechnology and Engineering*. One was the invited talk by Suyama about applications of DNA computing to genomic analysis. The other was by Basu *et al.*, who are trying to engineer signal processing in *E. coli* to make a sensor that detects a band at a particular molecular concentration [4].

The session on *Nucleic Acid Sequence Design* included many interesting talks about sequence design. In DNA8, interest moved towards secondary structure design, i.e., designing the sequences that fold into intended secondary structures [15].

In summary, although classical Adleman-style DNA computing is still actively pursued, it is no longer the main theme of the field, while applications to nanotechnology and biotechnology are becoming central issues. The following are some observations about DNA8 to support this view.

- The field of DNA nanotechnology treats both the construction and motion of structures.
- Even sequence design is geared towards DNA nanotechnology, as secondary structure design is becoming an important issue.
- Applications in genomic analysis are also becoming realistic. For example, Suyama proposed using his DNA computer for gene expression analysis

and SNP analysis. Cell engineering is no longer a dream. Weiss *et al.* are trying to engineer cells to make sensors. This is a movement towards *in vivo* molecular computing [41].

• DNA computing has flourished as a result of theoretical contributions related to the splicing system (or H-system) [28] and membrane system (or P-system) [27]. There were also some interesting and important talks in the theory session of DNA8. However, although new theoretical contributions are consistently being made, they are rather limited compared with the major contributions in the past, and the theory of DNA computing needs breakthroughs.

Lastly, I must mention that a number of chemists attended the conference, including Tomoji Kawai of Osaka University, who gave an invited talk about DNA nanotechnology, and Sen, who talked about electron transfer in DNA. Research on nanotechnology will involve increasing cooperation with chemistry and increasing numbers of chemists will likely develop an interest in the field.

3 Models for Molecular Computation and Their Analysis

In this section, I briefly summarize the models for molecular computation and analyse their computability and complexity. The analysis is intended to reveal the computational power of molecules and molecular reactions.

3.1 Computational Models

Adleman–Lipton Paradigm and its Refinements

I do not intend to explain the Adleman–Lipton paradigm in detail here [1, 21]. It is the first paradigm for DNA computing, and became the basis of many succeeding computational paradigms. It is a paradigm for solving a combinatorial search problem by 1) randomly generating solution candidates by hybridization of DNA, where each hybridized DNA molecule encodes a candidate, and 2) extracting solutions using data-parallel computation, where each experimental operation in a test tube is applied in parallel to all the molecules in the test tube.

I will mention two refinements of the paradigm. To solve 3-SAT problems, Suyama proposed a molecular algorithm in which solution candidates are not generated at once, but partial candidates are generated and gradually extended to form complete ones [47]. Once the partial candidates have been generated, those that can never be completed are immediately removed. This strategy reduces the number of candidates generated during computation. Suyama calls the strategy *dynamic programming*, because variables are ordered and each variable is processed according to the result of processing its preceding variables. Ogihara and Ray proposed a similar algorithm that they called *counting* [26].

Sakamoto *et al.* developed another refinement for 3-SAT problems [35]. Their machinery is called the *SAT Engine*, and it makes use of hairpin structures in DNA molecules (Fig. 1). In the SAT Engine, complementary literals are encoded by complementary nucleotide sequences in the sense of Watson and Crick. If a single-stranded DNA molecule contains two literals that are inconsistent with each other, i.e., a variable and its negation, then the molecule forms a hairpin. This means that inconsistent assignments correspond to molecules containing a hairpin, so a SAT problem can be solved by removing hairpin molecules and checking whether consistent assignments remain.

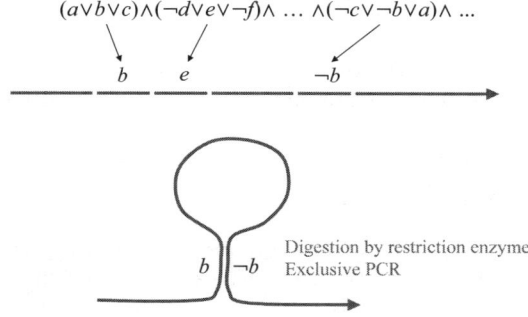

Fig. 1. Sakamoto's SAT Engine

Seeman–Winfree: Computing by Self-Assembly

Seeman, of the University of New York, is a central figure in DNA nanotechnology. He received the 1995 Feynman Prize in Nanotechnology at the Fourth Foresight Conference on Molecular Nanotechnology for his research on the synthesis of three-dimensional objects from DNA. Seeman has invented various DNA structures, while Winfree of the California Institute of Technology has proposed computational models based on self-assembly of those structures.

Among the structures that Seeman invented are *DNA tiles* [39]. *Double-crossover molecules* are examples of DNA tiles (Fig. 2, lower left). Each double-crossover molecule is made of two double-stranded molecules that exchange single strands at two points. Therefore, the tile is composed of four single-stranded DNA molecules that self-assemble.

Double-crossover molecules have four sticky ends, which each hybridize with the sticky end of another molecule (Fig. 2, right). Therefore, they can

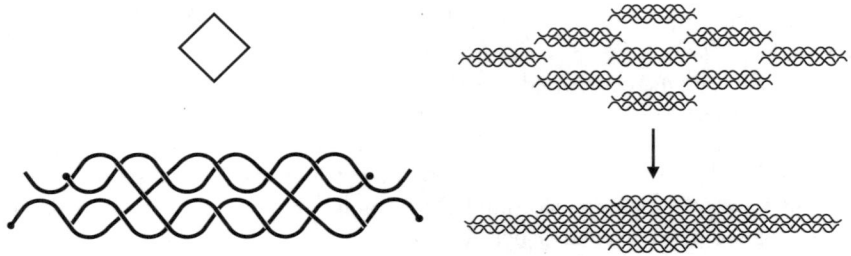

Fig. 2. Seeman's DNA tiles

self-assemble and form a planar structure [44]. This process of self-assembly corresponds to allowing square tiles with colored edges to hybridize only if adjacent edges are of the same color (Fig. 2, upper left).

As mentioned below, the process of tiling square tiles with colored edges is universal [45], because it can simulate execution of one-dimensional cellular automata.

Head: Splicing System

The *splicing system* introduced by Head, also known as the *H-system*, is a formal model of DNA recombination, i.e., digestion by restriction enzymes and concatenation by ligase [28]. It is a model that can be used to generate languages, by regarding nucleotide sequences as strings of the letters $\{A, G, C, T\}$.

Digestion and ligation are modelled using a set of splicing rules. A *splicing rule* is a string of the form $u_1 \# u_2 \$ u_3 \# u_4$. For the splicing rule $r = u_1 \# u_2 \$ u_3 \# u_4$, we define the relation $\vdash_r$ between pairs of strings as follows:

$$(x_1 u_1 u_2 x_2, y_1 u_3 u_4 y_2) \vdash_r (x_1 u_1 u_4 y_2, y_1 u_3 u_2 x_2)$$

Let R be a set of splicing rules, and A a set of initial strings called axioms. Then, the language L generated by R and A is inductively defined as follows:

- If $x \in A$, then $x \in L$.
- If $x, y \in L$, $r \in R$ and $(x, y) \vdash_r (z, w)$, then $z, w \in L$.

It is known that if R and A are finite, then L is regular.

Other Models

Other models for molecular computation include the *Sticker Model* [33], which is another enhancement of the Adleman–Lipton paradigm. There are also models for parallel computation of Boolean circuits, such as Ogihara and Ray's model [25]. Many models have been proposed for molecular state machines or finite automata, including those of Hagiya and Sakamoto [13, 19, 36] and Shapiro [6]. The last two will also be referred to at the end of this chapter.

3.2 Computability

Considering the computability of molecular computations, I mention two typical results here: one by Winfree concerning the computability of self-assembly of DNA, and the other by Head and his colleagues, who extended the splicing model in various ways to achieve universality of gene recombination.

The computational power of self-assembly by DNA molecules has been thoroughly investigated by Winfree [45]. He realized that languages generated by the hybridization of linear molecules are regular. Here, by linear molecules, I mean double-stranded DNA molecules with sticky ends that allow them to hybridize with each other. The set of all double-stranded molecules that can be generated by hybridization of copies of given linear molecules is regular, if we consider double-stranded molecules as strings of the letters $\{A, G, C, T\}$. Winfree pointed out that languages generated by linear, hairpin, and three-junction molecules are context-free. By hybridizing copies of those molecules with their sticky ends and ligating them together, we obtain single-stranded DNA molecules after denaturation. The set of these single strands is context-free as a language over $\{A, G, C, T\}$.

Finally, Winfree showed that languages generated by linear and double-crossover molecules are recursively enumerable. This beautiful result was obtained by simulating a one-dimensional cellular automaton using the two-dimensional self-assembly of DNA tiles. Note that one-dimensional cellular automata are Turing complete because they can simulate Turing machines.

Another important result concerning the computability of molecular computations was obtained by various extensions to the splicing model. As mentioned before, the original splicing system can only generate regular languages. Therefore, many theoreticians explored ways to extend the model to achieve universal computability [28]. Some extensions that are universal are:

- splicing circular molecules,
- using multiple tubes, and
- making rules time-varying.

3.3 Complexity

The computational complexity of molecular computation is measured in time and space, as are other kinds of computation. In molecular computing, time complexity has two aspects: the number of laboratory operations and the time taken for each operation. In particular, the latter is important from the viewpoint of analysing the computational power of molecular reactions, because the nature of a reaction determines the time required for obtaining a product, i.e., a computational result. (In some computational schemes, the time for a single operation grows exponentially as the size of a problem increases, while in some others, it is constant.) The space complexity of molecular computation corresponds to the number of required molecules and implies the degree of parallelism. The size or length of molecules should also be considered in order to accurately measure space complexity. In general, molecular computation gains time efficiency by sacrificing space efficiency, which is typical of the Adleman–Lipton paradigm. Therefore, it is important to analyse the trade-off between the two measures.

Some classical examples of analysing the complexity of molecular computation, in which time is measured only by the number of laboratory operations, include that of Reif, who showed that a non-deterministic Turing machine computation with input size n, space s, and time $2^{O(s)}$ can be executed in his PAM Model using $O(s)$ PA-Match steps and $O(s \log s)$ other PAM steps, employing aggregates of length $O(s)$ [29]. Beaver showed that polynomial-step molecular computers compute PSPACE [5]. Rooß and Wagner proved that the problems in $P^{NP} = \Delta_2^P$ can be solved in polynomial time using Lipton's model [30].

As for the complexity of self-assembly, Rothemund and Winfree proved that for any non-decreasing unbounded computable function $f(N)$, the number of tiles required for the self-assembly of an $N \times N$ square is bounded infinitely often by $f(N)$ [32]. Winfree et al. showed that the linear assembly of string tiles can generate the output languages of finite-visit Turing Machines [43].

No chemical reaction can be executed instantaneously. The efficiency of a chemical reaction depends on the yield of the reaction, which is related to its equilibrium constant, K, and the time to reach the equilibrium, which is related to the reaction constant k. For example, in the reaction $A \leftrightarrow B$, the concentration of B, denoted by $[B]$, is given as a function of elapsed time t as follows:

$$[B] = (K/(1 + K))(1 - e^{-(k+k_{-1})t})$$

In the equation, k and k_{-1} are the forward and backward reaction constants, respectively, and the equilibrium constant K is given by $K = k/k_{-1}$.

Note that no chemical reaction can escape error. A typical error in reactions involving DNA is mishybridization, which means that non-complementary nucleotide sequences hybridize. Since the probability of such error can never be zero, probabilistic analysis is essential in molecular computation. Some examples of probabilistic analysis of molecular computation follow. Karp et al. proved that the number of extract operations required to achieve error-resilient bit evaluation is $\Theta(\lceil \log_\epsilon \delta \rceil \times \lceil \log_\gamma \delta \rceil)$ [17]. Kurtz made a thermodynamic analysis of path formation in Adleman's experiment and showed that the time needed to form a Hamiltonian path is $\Omega(n^2)$ [20]. Winfree also made a thermodynamic analysis of DNA tiling in his PhD thesis [42].

4 Molecular Programming

Molecular programming aims at establishing systematic design principles for molecules and molecular systems with information-processing capabilities, and developing methods to ease their design and construction. In order to achieve this goal, we must solve many computational problems, which are also significant from the viewpoint of computer science.

Methods in molecular programming are roughly classified into those for designing molecules and those for designing molecular reactions. To design molecules, the problem of designing nucleotide sequences is an important issue in DNA computing. To design molecular reactions, models for DNA reactions, such as hybridization and denaturation, have been developed, as have methods for simulating those models using electronic computers. They are aimed at tuning reaction conditions to realize accurate and efficient reactions.

In this section, I briefly review the methods for sequence design used in DNA computing and those for simulating DNA reactions. I finally touch on molecular machines made of DNA, which are currently the main targets of molecular programming. They are molecules that change their state according to reactions with other molecules or environments.

4.1 Designing Molecules

Sequence Design

Sequence design has always been an important issue in DNA computing. It first aimed at designing a large library of nucleotide sequences that only hybridize with their complementary sequences, avoiding mishybridization. Research on sequence design began with defining a measure for evaluating the possibility of mishybridization in a library of nucleotide sequences. One well-established measure is the H-measure developed by Garzon *et al.*, which is computed using the Hamming distance, i.e., the number of mismatches, while considering frame shifts [10].

The genetic algorithm and related methods have been used to find libraries of nucleotide sequences with a maximum H-measure greater than a given threshold. Another solution to the problem was recently proposed by Arita [3] and elaborated on by Kobayashi [18]. Called the *template method*, it is an enhancement of the method originally proposed by Condon *et al.* [9].

A *template* is a sequence of zeros and ones. Either G or C is substituted for 1, and either A or T is substituted for 0. For example, from the template 011010, sequences such as $ACCTGA$, $TGCTCA$, and $TCGACA$ are obtained. Given a template of length n, a library of nucleotide sequences of size 2^n is obtained. Note that such a library satisfies the condition that all the sequences have the same GC content; therefore, their melting temperature is uniform.

In order to obtain a good library, template T should satisfy the condition that the following patterns always contain at least d mismatches with T with any frame shift:

$$T^R \quad TT^R \quad T^R T \quad TT \quad T^R T^R$$

In the pattern, T^R denotes the reverse of T. If $T = 110100$, then $T^R = 001011$. Using such a template, at least d mismatches are induced with shifting and reversal of sequences.

Here are some examples of templates. The following template has a length 6 and 2 mismatches:

$$110100$$

This is one of 2^6 such templates. For a length of 11 with 4 mismatches, we have

$$01110100100$$
$$01011100010$$
$$11000100101$$

For length 23 with 9 mismatches, we have

$$01111010110011001010000$$
$$10110011001010000011110$$
$$11100000101001100110101$$

Using a template of length n and with d mismatches, a library of DNA sequences is obtained by adopting an error-correcting code, such as BCH, Golay, Hamming, etc. If an n-bit code whose Hamming distance is d is used, then a library of DNA sequences is obtained by substituting G (or C) for 1 if the corresponding bit in a code word is 0 (or 1), and A or T for 0 if it is 0 (or 1). Given the conditions for the template, d mismatches are guaranteed with shifts and reverses. Otherwise, d mismatches are guaranteed with the code used.

Currently, sequence design that avoids both mishybridization with respect to the H-measure and unwanted secondary structures is being investigated [2].

Inverse Folding

Another direction in sequence design is designing a sequence that folds into a given secondary structure. This problem is called *inverse folding*, because it is the inverse of the problem of finding the secondary structure of a sequence with the minimum free energy. The inverse folding problem is to find a sequence whose minimum energy structure coincides with the given one.

The Vienna group has been working on the inverse folding problem and the folding problem [16], using dynamic programming as in the *mfold* of Zuker [49]. More accurately, the problem is formulated as follows. Let Ω be the target structure, and x be a sequence whose structure has a free energy of $E(x, \Omega)$. If $G(x)$ is the ensemble free energy of x, obtained using McCaskill's algorithm [23], then the probability p of x having the structure Ω is as follows:

$$p = \frac{e^{-E(x,\Omega)/RT}}{e^{-G(x)/RT}}$$

Therefore, the cost function $\Xi(x)$ of sequence x can be defined as

$$\Xi(x) = E(x,\Omega) - G(x) = -RT\ln p$$

The problem is then reduced to finding x, such that $\Xi(x)$ is minimized. The free energy of a secondary structure is computed using the nearest neighbour model, a standard model based on stacking energies [7].

4.2 Designing Molecular Reactions

In order make molecular reactions accurate and efficient, it is important to tune the reaction conditions such as temperature, salt concentration, and duration. The order of operations is also important. Simulating molecular reactions is considered a useful tool to find appropriate reaction conditions.

Molecular reactions should be simulated using good models. For example, for DNA hybridization, the nearest neighbour model based on stacking energies is widely used [7,37]. This model is also the basis for designing sequences that form target secondary structures. The staggering zipper model is an enhancement, in which hybridization consists of two steps. In the nucleation step, a nucleus of hybridization is formed between two strands, and in the zipping step, the nucleus is extended to form a complete hybrid.

Using such models, tools have been developed for simulating reactions among DNA molecules, such as e-PCR [38] and *the Virtual Nucleic Acid simulator* (VNA) [24].

VNA simulates reactions among molecules consisting of virtual bases [24]. A molecule in this simulator is a hybrid of virtual strands, such as

```
abcd
| |
CDEF
```

Each character is a virtual base, representing a nucleotide sequence of appropriate length, say 10 bases. Corresponding upper- and lowercase letters represent complementary sequences. The simulator supports the following reactions among such molecules: reactions, hybridization, denaturation, digestion, ligation, self-hybridization, and extension.

The simulator can verify the feasibility of algorithms for DNA computing, verify the validity of molecular biology experiments, such as PCR, and fit parameters in molecular biology experiments. Examples include Ogihara and Ray's computation of Boolean circuits [25], Winfree's construction of double-crossover molecules, and simulated PCR experiments.

VNA is implemented in Java and is executable as an applet. In order to simulate reactions among dynamically generated molecules, it combines combinatorial enumeration of molecules with continuous simulation based on differential equations. Interestingly, the concentrations of molecules are computed using differential equations to avoid a combinatorial explosion by setting a threshold on the concentration of a newly generated molecule. This also supports stochastic simulation.

Although it is still far from being able to simulate real reactions faithfully, because most of the reaction constants are not known, VNA has been used to examine the possibility of combining various reactions to find expected or unexpected reaction paths.

In general, it is becoming increasingly important to grasp both the static and dynamic behaviour of molecules, because research in DNA and molecular computing is moving towards designing molecules that move or change their states.

One of the next steps in this direction will be to estimate the energy barrier between two secondary structures of DNA or RNA. The Vienna group has already estimated energy barriers to design multi-state RNA molecules, and this is considered an important area of future research, particularly for designing molecular machines [8].

4.3 Molecular Machines

As mentioned in the previous section, there are many proposals and experiments for designing and implementing molecular machines out of DNA. Some of these include:

- Seeman's DNA motor using B-Z transition [22]
- Hagiya's whiplash PCR [13, 19, 36] (Fig. 3)
- Yurke's molecular tweezers [48]
- Seeman's PX-JX2 switch [46]
- Shapiro's DNA automata [6] (Fig. 4)

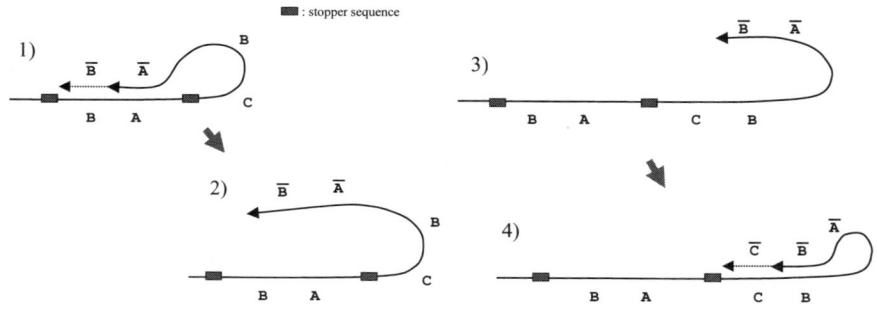

Fig. 3. Hagiya's Whiplash PCR

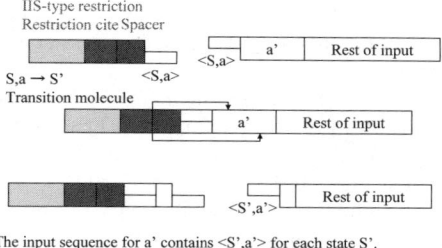

IIS-type restriction
Restriction cite Spacer

The input sequence for a' contains <S',a'> for each state S'.
The transition molecule cuts the input at the right place by the spacer.

Fig. 4. Shapiro's DNA Automata

5 Concluding Remarks

In this chapter, I summarized the research in DNA and molecular computing and research trends observed at DNA8, the recent international conference in this field, to explain the importance of research on molecular programming, i.e., designing molecules and molecular systems with information-processing capabilities.

In Japan, a research group on molecular programming funded as *Priority Area Research* by the Ministry of Education has recently been formed. In addition, a CREST project has been funded by the Japan Science and Technology Corporation to research *molecular addressing* for constructing molecular memory out of DNA and related molecules. It is also intended to explore new directions in molecular computing, including *optical molecular computing*.

The preparation of this chapter was supported by both of these projects.

References

1. Adleman, L.M.: Molecular computation of solutions to combinatorial problems, Science **266**, 1021–1024 (1994).
2. Andronescu, M., Dees, D., Slaybaugh, L., Zhao, Y., Condon, A.E., Cohen, B., Skiena, S.: Algorithms for testing that DNA word designs avoid unwanted secondary structure. In: *Proceedings of the Eighth International Meeting on DNA Based Computers* (2003) 92–104.
3. Arita, M., Kobayashi, S.: DNA sequence design using templates, New Generation Computing **20**, 263–277 (2002).
4. Basu, S., Karig, D., Weiss, R.: Engineering signal processing in cells: Towards molecular concentration band detection. In: *Proceedings of the Eighth International Meeting on DNA Based Computers* (2003) 80–89.
5. Beaver, D.: A universal molecular computer. In: *DNA Based Computers*. DIMACS Series in Discrete Mathematics and Theoretical Computer Science vol.**27** (1996) 29–36.

6. Benenson, Y., Paz-Elizur, T., Adar, R., Heina, E., Livneh, Z., Shapiro, E.: Programmable and autonomous computing machine made of biomolecules, Nature **414**, 430–434 (2001).

7. Cantor, C.R., Schimmel, P.R.: *Biophysical Chemistry, Part III: The behavior of biological macromolecules* (W.H. Freeman, San Francisco 1980).

8. Flamm, C., Hofacker, I.L., Maurer-Stroh, S., Stadler, P.F., Zehl, M.: Design of multistable RNA molecules, RNA **7**, 254–265 (2001).

9. Frutos, A.G., Liu, Q., Thiel, A.J., Sanner, A.M.W., Condon, A.E., Smith, L.M., Corn, R.M.: Demonstration of a word design strategy for DNA computing on surfaces, Nucleic Acids Research **25**(23), 4748–4757 (1997).

10. Garzon, M., Neathery, P., Deaton, R.J., Murphy, R.C., Franceschetti, D.R., Stevens, S.E. Jr.: A new metric for DNA computing. In: *Proceedings of 2nd Annual Genetic Programming Conference* (1997) 472–478.

11. Hagiya, M.: Perspectives on molecular computing, New Generation Computing **17**(2), 131–140 (1999).

12. Hagiya, M.: From molecular computing to molecular programming. In *DNA Computing, 6th International Workshop on DNA Based Computers, DNA 2000.* Lecture Notes in Computer Science, vol 2054 (Springer, Berlin Heidelberg New York 2001) 89–102.

13. Hagiya, M., Arita, M., Kiga, D., Sakamoto, K., Yokoyama, S.: Towards parallel evaluation and learning of Boolean μ-formulas with molecules. In: *DNA Based Computers III.* DIMACS Series in Discrete Mathematics and Theoretical Computer Science vol.48 (1999) 57–72.

14. Hagiya, M., Ohuchi, A. (Eds.): *Proceedings of the Eighth International Meeting on DNA Based Computers.* Lecture Notes in Computer Science, vol.2568 (Springer, Berlin Heidelberg New York 2003).

15. Heitsch, C.E., Condon, A.E., Hoos, H.H.: From RNA secondary structure to coding theory: A combinatorial approach. In: *Proceedings of the Eighth International Meeting on DNA Based Computers* (2003) 125–136.

16. Hofacker, I.L., Fontana, W., Stadler, P.F., Bonhoeffer, L.S., Tacker, M., Schuster, P.: Fast folding and comparison of RNA secondary structures, Monatshefte Chem. **125**, 167–188 (1994).

17. Karp, R., Kenyon, C., Waarts, O.: Error-resilient DNA computations. In: *Seventh ACM-SIAM Symposium on Discrete Algorithms* (1996) 458–467.

18. Kobayashi, S., Kondo, T., Arita, M.: On template method for DNA sequence design. In: *Proceedings of the Eighth International Meeting on DNA Based Computers* (2003) 115–124.

19. Komiya, K., Sakamoto, K., Gouzu, H., Yokoyama, S., Arita, M., Nishikawa, A., Hagiya, M.: Successive state transitions with I/O interface by molecules. In: *DNA Computing, 6th International Meeting on DNA Based Computers, DNA 2000.* Lecture Notes in Computer Science, vol.2054 (Springer, Berlin Heidelberg New York 2001) 17–26.

20. Kurtz, S.A., Mahaney, S.R., Royer, J.S., Simon, J.: Active transport in biological computing. In: *DNA Based Computers II.* DIMACS Series in Discrete Mathematics and Theoretical Computer Science, vol.44 (1999) 171–179.

21. Lipton, R.J.: DNA solution of hard computational problems, Science **268**, 542–545 (1995).

22. Mao, C., Sun, W., Shen, Z., Seeman, N.C.: A nanomechanical device based on the B-Z transition of DNA, Nature **397**, 144–146 (1999).

23. McCaskill, J.S.: The equilibrium partition function and base pair binding probabilities for RNA secondary structure, Biopolymers **29**, 1105–1119 (1990).
24. Nishikawa, A., Yamamura, M., Hagiya, M.: DNA computation simulator based on abstract bases, Soft Computing **5**(1), 25–38 (2001).
25. Ogihara, M., Ray, A.: DNA based self-propagating algorithm for solving bounded-fan-in Boolean circuits. In: *Genetic Programming'98* (1998) 725–730.
26. Ogihara, M., Ray, A.: DNA Based parallel computation by "counting". In: *DNA Based Computers III*. DIMACS Series in Discrete Mathematics and Theoretical Computer Science, vol.48 (1999) 255–264.
27. Păun, G.: *Membrane Computing – An Introduction* (Springer, Berlin Heidelberg New York 2002).
28. Păun, G., Rozenberg, G., Salomaa, A.: *DNA Computing* (Springer, Berlin Heidelberg New York 1998).
29. Reif, J.H.: Parallel molecular computation. In: *Seventh Annual ACM Symposium on Parallel Algorithms and Architectures* (1995) 213–223.
30. Rooß, D., Wagner, K.W.: On the power of DNA-computing, Information and Computation **131**, 95–109 (1996).
31. Rose, J.A., Hagiya, M., Deaton, R.J., Suyama, A.: A DNA based *in vitro* genetic program, Journal of Biological Physics **28**, (2003).
32. Rothemund, P.W.K., Winfree, E.: The program-size complexity of self-assembled squares. In: *Proceedings of the Thirty-Second Annual ACM Symposium on Theory of Computing* (2000) 459–468.
33. Roweis, S., Winfree, E., Burgoyne, R, Chelyapov, N.V., Goodman, M.F., Rothemund, P.W.K., Adleman, L.M.: A sticker based model for DNA computation. In: *DNA Based Computers II*. DIMACS Series in Discrete Mathematics and Theoretical Computer Science, vol.44 (1999) 1–29.
34. Sakakibara, Y., Suyama, A.: Intelligent DNA chips: Logical operation of gene expression profiles on DNA computers. In: *Genome Informatics 2000*. Genome Informatics Series, vol.11 (Universal Academy Press, Tokyo 2000) 33–42.
35. Sakamoto, K., Gouzu, H., Komiya, K., Kiga, D., Yokoyama, S., Yokomori, T., Hagiya, M.: Molecular computation by DNA hairpin formation, Science **288**, 1223–1226 (2000).
36. Sakamoto, K., Kiga, D., Komiya, K., Gouzu, H., Yokoyama, S., Ikeda, S., Sugiyama, H., Hagiya, M.: State transitions by molecules, BioSystems **52**, 81–91 (1999).
37. SantaLucia, J. Jr., Allawi, H.T., Seneviratne, P.A.: Improved nearest-neighbor parameters for predicting DNA duplex stability, Biochemistry **35**(11), 3555–3562 (1996).
38. Schuler, G.D.: Electronic PCR: Bridging the gap between genome mapping and genome sequencing, Trends in Biotechnology **16**(11), 456–459 (1998).
39. Seeman, N.C., et al: The perils of polynucleotides: The experimental gap between the design and assembly of unusual DNA structures. In: *DNA Based Computers II*. DIMACS Series in Discrete Mathematics and Theoretical Computer Science, vol 44 (1999) 215–233.
40. Suyama, A., Nishida, N., Kurata, K., Omagari, K.: Gene expression analysis by DNA computing. In: *Currents in Computational Molecular Biology* (2000) 12–13.
41. Weiss, R., Knight, T.F. Jr.: Engineered communications for microbial robotics. In: *DNA Computing, 6th International Meeting on DNA Based Computers,*

DNA 2000. Lecture Notes in Computer Science, vol.2054 (Springer, Berlin Heidelberg New York 2001) 1–16.

42. Winfree, E.: Simulations of computing by self-assembly. In: *Preliminary Proceedings, Fourth International Meeting on DNA Based Computers*, 1998 (University of Pennsylvania) 213–239. Also in Erik Winfree's PhD Thesis: *Algorithmic Self-Assembly of DNA* (California Institute of Technology, 1998).

43. Winfree, E., Eng, T., Rozenberg, G.: String tile models for DNA computing by self-assembly. In: *DNA Computing, 6th International Meeting on DNA Based Computers, DNA 2000*. Lecture Notes in Computer Science, vol.2054 (Springer, Berlin Heidelberg New York 2001) 63–88.

44. Winfree, E., Liu, F., Wenzler, L.A., Seeman, N.C.: Design and self-assembly of two-dimensional DNA crystals, Nature **394**, 539–544 (1998).

45. Winfree, E., Yang, X., Seeman, N.C.: Universal computation via self-assembly of DNA: some theory and experiments. In: *DNA Based Computers II*. DIMACS Series in Discrete Mathematics and Theoretical Computer Science, vol.44 (1999) 191–213.

46. Yan, H., Zhang, X., Shen, Z., Seeman, N.C.: A robust DNA mechanical device controlled by hybridization topology, Nature **415**, 62–65 (2002).

47. Yoshida, H., Suyama, A.: Solutions to 3-SAT by breadth first search. In: *DNA Based Computers V*. DIMACS Series in Discrete Mathematics and Theoretical Computer Science, vol.54 (1999) 9–22.

48. Yurke, B., Turberfield, A.J., Mills, A.P. Jr., Simmel, F.C., Neumann, J.L.: A DNA-fuelled molecular machine made of DNA, Nature **406**, 605–608 (2000).

49. Zuker, M., Steigler, P.: Optimal computer folding of large RNA sequences using thermodynamics and auxiliary information, Nucleic Acids Research **9**, 133–148 (1981).

Molecular Information Theory:
Solving the Mysteries of DNA

Sungchul Ji

Department of Pharmacology and Toxicology, Rutgers University, Piscataway, NJ
sji@eohis.rutgers.edu

Summary. DNA is the foundation stone of molecular information. Therefore, any theory that accounts for the basic molecular properties and behaviours of DNA inside the living cell can be viewed as constituting a *molecular information theory*. It is posited that the necessary and sufficient conditions for unraveling the workings of DNA are the successful construction of a *computer model of the living cell* that (i) takes into account not only *information* but also *free energy* transactions mediated by molecular machines, (ii) utilizes sequence-specific conformational strains of biopolymers (called *conformons*) as the immediate driving force for all teleonomic functions of molecular machines, and (iii) implements *cell language* in the form of fuzzy if–then rules at the levels of both molecular machines and the living cell. One important application of computer models of the cell is thought to be in the field of DNA microarray data analysis. It is suggested that, without computer models of the cell, it might be nearly impossible to extract meaningful biological information from microarray data.

1 Introduction

On the occasion of the 50*th* anniversary of the discovery of the DNA double helix by Watson and Crick in 1953, the February 17, 2003 issue of *Time* magazine featured a cover story entitled "Solving the Mysteries of DNA". It began with the following paragraph:

On February 28, 1953, Francis Crick walked into the Eagle pub in Cambridge, England, and announced that he and James Watson had "found the secret of life." In retrospect, the Watson and Crick's claim cited here must be viewed as only *partially true*, because, despite the many fundamental discoveries that the DNA double helix has led to during the past 50 years, we do not yet know exactly how the DNA molecule works and how the cell regulates the expression of tens of thousands of genes encoded in it [1]. Therefore it seems more accurate to say that what Watson and Crick had discovered was "*a* secret" rather than "*the* secret", since there may well be more secrets of life yet to be discovered. If this analysis is valid, the natural question that arises is: *What are the other secrets of life?* To characterize and discuss some of these secrets is the main objective of this contribution.

2 Physics of Molecular Machines and Conformons

Just as computers manipulate information on the *macroscopic* level, so molecular machines [2,3] (e.g., ion pumps, molecular motors, DNA topoisomerases, RNA polymerases) manipulate information on the molecular or *microscopic* level. What distinguishes computers and molecular machines is that the former is driven by energy supplied externally (i.e., electricity), whereas the latter is driven by the free energy generated internally through exergonic (i.e., free energy-releasing) chemical reactions that are catalyzed by molecular machines themselves. Consequently, the structures of molecular machines have evolved to catalyze not only *information* processing but also *free energy* processing simultaneously. This makes it absolutely necessary for molecular machines to be mechanically (i.e., conformationally) deformable in contrast to macroscopic machines which must be by and large mechanically rigid. Mechanical deformability of molecular machines is postulated to be essential for catalysis by, and storing free energy in, molecular machines. According to the conformon theory, the immediate causes for all goal-directed motions of molecular machines are conformational strains localized or trapped in sequence-specific loci in biopolymers [2,3]. Such sequence-specific conformational strains carrying both free energy (to do work) and genetic information (to control work) are known as *conformons* [2,3]. The first concrete evidence for conformons was provided by Benham in the early 1990s who referred to mechanical strains localized in sequence-specific sites in superhelical DNA molecules as SIDDs (stress-induced duplex destabilizations) [4–7]. It is clear that, by definition, SIDDs are indistinguishable from conformons resident in DNA and hence can be viewed as constituting a subset of conformons.

The living cell is an organized system or network of *molecular machines* whose activities are coordinated in space and time so as to accomplish a set of goals conducive to the survival of the cell. *Molecular machines* (MMs) are in many ways similar to macroscopic machines familiar to us, only reduced in size to molecular dimensions. This reduction in machine size entails dramatic changes in machine properties, including the emergence of Brownian motions (or *thermal fluctuations*) of MMs. Thermal motions are deemed essential for MMs to tap chemical free energy to power their goal-directed, irreversible motions without violating the laws of thermodynamics [2,3,8].

Well-known examples of MMs include those proteins that carry out various work functions such as muscle contraction, ion pumping across biomembranes, and cargo transport along molecular tracks in the cytosol. The laws of physics (Newtonian and quantum mechanics, statistical mechanics, and thermodynamics) dictate that, in order to carry out their functions, all molecular machines must possess the following properties, in addition to their ability to undergo Brownian motions:

I)   An MM must be able to exert force on its environment to effectuate a net displacement in space (e.g., so as to move the myosin head along the actin filament) or in time (e.g., so as to transmit genetic information from one generation to the next by encoding it in, and decoding it from, DNA). The generating force entails storing energy within MMs that does not equilibrate with thermal energies of the environment.

II)   Any directional motions of an MM must be accompanied by the irreversible progression of free-energy releasing (exergonic) chemical reactions (e.g., ATP hydrolysis) or physical processes (e.g., down-hill movement of ions across biomembranes).

III) Processes I) and II) must be coupled within an MM proper in such a way that process I) cannot occur without process II), or vice versa.

The *conformon theory* [2, 3] postulates that MMs accomplish the above three processes via the following mechanisms:

1) An MM exists in at least two stable conformational states, say A and B:

(1) MM(A) $\leftrightarrow$ MM(B)

2) An MM catalyzes at least one exergonic chemical reaction when it exists in a conformational state intermediate between A and B, to be designated as MM(A&B). At this state, the chemical subsystem most likely exists in a state intermediate between a and b, denoted by a&b, where a and b are the reactant (or reactant system) and the product (or product system), respectively, both together forming the chemical subsystem of the enzyme-substrate complex. We can represent these processes symbolically as follows:

(2) MM(A,a) $\leftrightarrow$ MM(A&B, a&b) $\leftrightarrow$ MM(B,b)

where $\leftrightarrow$ indicates reversibility.

3) If the Gibbs free energy difference ΔG between MM(B) and MM(A) is less than the free energy difference between b and a in absolute magnitudes, we can represent it as:

(3) $\Delta|G(B - A)| \quad < \quad \Delta|G(b - a)|$

where the difference is calculated by subtracting the initial value from the final one. Under the condition where (3) obtains, Reaction (2) can proceed from left to right spontaneously because the total Gibbs free energy change accompanying (2) is negative.

We can treat an MM as a set having two elements, X and Y, the former indicating the conformational state of MM and the latter the state of the chemical subsystem bound to and catalyzed by MM. The progression of the action of MM (X,Y) can be represented using the geometric language of the fuzzy set theory described by Kosko [9], in the form of a series of sets MM(X,Y), each having different numerical values for X and Y. We may represent this idea symbolically as:

(4) Molecular Machine in Action $= MM(X, Y | X, Y \to [0, 1])$

where "|" means "such that", and "$\to [0, 1]$" means "assumes the value of a positive real number between 0 and 1". MM(X,Y) is not a crisp set but fuzzy, because the values of X and Y are not crisp such as 0 or 1 but can vary more or less continuously between 0 and 1. This is especially true for the value of X, due to the fact that a protein can exist in thousands of conformational substates [10] or *microconformations*, each of which requiring one number between 0 and 1 to be designated. We can indicate the fuzziness of the set MM(X,Y) using the "fuzzy hypercube" of Kosko as shown in Fig. 1, where X can assume any positive real numbers between 0 and 1, each number representing one conformational substate of Frauenfelder [10].

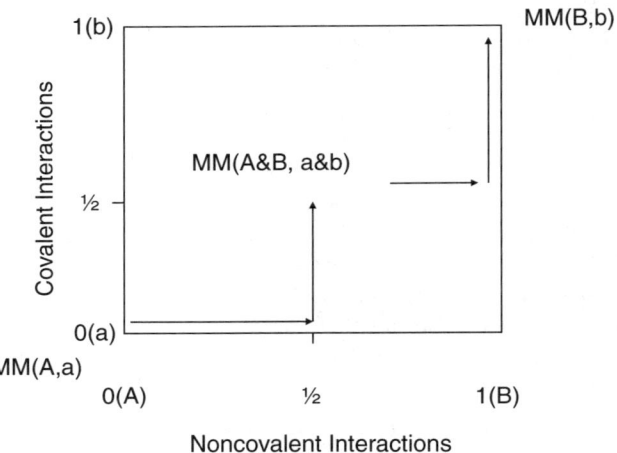

Fig. 1. Molecular machine fuzzy cube

The MM fuzzy cube is adopted from the concept of the fuzzy cube described in [9]. The X-axis represents the conformational states (involving noncovalent interactions) of an MM, and the Y-axis, the chemical reaction coordinates reflecting the progression of the chemical reaction, a↔b, catalyzed by the MM (involving covalent interactions). The two stable conformational states of the MM, namely, A and B, are indicated as 0 and 1, respectively, on the X-axis, and the stable chemical species, a and b, are indicated as 0 and 1, respectively, on the Y-axis. Both the chemical subsystem and protein subsystem are shown to possess an intermediate state indicated by the numerical value of 1/2. When both the chemical and protein subsystems are in the 1/2 state, the MM reaches the so-called "transition state" (also called the Franck–Condon state [2,3]), symbolized by MM(A&B, a&b). The sets at 1/2 indicate that the protein subsystem is at the "A and B" (i.e., "A and not-A") state and the chemical subsystem at the "a and b" (i.e., "a and not-a") state. Such "A and not-A" states are characteristic of fuzzy sets and can be expressed in terms of the so-called "fuzzy entropy" introduced by Kosko [9].

In Fig. 1, the progression of one half cycle of MM(X,Y) is depicted as a zigzag path indicated by the four arrows. Notice that the path changes abruptly at the coordinate $(1/2,0)$ and $(1,1/2)$, and this is because the conformational coordinates X vary more or less continuously (giving rise to the two horizontal arrows) while the chemical coordinates Y vary abruptly (as the two vertical arrows show), due to the large free energy barriers associated with the electronic transitions responsible for the chemical reaction a$\rightarrow$ b. The net flow of MM states from MM(A,a) to MM(B,b) through the intermediate state MM(A&B, a&b) or, equivalently, from MM(0,0) to MM(1,1) through MM(1/2,1/2), can be driven by the net free energy change given by (3). It should be pointed out that, in the final state, M(1,1), a part of the chemical energy "released" from the exergonic reaction, a$\rightarrow$ b, is postulated to have been converted into and stored as the mechanical energy of sequence-specific conformational strains, called conformons [2, 3], entrapped in specific loci within MM(1,1). It is these conformons that serve as the source of energy to generate *directed* molecular forces exerted by MM(B,b) on its environment, as required by property I) described above.

It is proposed here that the conformon-driven action path shown in Fig. 1 can be viewed as a molecular analog of the "if–then rule" of fuzzy logic [9] pp.158-171. Thus, the conformon-driven action path can be expressed as:

(5) "If X is A and Y is a, then X is B and Y is b."

If this conjecture is right, we can view MMs as molecular systems capable of implementing fuzzy if–then rules and hence as "fuzzy molecular machines".

3 The Living Cell as a *Fuzzy Molecular System*

The living cell can be viewed as an organized system of MMs (or motors), each implementing a set of if–then rules. Hence, the cell can be thought of as a giant or super-molecular machine implementing a system of M if–then rules, where M is the number of active enzymes in the cell at any given time. The progression of the action of the cell, then, can be represented geometrically in terms of a 2M-dimensional molecular machine hypercube [9], 2M because each enzyme undergoes conformational transitions and also catalyzes its "conjugate" chemical reaction as shown in Fig. 1.

About 5 years before coming into contact with the fuzzy set theory of Zadeh and his concept of if–then rules [9, 11, 12], the present author came to the conclusion, based on then available experimental data on *apoptosis* (also called *programmed cell death*), that the living cell has evolved to obey a system of *if–then* statements of the following type [13, 14]:

(6) *"If you are in cell state X and receive signal Y, then do Z."*

This if–then statement or rule was postulated to be encoded in the cellular genome as an organized set of genes that can be represented as:

(7) $<x, y, z>$

where x is the set of genes coding for enzymes that determine cell state X; y is the set of genes coding for the receptor for signal Y; z is the set of genes coding for the enzymes that execute the consequent of the if–then rule, Z; and finally $<...>$ represents the set of genes called "spatiotemporal genes" [15,16] that controls the timing of the expression of structural genes. Interestingly, indirect experimental evidence exists indicating that spatiotemporal genes are predominantly located in the so-called noncoding regions of DNA [17].

The *molecular* if–then rule indicated in (6) and (7), although derived from the phenomenon of apoptosis, is thought to apply to other gene-directed cell functions, such as chemotaxis, cell division, cell differentiation, cell growth, etc. Therefore, we can represent the living cell as a system of M if–then rules, where M is the number of active enzymes inside the cell at any given time.

The system of if–then rules constituting the living cell does not reveal any causal relation among the M component if–then rules. This deficiency can be remedied by employing the diagrammatic scheme known as a *fuzzy cognitive map* (FCM) [9]. In this scheme, each if–then rule is represented as an arrow connecting two nodes, one representing the conditional and the other the consequent.

The fuzzy-logic-based computational tools such as *fuzzy systems*, *fuzzy approximation theorem* (FAT), and *fuzzy cognitive maps* (FCMs) provide versatile methodologies to model the complex information and energy transactions that go on inside the living cell mediated by molecular machines.

4 A Cell Model-Based Approach to Microarray Data Analysis

It is possible that many of the contemporary biological problems cannot be solved without using theoretical models of the cell, just as the atomic spectra measured throughout the 19th century could not be correctly interpreted until Niels Bohr succeeded in formulating the theoretical model of the atom in 1913. Based on a conjectured analogy between 20th century atomic physics and 21st century molecular/cell biology, it is predicted here that the DNA microarray data that have accumulated in the literature in the past decade or so cannot be correctly interpreted without utilizing realistic computer models of the living cell.

The advent of the microarray technique in molecular biology in 1995 [18–24] may mark an important turning point in the history of biology, comparable to the discovery of the DNA double helix in 1953. A microarray refers to a microscopic slide (or its equivalent), about 2 cm by 2 cm in dimension, divided typically into 10,000 squares, each of which contains a fragment of DNA several hundred nucleotides long (called complementary DNA, cDNA) that is complementary to the stretches of the genome encoding an mRNA molecule [22]. Therefore, using one microarray, it is possible to measure the levels of 10,000 mRNA molecules in a biological sample simultaneously. Before the development of the microarray technique, it was possible to study only one or a few gene expressions at a time.

Due to the instability of mRNA, it is necessary to convert mRNA in a biological sample into its more stable complementary DNA fragments using reverse transcriptase and measure these DNA fragments with the microarrays as indirect measures of the original levels of mRNA molecules [22]. The instability of mRNA implies that the level of mRNA inside the cell at any given time, R_{cell}, is determined by two rate processes: the rate of synthesis V_{syn}, and the rate of degradation or hydrolysis V_{hyd} [25–27]. Therefore, we can write the rate of change of the mRNA pool in the cell as

(8) $dR_{cell}/dt = k_1 - k_2 = V_{syn} - V_{hyd}$ or

(9) $R_{cell} = (k_1 - k_2)t = (V_{syn} - V_{hyd})t,$

where R_{cell} is the intracellular level of mRNA at time t above the control level at $t = 0$. It is assumed in (8) that the rates of synthesis and hydrolysis of mRNA are zero-order with respect of mRNA concentrations inside the cell. The average change in R_{cell} over an observational period Δt is given by

(10) $\Delta R_{cell}/\Delta t = k_1 - k_2 = V_{syn} - V_{hyd}$

The stability of mRNA inside the cell varies widely depending on conditions, and the half-lives of mRNA can range from a few minutes to over 24 hours [26]. From Eq. (10), we can readily derive the relation between a change in the cellular levels of mRNA, measured over time period Δt, i.e., $\Delta R_{cell}/\Delta t$, and the changes in the rates of synthesis, V_{syn}, and hydrolysis of mRNAs, V_{hyd}, induced by control mechanisms of the cell other than by mRNA levels. The changes in the rates of mRNA synthesis or hydrolysis may be due to any one or more of the following causes: (i) changes in the levels of enzymes involved resulting from alterations in gene expressions, (ii) activation or inhibition of enzymes by ligand binding, and (iii) changes in the patterns of couplings among different enzymes due to cell-state transitions.

In order for there to be a one-to-one correspondence between mRNA levels and gene expression, the V_{hyd} term in Eq. (10) must be identically zero. Only under such a condition would there be a simple relation between ΔR_{cell} and V_{syn}. If this is not the case, the relation between mRNA level, i.e., ΔR_{cell}, and gene expression, i.e., V_{syn}, can be quite complex, as shown in Table 1.

As evident in Table 1, the mRNA levels can increase (as measured by microarrays) even when the gene expression (i.e., mRNA synthesis) is decreased (as seen in Mechanism 3), which is counter-intuitive. Similarly, the mRNA

Table 1. The multiplicity of the correlations between mRNA levels in the cell and the rates of mRNA synthesis and hydrolysis, predicted on the basis of a simple kinetic assumption, Eq. (10). ↑ = Increase; **0** = No change; ↓ = Decrease. Multiple arrows reflect the relative magnitudes of changes.

R_{cell}	↑	↑	↑	0	0	0	↓	↓	↓
V_{syn}	↑	0	↓	↑	0	↓	↑	0	↓
V_{hyd}	0	↓	↓↓↓	↑	0	↓	↑↑↑	↑	0
Mechanisms	1	2	3	4	5	6	7	8	9

level can decrease even when the gene expression rate increased (see Mechanism 7). There are a total of 9 classes of possible kinetic mechanisms that can be predicted to occur inside the cell based on the simple kinetic assumptions made in Eq. (10). Therefore, the intracellular level of mRNA levels at any given time may be represented by a linear combination of the 9 distinct classes of mechanisms shown in Table 1:

$$(11) \quad R_{cell} = \sum_{i=1}^{9} a_i M_i,$$

where a_i is the coefficient indicating the degree of contribution of the ith mechanism, M_i, with the index i running from 1 to 9, the total number of distinct kinetic mechanisms. It is interesting to note that Eq. (11) is identical in form with Eq. (2) in [21]. This suggests that the small number of so-called "characteristic modes" underlying the microarray data analysed in [21] using singular value decomposition (SVD) may be theoretically related to the kinetic mechanisms identified in Table 1.

It is also possible to view Eq. (11) as a *fuzzy system* consisting of 9 if–then rules, each of which can be expressed as:

$$(12) \quad \text{If } V_{syn} = A, \text{ then } V_{hyd} = B,$$

where A has the value "increase", "no change", or "decrease", and B has the value "no change", "normal increase", "large increase", "normal decrease", or "large decrease". In order for the if–then rule, Expression (12), to hold, there must exist in the cell some mechanisms which couple and coordinate the rates of mRNA synthesis and hydrolysis. Otherwise, the if–then rules cannot be activated in the cell.

5 Conclusions

Despite much detailed experimental knowledge that has accumulated in the molecular and cell biology of gene expression during the past five decades, we still do not know how the cell expresses the right set of genes at the right times and right places for the right durations under given environmental conditions. To understand how the cell accomplishes such miraculous molecular feats, it is suggested here that a computer model of the living cell is needed – a computer model based not on a traditional physics approach utilizing differential equations but rather on a fuzzy-logic-based approach employing fuzzy if–then rules (as discussed in Sect. 2 and 3). According to the *fuzzy approximation theorem* (FAT) [9, 11], the behaviour or the function of any system capable of input and output processes can be approximated by a system of fuzzy if–then rules, i.e., as a *fuzzy system*. Since the living cell has been shown to obey molecular if–then rules [3], it should be possible, based on FAT, to represent the molecular behaviours of the living cell in terms of a system of inter-connected fuzzy if–then rules that may be implemented on a computer. Important applications of such models of the living cell may be found in the general field of *bionomics*, including *transcriptomics* (described above), *proteomics*, *metabonomics*, and *phenomics* (study of phenotypes).

References

1. W.W. Gibbs. The Unseen Genome: Gems among the Junk. *Sci. Am.*, Vol. 289(5), pp.48-53, 2003.
2. S. Ji. Energy and Negentropy in Enzymic Catalysis. *Ann. N.Y. Acad. Sci.*, Vol. 227, pp.419-437, 1974.
3. S. Ji. Free energy and information contents of *Conformons* in proteins and DNA. *BioSystems,* Vol. 54, pp.107-130, 2000.
4. C.J. Benham. Energetics of the strand separation transition in superhelical DNA. *J. Mol. Biol.,* Vol. 225, pp.835-847, 1992.
5. C.J. Benham. Sites of predicted stress-induced DNA duplex destabilization occur preferentially at regulatory sites. *Proc. Natl. Acad. Sci. USA,* Vol. 90, pp.2999-3003, 1993.
6. C.J. Benham. Duplex destabilization in superhelical DNA is predicted to occur at specific transcriptional regulatory regions. *J. Mol. Biol.,* Vol. 255, pp.425-434, 1996.
7. C.J. Benham. Computation of DNA structural variability: a new predictor of DNA regulatory regions. *CABIOS,* Vol. 12(5), pp.375-381, 1996.
8. R.D. Astumian. The role of thermal activation in motion and force generation by molecular motors. *Phil. Trans. R. Soc. Lond. B,* Vol. 355, pp.511-522, 2000.
9. B. Kosko. *Fuzzy Thinking: The New Science of Fuzzy Logic,* Hyperion, New York, 1993.
10. G.U. Nienhaus, J.D. Muller, B.H. McMahon, H. Frauenfelder. Exploring the conformational energy landscape of proteins. *Physica D,* Vol. 107, pp.297-311, 1997.
11. B. Kosko. Fuzzy Systems as Universal Approximators. *IEEE Trans. Comput.,* Vol. 43(11), pp.1329-1333, 1994.
12. K. Kilic, B.A. Sproule, I.B. Turksen, C.A. Naranjo. Fuzzy system modeling in pharmacology: an improved algorithm. *Fuzzy Sets Syst.,* Vol. 130, pp.253-264, 2002.

13. S. Ji. A Cell Linguistic Analysis of Apoptosis. *Comments Toxicol.*, Vol. 5(6), pp.571-585, 1997.

14. S. Ji. Isomorphism between cell and human languages: molecular biological, bioinformatic and linguistic implications. *BioSystems*, Vol. 4, pp.17-39, 1997.

15. S. Ji. Biocybernetics: A Machine Theory of Biology. In: *Molecular Theories of Cell Life and Death*, Rutgers University Press, New Brunswick, pp.1-237, 1991.

16. S. Ji. The Bhopalator: An Information/Energy Dual Model of the Living Cell (II). *Fundam. Inf.*, Vol. 49(1-3), pp.147-165, 2002.

17. S. Ji. Microsemiotics of DNA. *Semiotica*, Vol. 138, pp.15-42, 2002.

18. M. Schena, D. Shalon, R.W. Davis, P.O. Brown. Quantitative Monitoring of Gene Expression Patterns with a Complementary DNA Microarray. *Science*, Vol. 270, pp.467-470, 1995.

19. A.C. Pease, D. Solas, E.J. Sullivan, M.T. Cronin, C.P. Holmes, S.P. Fodor. Light-generated Oligonucleotide arrays for Rapid DNA Sequence Analysis. *Proc. Natl. Acad. Sci. USA*, Vol. 91, pp.5022-5026, 1994.

20. M.B. Eisen, P.T. Spellman, P.O. Brown, D. Botstein. Cluster analysis and display of genome-wide expression patterns. *Proc. Natl. Acad. Sci. USA*, Vol. 95, pp.14863-14868, 1998.

21. N.S. Holter, M. Mitra, A. Maritan, M. Cieplak, J.R. Banavar, N.V. Fedoroff. Fundamental patterns underlying gene expression profiles: simplicity from complexity. *Proc. Natl. Acad. Sci. USA*, Vol. 97(15), pp.8409-8414, 2000.

22. S.J. Watson, U. Akil. Gene Chips and Arrays Revealed: A Primer on Their Power and Their Uses. *Biol. Psychiatry*, Vol. 45, pp.533-543, 1999.

23. U. Alon, N. Barkai, D.A. Notterman, K. Gish, S. Ybarra, D. Mack, A.J. Levine. Broad patterns of gene expression revealed by clustering analysis of tumor and normal colon tissues probed by oligonucleotide arrays. *Proc. Natl. Acad. Sci. USA*, Vol. 96, pp.6745-6750, 1999.

24. K.P. White, S.A. Rofkin, P. Hurban, D.S. Hogness. Microarray Analysis of *Drosophila* Development During Metamorphosis. *Science*, Vol. 286, pp.2179-2184, 1999.

25. D.G. Hoel. A Simple Two-Compartmental Model Applicable to Enzyme Regulation. *J. Biol. Chem.*, Vol. 245, pp.5811-5812, 1970.

26. D.J. Shapiro, J.E. Blume, D.A. Nielsen. Regulation of Messenger RNA Stability in Eukaryotic Cells. *BioEssays*, Vol. 6(5), pp.221-226, 1987.

27. J.L. Hargrove, F.H. Schmidt. The role of mRNA and protein stability in gene expression. *FASEB J.*, Vol. 3, pp. 2360-2370, 1989.

Formal Modelling of *C. elegans* Development. A Scenario-Based Approach *

Na'aman Kam[1], David Harel[1], Hillel Kugler[1], Rami Marelly[1], Amir Pnueli[1], Jane Albert Hubbard[2], and Michael J. Stern[3]

[1] Dept. of Computer Science and Applied Mathematics, The Weizmann Institute of Science, Rehovot 76100, Israel
  {kam,dharel,kugler,rami,amir}@wisdom.weizmann.ac.il
[2] Dept. of Biology, New York University, New York, NY
  jane.hubbard@nyu.edu
[3] Dept. of Genetics, Yale University School of Medicine, New Haven, CT
  michael.stern@yale.edu

Summary. We present preliminary results of a new approach to the formal modelling of biological phenomena. The approach stems from the conceptual compatibility of the methods and logic of data collection and analysis in the field of developmental genetics with the languages, methods, and tools of scenario-based reactive system design. In particular, we use the recently developed methodology consisting of the language of live sequence charts with the play-in/play-out process, to model the well-characterized process of cell fate acquisition during *C. elegans* vulval development.

1 Introduction

Our understanding of biology has become sufficiently complex that it is increasingly difficult to integrate all the relevant facts using abstract reasoning alone. This is exacerbated by current high-throughput technologies that spew data at ever-increasing rates. While bioinformatic approaches to handling this mass of data have generated databases that ease the storage and accessibility of the data, rigorous modelling approaches are necessary to integrate these data into useable models that can exploit and analyse the available information. There are many current efforts aimed at biological modelling, and it is likely that different approaches will be appropriate for various types of biological information and for various research objectives [2, 21]. Here, we present a novel approach to modelling biological phenomena. It utilizes in a direct and powerful way the mechanisms by which raw biological data are amassed, and

---

* An early version of this chapter appeared in [14].

smoothly captures that data within tools designed by computer scientists for the design and analysis of complex reactive systems.

A considerable quantity of biological data is collected and reported in a form that can be called "condition–result" data. The gathering is usually carried out by initializing an experiment that is triggered by a certain set of circumstances (conditions), following which an observation is made and the results recorded. The condition is most often a perturbation, such as mutating genes or exposing cells to an altered environment. For example, genetic data often first emerge as phenotypic assessments (anatomical or behavioural outputs) that are compared between a mutant background and a defined "wild-type". Another example includes observations of the effects of anatomical manipulations (e.g., cell destruction or tissue transplantation) on the behaviour of the remaining structures. These types of experiments test specific hypotheses about the nature of the system that is perturbed. Many inferences about how biological systems function have been made from such experimental results, and our consequent understanding based on these logical inferences is becoming increasingly, even profoundly, complex.

One feature of these types of experiments is that they do not necessitate an understanding of the particular molecular mechanisms underlying the events. For example, much information can be ascertained about a gene's function by observing the consequences of loss of that function before the biochemical nature or activity of the gene product is known. Moreover, even when the biochemical activity is known, the functional significance of that activity in the context of a biological system is often deduced at the level of phenotypic output. Naturally, with knowledge of molecular mechanisms, increasingly sophisticated inferences can be made and more detailed hypotheses tested, but the outputs, be they a certain cell fate acquisition, changes in gene expression patterns, etc., are often recorded and analysed at the level of a phenotypic result. Thus, a large proportion of biological data is reported as stories, or "scenarios", that document the results of experiments conducted under specific conditions. The challenge of modelling these aspects of biology is to be able to translate such "condition–result" phenomena from the "scenario"-based natural language format into a meaningful and rigorous mathematical language. Such a translation process will allow these data to be integrated more comprehensively by the application of high-level computer-assisted analysis. In order for it to be useful, the model must be rigorous and formal, and thus amenable to verification and testing.

We have found that modelling methodologies originating in computer science and software engineering, and created for the purpose of designing complex *reactive systems*, are conceptually well suited to model this type of condition–result biological data. Reactive systems are those whose complexity stems not necessarily from complicated computation but from complicated reactivity over time. They are most often highly concurrent and time-intensive, and exhibit hybrid behaviour that is predominantly discrete in nature but has continuous aspects as well. The structure of a reactive system consists of many interacting components, in which control of the behaviour of the system is highly distributed among the components. Very often the structure itself is dynamic, with its components being repeatedly created and destroyed during the system's life span.

The most widely used frameworks for developing models of such systems feature *visual formalisms*, which are both graphically intuitive and mathematically rigorous. These are supported by powerful tools that enable full model executability and analysis, and are linkable to graphical user interfaces (GUIs) of the system. This enables realistic simulation prior to actual implementation. At present, such languages and tools (often based on the *object-oriented* paradigm) are being strengthened by verification modules, making it possible not only to execute and simulate the system models (test and observe), but also to verify dynamic properties thereof (prove).

A central premise of this chapter is that many kinds of biological systems exhibit characteristics that are remarkably similar to those of reactive systems. The similarities apply to many different levels of biological analysis, including those dealing with molecular, cellular, organ-based, whole organism, or even population biology phenomena. Once viewed in this light, the dramatic concurrency of events, the chain-reactions, the time-dependent patterns, and the event-driven discrete nature of their behaviours, are readily apparent. Consequently, we believe that biological systems can be productively modelled as reactive systems, using languages and tools developed for the construction of human-made systems.

In previous work on T cell activation [13] and T cell behaviour in the thymus [8], the feasibility of this thesis was addressed on a small scale, and the results have been very encouraging. In particular, that work demonstrated the adequacy of object-oriented analysis to modelling biological systems, and showed the applicability of the *visual formalism* of statecharts for representing their behaviour.

In our current work, we not only begin to tackle a more complex system, but also incorporate two levels of data into our model. One of these levels, shared also by the models from the previous work, represents the "rules" of the behaviour of the system. These rules are based on abstractions and inferences from various sets of primary data. The second level of data being incorporated into our model, is new, and it includes the "observations" that comprise the primary data themselves. The set of observations utilized is a crucial component of the model, allowing both execution and verification, including consistency checks between the observations and the inferred rules. To accomplish this, we also depart in a significant way from the intra-object, state-based, statechart approach used in the previous work, and instead use a more recently developed inter-object, scenario-based approach to reactive system specification. The language we use to formalize the data is called *live sequence charts* (LSCs) [4], and capturing the data and analysing them is carried out using the *play-in/play-out* methodology, supported by the *Play-Engine* tool [11, 12].

1.1 The Biological System

Our current effort is focused on a means to formalize and analyse primary data such that the consistency of inferences made from these data can be tested as part of the model. We have chosen the development of the nematode *Caenorhabditis elegans* as a subject since this organism is extremely well defined at the anatomical, genetic, and molecular level. Specifically, the entire cell lineage of *C. elegans* has been traced, many genetic mutants have been described, and the entire genome is sequenced [3, 18]. Moreover, virtually all of the researchers working on this organism use conceptually similar methodologies and logic in their genetic experiments, and use a uniform substrate wild-type strain (N2). These circumstances make feasible computer-assisted integration of the primary data. Finally, genome-wide efforts to define the function, expression levels, expression patterns, and interactions of all genes are underway [5, 6, 15, 17].

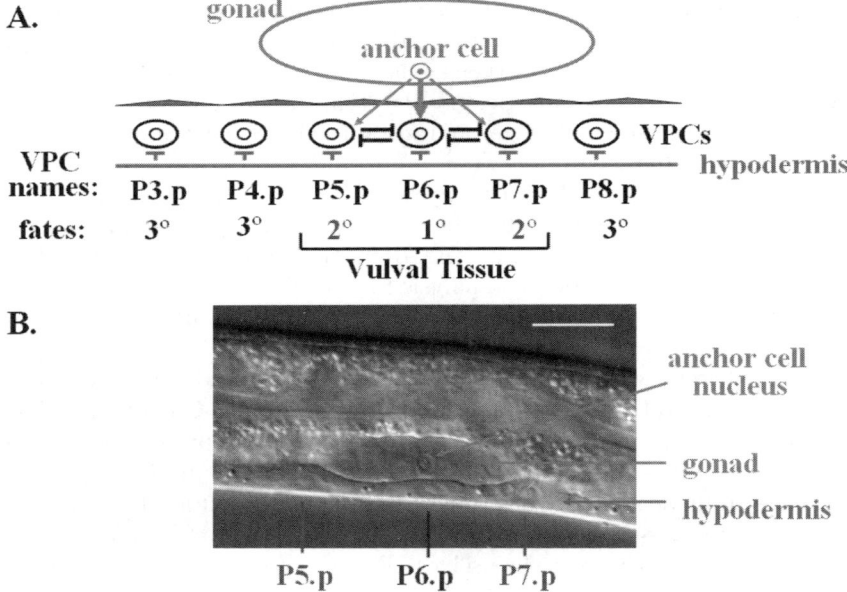

Fig. 1. Vulval cell fate determination. (A) Schematic representation of cellular interactions that determine vulval fates. Signals (arrows and T-bars) are color-coded based on their source. The anchor cell promotes vulval fates. VPC-VPC interactions inhibit adjacent cells from acquiring a 1° fate (and instead promote a 2° fate). The ventral hypodermis inhibits vulval (1° and 2°) fates. (B) Differential interference contrast (DIC) photomicrograph of the central body region of a live *C. elegans* L3-stage worm. White bar ~ 15 *μm*.

As a specific test-case, we have begun to model *C. elegans* vulval development (see [20] for a recent review). The vulva is a structure through which eggs are laid. This structure derives from three cells within a set of six cells with an equivalent set of multiple developmental potentials (Fig. 1). These six cells are named P3.p, P4.p, P5.p, P6.p, P7.p, and P8.p (collectively termed P(3–8).p). Due to their potential to participate in the formation of the vulva, they are also known as the *vulval precursor cells* (VPCs). Each has the potential to acquire either a non-vulval fate (a 3° fate) or one of two vulval cell fates (a 1° or a 2° cell fate; see Fig. 1). During normal development, after a series of cell divisions in a characteristic pattern, a vulva consisting of 22 nuclei is formed. Vulval development was one of the first areas to which considerable effort was applied to achieve a molecular understanding of how cells acquire their particular fates. This system, though limited in cell number, is quite complex and ultimately integrates at least four different molecular signalling pathways in three different interacting tissue types. Cell fate acquisition in development also occurs in the context of cell cycle control and the general global controls on the growth and development of the organism. Hence, vulval development indeed represents a rather complex reactive system.

Here, we describe our progress applying the visual formalism of LSCs and the play-in/play-out methodology to the representation of biological scenarios from one particular study, the data published in a paper by Sternberg and Horvitz in 1986 [19]. LSCs appear to be highly accessible to biologists, while retaining mathematical rigor. Designed specifically to capture scenario-based behaviour, the structure of LSCs fits extremely well into our framework of condition–result data. In this chapter, data are reported regarding a series of experiments in which cells are destroyed using a laser microbeam (ablation). Since signals among the VPCs and between the VPCs and adjacent tissues cooperate to determine the fates of these cells, these experiments test the nature of these interactions and the relative potential of the VPCs to adopt vulval (versus non-vulval) fates and, among those that adopt a vulval fate, the specific type of vulval fate. These data are used to infer specific properties of the events that occur in the wild-type situation and to generate a "model" for how the unperturbed system works based on the behaviour of the perturbed system. We present a subset of this test-case below with special emphasis on the LSC language and the play-in/play-out methodology. In addition, we present our solutions to particular challenges posed by the nature of biological data themselves and its formalization from the natural language used by *C. elegans* biologists. In representing even a small and simple set of experimental results, we have addressed a number of issues that provide evidence for the flexibility and potential of this modelling approach.

1.2 The Modelling Methodology

We are adopting an inter-object, scenario-based modelling approach, using the language of *live sequence charts* (LSCs) [4] and the *play-in/play-out* methodology [11, 12], both supported by the *Play-Engine* modelling tool [11]. The decision to take this approach, rather than the statechart-based one, emerged from the consideration of how to best represent the *C. elegans* data formally, and how to best carry out the formalization process.

LSCs constitute a visual formalism for specifying sequences of events and message passing activity between objects. They can specify scenarios of behaviour that cut across object boundaries and exhibit a variety of modalities, such as scenarios that can occur, ones that must occur, ones that may not occur (called *anti-scenarios*), ones that must follow others, ones that overlap with others, and more. Technically, there are two types of LSCs, universal and existential. The elements of LSCs (events, messages, guarding conditions, etc.) can be either mandatory (called *hot* in LSC terminology) or provisional (called *cold*). Universal charts are the more important ones for modelling, and comprise a *prechart* and a *main* chart, the former triggering the execution of the latter. Thus, a universal LSC states that whenever the scenario in the prechart occurs (e.g., the user has flipped a switch), the scenario in the main chart must follow it (e.g., the light goes on). Thus, the relation between the prechart and the chart body can be viewed as a dynamic condition–result: if and when the former occurs, the system is obligated to satisfy the latter.

Play-in/play-out is a recently developed process for modelling in LSCs, with which one can conveniently capture inter-object scenario-based behaviour, execute it, and simulate the modelled system in full. The play-in part of the method enables people who are unfamiliar with LSCs to specify system behaviour using a high-level, intuitive, and user-friendly mechanism. The process asks that the user first build the graphical user interface (GUI) of the system, with no behaviour built into it. The user then "plays" the GUI by clicking the graphical control elements (in electronic systems these might be buttons, knobs, and so on) in an intuitive manner, in this way giving the engine sequences of events and actions, and teaching it how the system should respond to them. As this is being done, the Play-Engine continuously constructs the corresponding LSCs automatically.

While play-in is the analog of writing programs, play-out is the analog of running them. Here the user simply plays the GUI as he/she would have done when executing the real system, also by clicking buttons and rotating knobs, and so on, but limiting him/herself to end-user and external environment actions. As this is going on, the Play-Engine interacts with the GUI and uses it to reflect the system state at any given moment. The scenarios played in using any number of LSCs are all taken into account during play-out, so that the user gets the full effect of the system with all its modelled behaviours operating correctly in tandem. All specified ramifications entailed by an occurring event or action will immediately be carried out by the engine automatically, regardless of where in the LSCs it was originally specified. Also, any violations of constraints (e.g., playing out an anti-scenario), or contradictions between scenarios, will be detected if attempted. This kind of sweeping integration of the specified condition–result style behaviour is most fitting for biological

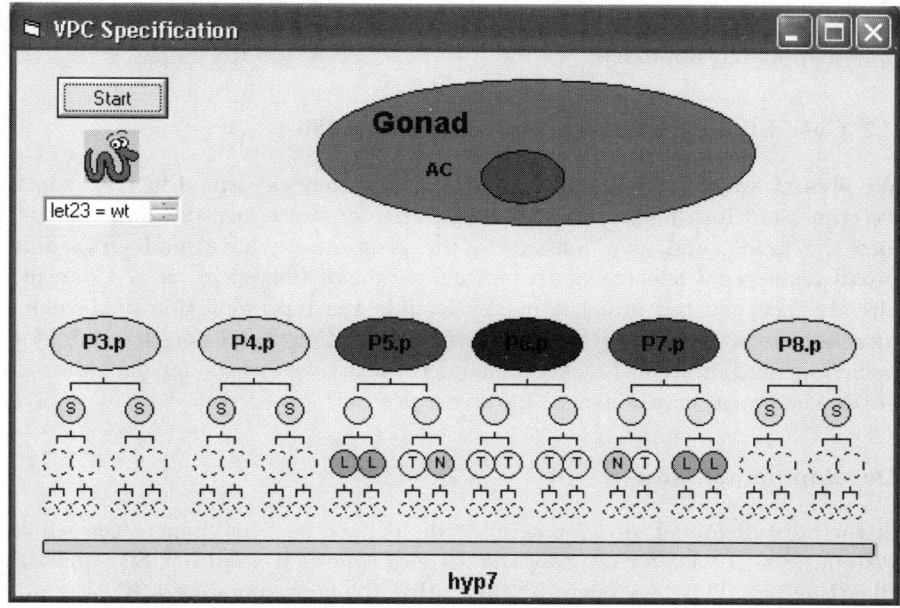

Fig. 2. The graphical user interface

systems, where it can often be very difficult to connect the many pieces of behavioural data that are continuously discovered or refined.

2 Results

2.1 The GUI of Vulval Fate Determination

A critical aspect of the GUI is that it can be designed by the user to reflect his/her view of the system. In our case, the GUI (Fig. 2) is a simplified representation of the actual anatomical situation under study (Fig. 1). The six VPCs and the interacting adjacent tissues, which include the gonad and the ventral hypodermis, are represented. This GUI represents the developmental decisions that occur during a discrete window of time, and we developed it the way we did in order for it to best capture that particular stage in the organism's development. It can be expanded to capture more parts and stages in the development process, and also to include inputs from other GUIs (possibly constructed by other people working on a growing distributed model) that represent other connected developmental vignettes. The beauty of this kind of GUI is that it can be made to directly reflect the way biologists represent their system's anatomy, and the particular GUI in Fig. 2 is intuitive for anyone working on vulval development. However, the more important aspect of the GUI is that it allows the model to "come alive" in the context of the play-in/play-out approach.

2.2 Case Study: Sternberg and Horvitz, 1986

We present several examples of scenarios that were described in [19], which we translated into LSCs. These examples demonstrate how data are entered into the model and give context to the progress we have made regarding broad conceptual aspects of the formalization of biological data. Concepts already incorporated into the model include the representation of developmental time, symbolic instances, default assumptions, and non-deterministic behaviour. Examples of LSCs are included that represent the formalization of both primary observations and inferred rules.

Developmental time

In the current model, developmental time drives the behaviour of the whole system. (See [10, 11] for the way the issue of time is treated in LSCs and the Play-Engine.) Thus, we begin by presenting the corresponding LSC (Fig. 3). We decided to concentrate on post-embryonic development (after the embryo has hatched from the egg), but earlier developmental stages can easily be incorporated as well. Post-embryonic development is divided into four larval stages (denoted L1 to L4), each of which ends in a molt. The time units represent hours from hatching (the beginning of L1). This LSC is activated when the user clicks the start button on the GUI. The time at which this operation was performed is stored in a variable called T, and all other events in this LSC refer to this time point. Thus, for example, the L1 phase lasts 16 hours, the L2 phase lasts 9 hours, and the Pn.p cells divide 29-30 hours after hatching, which is about 4 hours after entry into the L3 phase[1].

Representing the experimental setup in the prechart

The experiment represented by the LSC in Fig. 4 was devised to establish the ground-state fate of the P(3,4,8).p cells, that is, to see what cell fate these cells would acquire if they were not influenced by any known interactions from adjacent cells. At the time of this publication, the influence of the hypodermis on vulval fate specification was not known (it is also irrelevant under these specific conditions). The cells known to influence VPC cell fate were removed by laser ablation, including the three VPCs that are normally induced to form the vulva (P5.p, P6.p, and P7.p). The fates of the remaining VPCs were observed and recorded. The results of this set of experiments were reported in

---

[1] The time delays in this chapter refer to standard growing conditions at $20°C$; in a previous version [14], we provided some different figures, referring to developmental time at $25°C$.

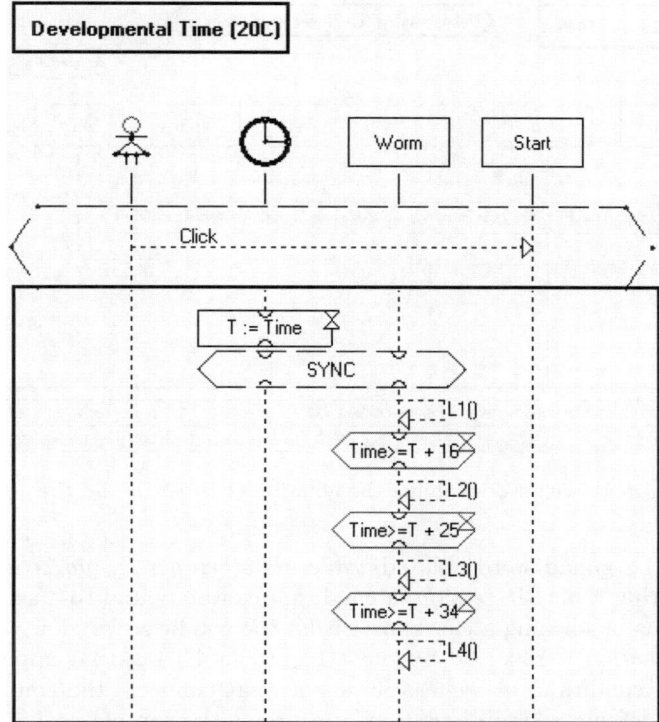

Fig. 3. An LSC depicting developmental time

the following statement (which is also attached as a note to the LSC, including a hyperlink to the paper abstract in Medline):

"We ablated P(5-7).p and the gonad in five L1 hermaphrodites; in these animals P(3,4,8).p each adopted a tertiary fate" [19].

In this experiment, the cell ablations took place during the first larval stage (L1). This fact is stated in the prechart by placing the condition before the L2 event: the prechart will be satisfied only if by the time the L2 event occurs (determined in the developmental time LSC) the specified cells will already be ablated.

Experimental setups, such as the above one, implicitly assume that besides the reported perturbations all other elements of the system remained intact. We represent such implicit assumptions in a set of *default assumptions*, which includes all the properties of the system for which we would like to set some default value (e.g., genes are not mutated, or "wild type"). If a given condition within the LSC includes assigning a non-default value to some property, this requirement will override the default assumption. Thus, the condition that appears in the prechart of Fig. 4 corresponds to a situation in which P(5-

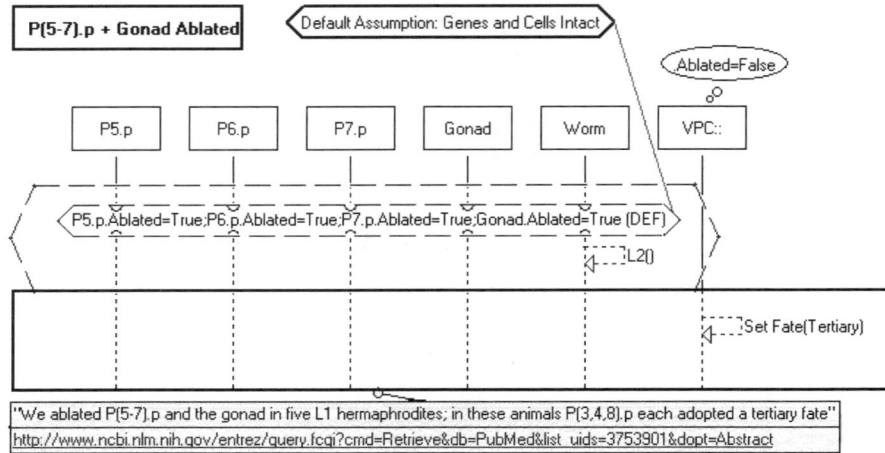

Fig. 4. A scenario involving the ablation of P(5-7).p and the gonad

7).p and the gonad were ablated *while all other genes and cells remained intact.* During a play-in session, a condition can be bound to a set of default assumptions by clicking a checkbox within the condition window. Within the LSC, this action is reflected by the *DEF* (default) tag that appears at the end of the condition, as well as by a line that connects the condition with the set of default assumptions that appears at the top of the LSC (the line appears only when the mouse cursor is placed over the condition).

Symbolic instances

The LSC in Fig. 4 illustrates the concept of symbolic instances. (See [11] or [16] for how symbolic instances are dealt with in LSCs and the Play-Engine.) In this experiment, the behaviour of all remaining VPCs is similar: they all adopt tertiary fates. This is expressed in the LSC by making a general statement that applies to a symbolic instance of the VPC class. This depiction indicates that during execution, all instances of the VPC class that were not ablated will be bound to this LSC and execute a tertiary fate.

Non-deterministic behaviour

Part of the experiment designed to assess the ground state of the VPCs involved ablating the cells that normally adopt vulval fates [P(5-7).p] but leaving the gonad intact. Three animals were observed after performing these ablation conditions; the fates of the VPCs are shown below (reported in [19], Table 3, line 3):

Pn.p cells:	P3.p	P.4p	P.5p	P.6p	P.7p	P.8p
Fates:	2° or 3°	1° or 2°	ablated	ablated	ablated	1° or 2°

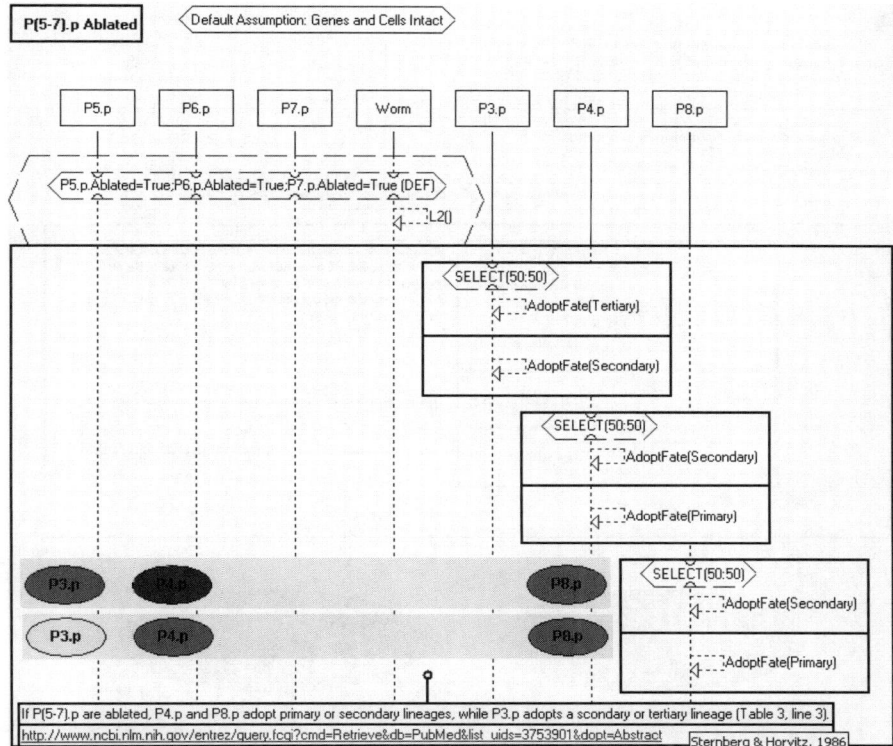

Fig. 5. Representing non-deterministic events. The results of two consecutive executions of this LSC are shown. P3.p and P4.p adopted different fates on each of these executions

As is often the case in biological experiments, the results of the conditions of this experiment were not deterministic: P3.p adopted either a 2° or a 3° fate, while P4.p and P8.p adopted either a 1° or a 2° fate. Each of these fates is recognizable by a specific pattern of cell divisions and cellular morphologies. Non-determinism is represented in the LSC that depicts this experiment (Fig. 5) using selection boxes within the main chart. This representation enables the model to execute non-deterministic behaviour of objects using a selected list of outcomes at prescribed frequencies.

Another example for the usage of non-determinism involves cell movements in the context of VPC ablations [19]:

> "In two unc-84 animals, where P4 was the only Pn cell that entered the ventral cord, P4.p moved towards the anchor cell to occupy the position normally occupied by P5.p or P6.p. In these animals, the isolated P4.p executed a 2° fate (LLTN) or a 1° fate (LTTT), respectively" (Fig. 6).

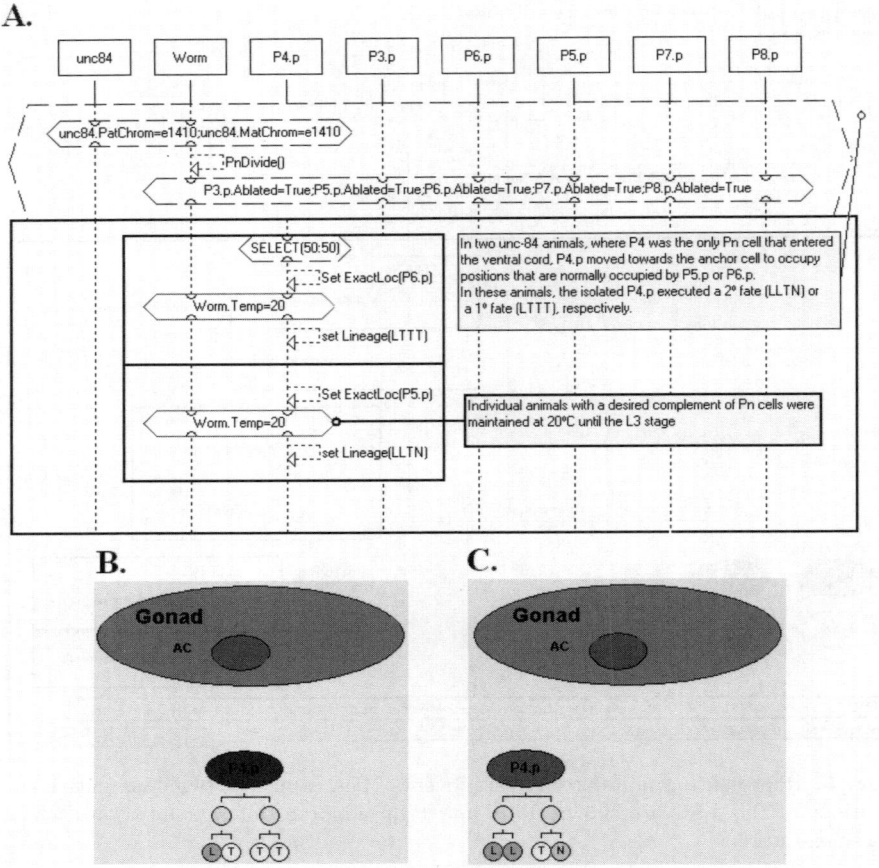

Fig. 6. Non-determinism in cell movement. In *unc-84* (e1410) mutants, isolated VPCs can be depicted, as a consequence of a nuclear migration failure in all other VPCs (A). In animals where P4.p was the isolated VPC, it migrated towards the AC to occupy either P6.p's or P5.p's location. Accordingly, P4.p can execute a primary lineage (B) or a secondary lineage (C)

In the above experiment, the isolated P4.p acquired its final position in a non-deterministic fashion. The lineage executed by such a VPC is believed to result from its relative position to the anchor cell. Note that in the corresponding LSC (Fig. 6A) we specified the pure experimental observation, therefore not relating explicitly the observed cell lineage to the spatial location of the VPC. Such biological inferences can be added later on as behavioural rules (see below).

Representing genes as internal objects

The LSC shown in Fig. 6A involves a homozygotic mutation in the *unc-84* gene. Such a mutation implies that both the maternal chromosome and the paternal chromosome carry the same *unc-84* allele (in this case, *e1410*). We decided to represent genes as internal objects. These differ from GUI objects by not having a graphical representation in the GUI itself. Internal objects can be accessed via an internal object map, as shown in Fig. 7.

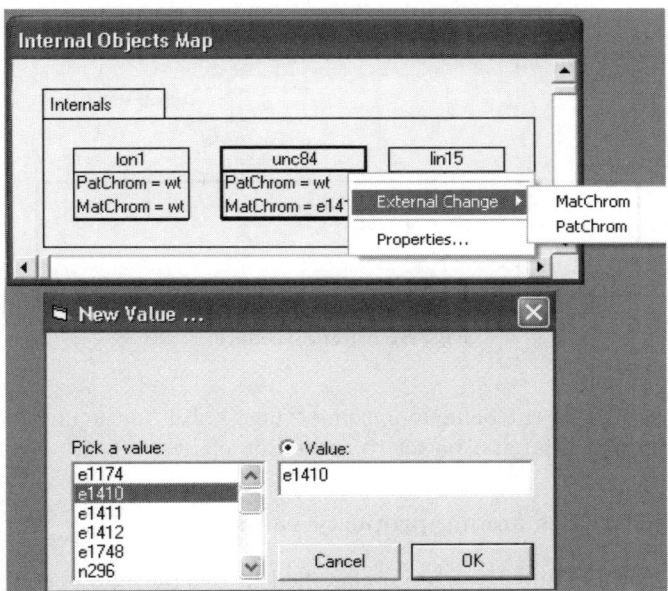

Fig. 7. Genes as internal objects

Representing general rules as anti-scenarios

In addition to the input of raw data from specific experiments, general rules of system behaviour can also be included as LSCs. Some rules can be represented most easily by what are known as *anti-scenarios*. For example, to represent the statement:

> *"The anchor cell is required for Pn.p cells to adopt vulval (1° or 2°) as opposed to non-vulval (3°) fates"* [19],

an LSC representing the following anti-scenario can be used:

> *"it cannot be the case that a VPC will adopt a vulval fate in the absence of the anchor cell"* (Fig. 8).

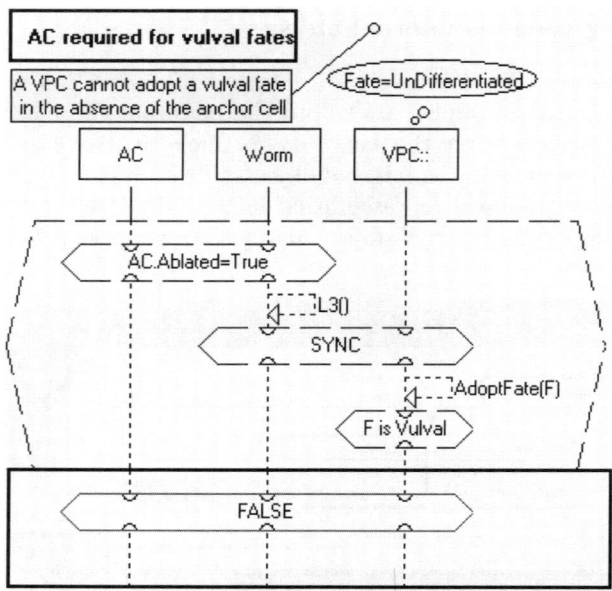

Fig. 8. An anti-scenario

The ability to represent behaviour using "must" and "must not" statements lends greater potential and power to modelling efforts.

Setting hierarchies among property values

VPC fates can be, and have been, classified in several different ways. These classifications can be arranged in a hierarchy that reflects a higher level interpretation of the biology or just the level at which the data were obtained. For example, a VPC fate can be classified either as "vulval" or, more specifically, as "primary" or "secondary", the first inclusive of, but less specific than, the other two. In some cases, we may wish to refer only to the higher level description; either because we do not know the details at the more concrete level (e.g., we know that the VPC went through a second round of cell division, implying that it adopted a vulval fate, but the exact pattern of division was not observed), or because this is not relevant to the specified scenario. For such cases, hierarchical types can be defined, allowing the user to arrange values at several levels. In the VPC fate example, when the user clicks a VPC and chooses the *Fate* property, the various values that can be assigned to this property are arranged in a tree-like structure (see Fig. 9). The user can then pick the desired value from any of the tree's levels. This representation is not merely cosmetic, but has a semantic significance that influences the dynamic execution of the model. Using this hierarchy, one can specify a condition that

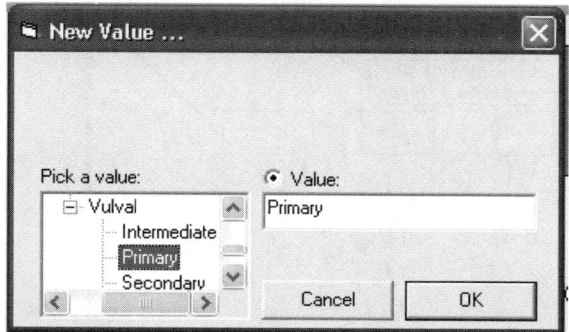

Fig. 9. Hierarchy among property values

says *Fate* is *Vulval*, and the condition will hold if *Fate* evaluates to *Vulval*, *Intermediate*, *Primary*, or *Secondary* (see Fig. 8).

System analysis: The play-out mechanism

The play-out mechanism can be used to test the system in various ways:

1. *Detecting inconsistencies among LSCs.* In its anti-scenario guise, an LSC representing, for example, a scenario in which a VPC adopts a vulval fate in the absence of the anchor cell would cause the Play-Engine to announce that a "hot" condition was violated. Certain mutations can obviate the requirement of the anchor cell for vulval cell fates. For example, if the hypodermal inhibitory mechanism (see Fig. 1) is compromised by mutation, vulval fates can occur even in the absence of the gonad (Fig. 10). Executing a play-out session in which an "experiment" of this type was

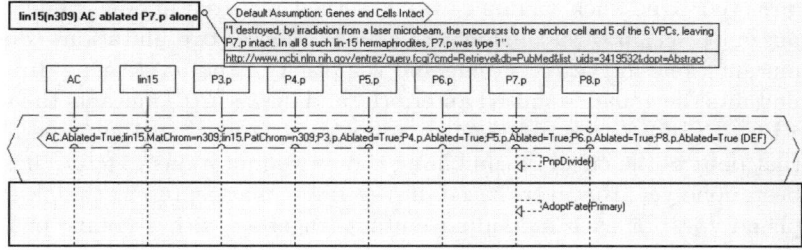

Fig. 10. A VPC can adopt a vulval fate in the absence of the anchor cell

performed resulted in a violation of the "old" rule (Fig. 11). Thus, this modelling approach can help integrate new results into the framework of

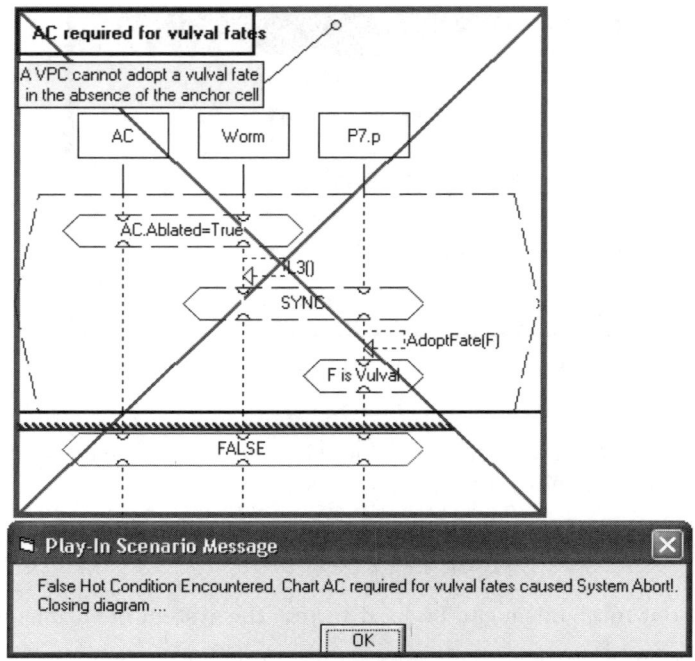

Fig. 11. Violation of an anti-scenario

existing data, pointing out inconsistencies that might have been ignored by the experimentalist.

2. *Predictability.* The Play-Engine can juxtapose LSCs generated either from specific information results or from general biological rules, which are inferred from many data sets. This juxtaposition has the potential to simulate novel scenarios and to highlight behaviours that were not previously observed. One such example was observed over our model while playing out a scenario that involved a combination of two mutations (double mutant), one in the *dig-1* gene and the other in the *lin-12* gene. In *dig-1* mutants the gonad is shifted anteriorly, and in *lin-12(0)* mutants the three VPCs P(5-7).p were observed to adopt a 1° fate. An actual experiment that detects the effect of combining these two mutations together has not been done yet. However, based on the LSCs that represent the observed phenotypes for each individual mutant, together with a couple of LSCs that represent deduced rules regarding signalling mechanisms involved in vulval induction, the Play-Engine predicted that such an experiment would result in P(4-6).p adopting a 1° fate (Fig. 12). Such a play-out session illustrates that the Play-Engine can execute scenarios that were not played in explicitly. (For further explanations, including a demo of this execution, see [1].)

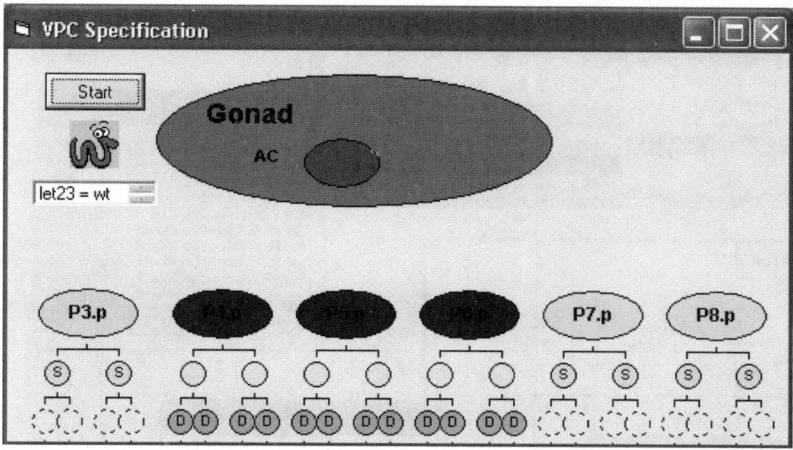

Fig. 12. Predicting the VPC lineages of a *lin-12(0);dig-1* double mutant. *D* denotes a nucleus that divides, but the division axis is not known

3. *Query the system for scenarios that satisfy a given behaviour.* Play-out has been extended with a powerful new module, called *smart play-out* [9, 11], which utilizes tools from program verification in order to analyse an LSC model in much richer ways than just executing it. Many of its uses come from asking the system to find a way to do something on its own, or to ask it "Is this possible?" kinds of questions. For example, we can ask the system to automatically figure out if there is some possible way of satisfying a particular scenario, and to then run the resulting discovered sequence on its own. An example of such a test is depicted in Fig. 13, in which an existential LSC is used to query the system for scenarios that result in P7.p adopting a 3° fate. We instruct the smart play-out module to find a way to initialize the system with any set of mutations, cell ablations, and environmental conditions from which there exists some behavioural pathway that satisfies the requested scenario (Fig. 13A). This corresponds to identifying an experimental setup that can lead to the requested result. If the system succeeds, it announces the initial configuration that was used for satisfying the test (Fig. 13B), and executes the corresponding LSCs. While executing the solution, the system monitors the existential chart that specifies the test (Fig. 13C). Figure 13.D shows the results of such a successful run, in which P7.p adopted a tertiary fate as a consequence of compromising the inductive signalling pathway by anchor cell ablation. In this specific example, the smart play-out module chose to execute a scenario in which, besides the anchor cell, 5 VPCs were ablated as well. Under these experimental conditions, the ablation of the VPCs has no influence on the fate adopted by P7.p; yet, this is a valid scenario that results in P7.p adopting a tertiary fate. The Play-Engine enables the user

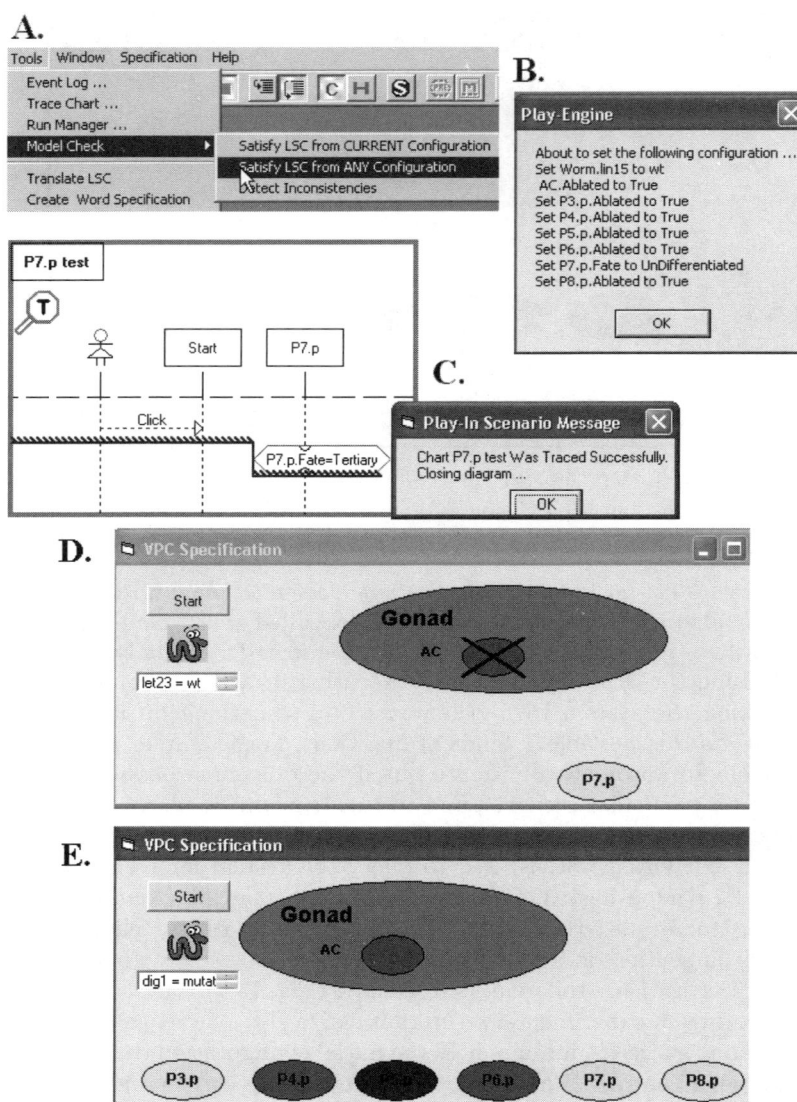

Fig. 13. Using smart play-out for satisfying the test "can P7.p adopt a 3° fate?". The smart play-out module is instructed to find a way to initialize the system with any configuration from which there exists a possibility to satisfy the test (A). Solution 1: A scenario that involves compromising the inductive signal by anchor cell ablation (B–D). Solution 2: A scenario that involves an anterior shift of the gonad (E).

to identify additional runs that satisfy a given test, if such runs exist (see [11] for further details). Figure 13.E shows another scenario depicted by the smart play-out module to satisfy the same test. In *dig-1* animals, an anterior displacement of the gonad can result in a shift in the pattern of fates adopted by the VPCs, leading to P7.p adopting a tertiary fate. (For further explanations, including a demo of applying this test to the model, see [1].)

Thus, play-out, with its smart play-out enrichment, makes it possible to detect, or predict, the outcome of combinations of conditions not previously tested, and to query the system for many different kinds of desired (or undesired) outcomes.

3 Discussion

In previous sections, we illustrated the capability of our modelling approach to facilitate the transformation of biological data into formal computer-analysable statements. In particular, our modelling approach provides solutions for the following challenges involved in this formalization process:

Incorporating raw biological data

One of the unique features of our approach is the incorporation of primary data, rather than modelling the general rules that are derived or inferred from the actual data. One challenge to modelling primary data is the integration of information obtained in different ways and using different approaches. The assessment of results can vary with the individual experimentalist and changes over time as new information about the system and technologies to evaluate it become available. The variety of data collection methods can be integrated by incorporating LSCs that link the different types of data together. We can then use the model to validate equivalence statements against the data, thus enhancing our understanding of the biological system.

Importantly, primary data, once entered as LSCs, can be used even when the general rules of behaviour are modified with time. Primary data scenarios can be derived from many different laboratories and many different types of experimentation, and can be used to detect conflicts within the primary data, highlighting important aspects to be later tested more rigorously.

Rules as a second level

A computer-assisted approach to a formalized list of primary data is useless unless it can be tested for overall consistency and logical conflict, in ways that are better than can be done by human reasoning alone. This possibility is also built into our approach. A logical system is routinely used by experimentalists to infer abstractions about the real system from condition–result types of data. In our model, these abstractions can be formalized as "rules". For example, they can be entered in the form of anti-scenarios (as shown above). Thus, it may be possible to formalize a system that is already in use in a non-formal way so that the rigorous tools of computer-based verification can be brought to bear on the data. Moreover, having both primary data and associated "rules", allows novel scenarios to be obtained from the input data, as illustrated above.

Non-deterministic events

One of the hallmarks of biological data is non-determinism. While some non-determinism likely derives from the paucity of our knowledge of the important details of the systems of interest, the very nature of biological systems contains some inherent non-determinism. For example, stochastic events, threshold phenomena, and feedback systems all operate at the molecular and cellular levels and it is crucial that methods and tools for their modelling support explicit choice-based, or probability-based, branching of behaviour.

A behavioural database

At the most basic level, a model of the type we propose forms a behavioural database that can be used for retrieving dynamic data. Existing biological databases can be used to retrieve the sequence of a given gene, or even to find its homologues, but there is no current mechanism for asking questions such as "what will happen if gene X is mutated together with gene Y", or "what will be the phenotypic outcome of ablating cell A in a specific genetic background".

Detecting inconsistencies

New inferences that are made can be checked for consistency against existing observations and rules. As illustrated in Sect. 2.2 above, "old" rules that are contradicted by new experiments can be detected as well.

Predictions

The play-out approach offers the additional possibility of asking the system for the predicted result of an experiment that has not yet been performed. Such predictions can then be tested in the lab. If the model predicts an outcome different from the actual outcome, there are likely factors that have not been included in the model that are relevant to the experimental system. Identifying such putative factors can then lead to new experiments that will aim at filling in these gaps. Alternatively, such a mismatch between prediction and experiment can result from improper rules or interpretations that currently constitute the model. Thus, the kind of model analysis we propose can assist in improving, correcting, and sharpening our understanding of the biological system.

Explaining surprising results

The smart play-out mechanism can be used to answer questions such as: "Is there any scenario in which a given behaviour will be observed?" Such tests can be applied to explaining surprising experimental outcomes that were observed in the lab (based on the current data, is there any scenario that can produce this surprising result?), or to test our understanding of the system (is there indeed no scenario, in which, although the product of gene X was eliminated, a given tissue T will still develop normally?).

4 Current Status and Future Directions

To date, we have formalized as LSCs only a small set of data pertaining to VPC specification. These have served to highlight some of the critical issues that need to be addressed in modelling this system in its entirety. Several of these issues have already been addressed, as described in this chapter. In addition to the conceptual advances this small data set has prompted, these LSCs reveal much of the promise and feasibility of applying LSCs and play-in/play-out to the modelling of biology. One of the strengths of the entire approach is the ability to execute a model even with incomplete data sets. As we continue to fill out our test-case model, including the integration of different signal-transduction inputs into the determination of vulval fates, we are also considering the expansion of the methodology in several areas:

Expanding the experimental repertoire

Although our current efforts are concentrated on condition–result data from genetic and anatomical manipulations, there is no a priori reason why the same methods should not be applied to other types of data, representing other levels of biological inquiry. These include biochemical data (such as signal-transduction pathways, protein-protein interactions, etc.) and gene expression data (microarray data, anatomical expression pattern information, etc.).

Distributed play-in/play-out

It is already possible to connect separate GUIs and Play-Engines to each other. This will enable multiple labs to participate in the modelling effort.

Each lab can design its own GUI to represent its sub-system of interest, and then play in the relevant scenarios. Connecting Play-Engines to each other can facilitate a distributed play-out mechanism, in which an event that occurs in an LSC that is being executed on one computer will activate an LSC that belongs to a specification running on another computer. The Play-Engine development team is also working on tools to connect the Play-Engine to other environments as well, e.g., tools that enable statechart-based modelling, such as Rhapsody, and ones that deal with the continuous aspects of systems. These too will significantly expand the possibilities of modelling biological systems.

Another possibility is to connect to the *C. elegans* database, Wormbase, (or its equivalent in other systems), so that just as we can obtain the sequence and homologues, etc., for each gene, we would be able to retrieve via Wormbase the LSCs in which it participates.

The long-term goal

In beginning to model C. elegans biology using these methodologies, we have approached this task with the long-term goal of modelling as much of C. elegans biology as is possible, and to represent it as realistically as is possible. The dream is to construct a full 4-dimensional model of a multi-cellular animal, which is true to all known facts about it, and which is easily extendable as new facts are discovered (see [7] for a more detailed discussion of this). The expandability and flexibility of our tools make us hopeful that this work represents the initial step towards that goal.

Acknowledgments

We would like to thank Anat Maoz and Yael Kfir for their help in modelling cell movements and cell divisions, as depicted in Sect. 2, Figs. 2, 6, and 12.

References

1. *http://www.wisdom.weizmann.ac.il/mathusers/kam/CelegansModel/Demos.htm.*
2. J. Bower and H. Bolouri (Eds.). *Computational Modeling of Genetic and Biochemical Networks.* MIT Press, Cambridge MA, 2001.
3. C. elegans Sequencing Consortium. Genome sequence of the nematode C. elegans: a platform for investigating biology. *Science*, 282(5396):2012–8, 1998.
4. W. Damm and D. Harel. LSCs: Breathing life into message sequence charts. *Formal Methods in System Design*, 19(1):45–80, 2001.
5. A.G. Fraser, R.S. Kamath, P. Zipperlen, M. Martinez-Campos, M. Sohrmann, and J. Ahringer. Functional genomic analysis of C. elegans chromosome I by systematic RNA interference. *Nature*, 408(6810):325–30, 2000.

6. P. Gonczy, C. Echeverri, K. Oegema, A. Coulson, S.J. Jones, R.R. Copley, J. Duperon, J. Oegema, M. Brehm, E. Cassin, E. Hannak, M. Kirkham, S. Pichler, K. Flohrs, A. Goessen, S. Leidel, A.M. Alleaume, C. Martin, N. Ozlu, P. Bork, and A.A. Hyman. Functional genomic analysis of cell division in C. elegans using RNAi of genes on chromosome III. *Nature*, 408(6810):331–6, 2000.

7. D. Harel. A grand challenge for computing: Full reactive modeling of a multicellular animal. Position paper. UK Workshop on Grand Challenges in Computing Research, 2002.

8. D. Harel, S. Efroni, and I.R. Cohen. Reactive animation. In *Proc. 1st Int. Symposium on Formal Methods for Components and Objects (FMCO 2002)*, Lecture Notes in Computer Science. Springer-Verlag, 2003.

9. D. Harel, H. Kugler, R. Marelly, and A. Pnueli. Smart play-out of behavioral requirements. In *Proc. 4th International Conference On Formal Methods in Computer-Aided Design (FMCAD'02)*, 378–98, 2002.

10. D. Harel and R. Marelly. Playing with time: On the specification and execution of time-enriched LSCs. In *Proc. 10th IEEE/ACM International Symposium on Modeling, Analysis and Simulation of Computer and Telecommunication Systems (MASCOTS'02)*, 2002.

11. D. Harel and R. Marelly. *Come, Let's Play: Scenario-Based Programming Using LSCs and the Play-Engine*. Springer-Verlag, 2003.

12. D. Harel and R. Marelly. Specifying and executing behavioral requirements: the play-in/play-out approach. *Software and Systems Modeling*, 2(2):82–107, 2003.

13. N. Kam, I.R. Cohen, and D. Harel. The immune system as a reactive system: Modeling T cell activation with statecharts. In *Visual Languages and Formal Methods*, Human-Centric Computing, IEEE, 15–22, 2001.

14. N. Kam, D. Harel, H. Kugler, R. Marelly, A. Pnueli, E.J.A. Hubbard, and M.J. Stern. Formal modeling of C. elegans development: A scenario based approach. In C. Priami (Ed.). *1st Workshop on Computational Methods in Systems Biology*, Lecture Notes in Computer Science 2602, Springer-Verlag, 4–20, 2003.

15. I. Maeda, Y. Kohara, M. Yamamoto, and A. Sugimoto. Large-scale analysis of gene function in Caenorhabditis elegans by high-throughput RNAi. *Curr. Biol.*, 11(3):171–6, 2001.

16. R. Marelly, D. Harel, and H. Kugler. Multiple instances and symbolic variables in executable sequence charts. In *Proc. 17th Ann. ACM Conf. on Object-Oriented Programming, Systems, Languages and Applications (OOPSLA'02)*, 83–100, 2002.

17. F. Piano, A.J. Schetter, M. Mangone, L. Stein, and K.J. Kemphues. RNAi analysis of genes expressed in the ovary of Caenorhabditis elegans. *Curr. Biol.*, 10(24):1619–22, 2000.

18. D.L. Riddle, T. Blumenthal, B.J. Meyer, and J.R. Priess (Eds.). *C. elegans II*. Cold Spring Harbor Laboratory Press, Plainview NY, 1997.

19. P.W. Sternberg and H.R. Horvitz. Pattern formation during vulval development in C. elegans. *Cell*, 44(5):761–72, 1986.

20. M. Wang and P.W. Sternberg. Pattern formation during C. elegans vulval induction. *Curr. Top. Dev. Biol.*, 51:189–220, 2001.

21. A.S. Wilkins (Ed.). *Modelling complex biological systems: a special issue. BioEssays*, 24, Wiley Periodicals, 2002.

P Systems with Symport/Antiport Rules. A Survey

Carlos Martín-Vide[1] and Gheorghe Păun[1,2]

[1] Research Group on Mathematical Linguistics, Rovira i Virgili University,
Pl. Imperial Tárraco 1, Tarragona, Spain
cmv@astor.urv.es
[2] Institute of Mathematics of the Romanian Academy, Bucureşti
gpaun@imar.ro

Summary. After briefly presenting the basic ideas and types of results of membrane computing, we introduce a widely investigated class of P systems, with a direct biological motivation, the symport/antiport P systems. We recall the generative variants (including the case when the result of a computation is obtained by means of the trace of a specified object in its movement through membranes), as well as the automata-like variants. The central results about the computing power of these systems are recalled, and in this context several open problems are mentioned.

1 Introduction to Membrane Computing

Membrane computing is a branch of natural computing which aims to devise distributed parallel computing models that are abstracted from the living cell structure and functioning [24]. We refer to [25] and to the web page http://psystems.disco.unimib.it for details and for references. The computing devices investigated in this framework are called *membrane systems* or *P systems*.

The basic ingredients of such a system are the *membrane structure* and the sets of *evolution rules* which process *multisets* of *objects* placed in the compartments of the membrane structure. A membrane structure is a hierarchically arranged set of membranes, as suggested in Fig. 1, where we distinguish the external membrane (corresponding to the plasma membrane and usually called the *skin* membrane) and several internal membranes (corresponding to the membranes present in a cell, around the nucleus, in Golgi apparatus, vesicles, etc.); a membrane without any other membrane inside is said to be *elementary*. Each membrane determines a compartment, also called *region*, the space delimited from above by it and from below by the membranes placed directly inside, if any exist. The correspondence membrane-region is one-to-one, that is why we sometimes use these terms interchangeably; also, we identify by the same label a membrane and its associated region. (A membrane structure can be described by a rooted tree, in the obvious way.)

In the basic variant of P systems, each region contains a multiset of symbol-objects, which correspond to the chemicals present in a solution in a cell compartment; these chemicals are considered here as unstructured, that is why we describe them by symbols from a given alphabet. (A strong branch of membrane computing deals with structured objects, described by strings and processed by string-transforming rules, such as rewriting rules, splicing rules, and so on, but in this chapter we do not consider this case.)

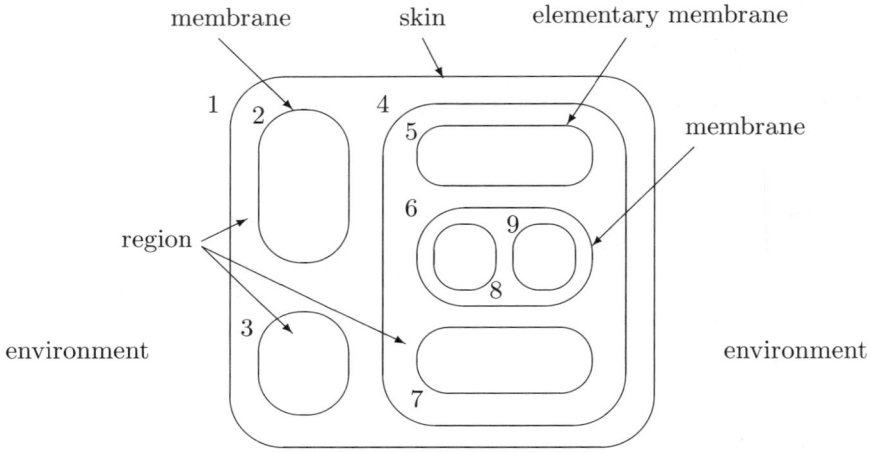

Fig. 1. A membrane structure

The objects evolve by means of evolution rules, which are also localized, associated with the regions of the membrane structure. The rules correspond to the chemical reactions possible in the compartments of a cell. The typical form of such a rule is $aad \rightarrow (a, here)(b, out)(b, in)$, with the following meaning: two copies of object a and one copy of object d react and the reaction produces one copy of a and two copies of b; the new copy of a remains in the same region (indication *here*), one of the copies of b exits the compartment (indication *out*), and the other enters one of the directly inner membranes (indication *in*). We say that the objects a, b, b are *communicated* as indicated by the commands associated with them in the right hand member of the rule. When an object exits a compartment, it will go to the surrounding compartment; in the case of the skin membrane this is the environment, hence the object is "lost": it never comes back into the system. If no inner membrane exists (that is, the rule is associated with an elementary membrane), then the indication *in* cannot be followed, and the rule cannot be applied. The communication of objects through membranes reminds us of the fact that the biological membranes contain various (protein) channels through which the molecules can pass (in a passive way, due to concentration difference, or in an active way, with a consumption of energy), in a rather selective manner.

A rule as above, with several objects in its left hand member, is said to be *cooperative*; a particular case is that of *catalytic* rules, of the form $ca \rightarrow cu$, where c is a catalyst, always appearing only in such rules, never changing, and a is a symbol-object. A rule of the form $a \rightarrow u$ is called *non-cooperative*.

The rules associated with a compartment are applied to the objects from that compartment, in a *maximally parallel way*: all objects which can evolve by means of local rules should do it (we assign objects to rules, until no further assignment is possible). The used objects are "consumed"; the newly produced objects are placed in the compartments of the membrane structure according to the communication commands assigned to them. The rules to be used and the objects to evolve are chosen in a nondeterministic manner. It is also possible to have a priority relation among rules (corresponding to the biochemical observation that certain reactions are more likely to appear – are more active – than others), and in this way the nondeterminism is decreased. In turn, all compartments of the system evolve at the same time, synchronously (a common clock is assumed for all membranes). That is, we have two levels of parallelism, one at the level of compartments and one at the level of the whole "cell".

The rules can also have other effects than changing the multisets of objects, for instance, they can control the membrane permeability (this corresponds to the fact that the protein channels from cell membranes can sometimes be closed, e.g., when an undesirable substance should be kept isolated, and they are reopened when the "poison" vanishes). If a membrane is impermeable, then no rule asking for passing an object through it can be used. In this way, the processes taking place in a membrane system can be controlled ("programmed"). In particular, membranes can be *dissolved* (all objects and membranes from a dissolved membrane are left free in the surrounding compartment – the skin membrane is never dissolved, because this destroys the "computer" itself; the rules of the dissolved membrane are "lost" – they are supposed to be specific to the reaction conditions from the former compartment, hence they cannot be applied in the upper compartment, which has its own rules), and *divided* (as in biology, when a membrane is divided, its content is replicated in the newly obtained membranes). Some other possibilities exist to control the communication of objects and the use of rules, but we do not enter into such detail here.

A membrane structure and the multisets of objects from its compartments identify a *configuration* of a P system. By a nondeterministic, maximally parallel use of rules as suggested above we pass to another configuration; such a step is called a *transition*. A sequence of transitions constitutes a *computation*. A computation is successful if it halts, that is, it reaches a configuration where no rule can be applied to the existing objects. With a halting computation we can associate a *result* in various ways. The simplest possibility is to count the objects present in the halting configuration in a specified elementary membrane; this is called *internal output*. We can also count the objects which leave the system during the computation, and this is called *external output*. In both cases the result is a number. If we distinguish among different objects, then we can have as a result a vector of natural numbers. The objects which leave the system can also be arranged in a sequence according to the moments when they exit the skin membrane, and in this case the result of a computation is a string.

This last possibility is worth emphasizing, because of the qualitative difference between the data structure used inside the system (multisets of objects, hence numbers) and the data structure of the result, which is a string, and contains positional information, a syntax.

Because of the nondeterminism of the application of rules, starting from an initial configuration, we can get several successful computations, hence several results. Thus, a P system *computes* (one also can say *generates*) a set of numbers, or a set of vectors of numbers, or, in the case of the external output, a language.

From a computability point of view, it is quite interesting that many types of P systems, of rather restricted forms, are computationally universal; they can compute, for instance, all Turing computable sets of natural numbers. This is true even for systems with simple rules (catalytic), with a very reduced number of membranes (most of the universality results recalled in [25] refer to systems with less that five membranes).

The computational power is only one criterion for assessing the quality of new computing machinery; from a practical point of view at least equally important is the *efficiency* of the new device. The P systems display a high level of parallelism. Moreover, at the mathematical level, rules of the form $a \rightarrow aa$ are allowed and by iterating such rules we can produce an exponential number of objects in a linear time. The parallelism and the possibility to produce an exponential working space are standard ways to speedup computations. In the general framework of P systems with symbol-objects (and without membrane division or membrane creation) these ingredients do not suffice in order to solve computationally hard problems (e.g., **NP**-complete problems) in a polynomial time: in [32] (see also [31]) it is proved that any deterministic P system can be simulated by a deterministic Turing machine with a linear slowdown.

However, pleasantly enough, if additional features are considered, either able to provide an enhanced parallelism (for instance, by membrane division, which may produce exponentially many membranes in a linear time), or a better structure of the multisets of objects (by membrane creation), then **NP**-complete problems can be solved in a polynomial (often, linear) time. The procedure is as follows (it is consistent with standard computational complexity requirements, see, e.g., [20]). Given a decision problem, we construct in a uniform manner (hence in polynomial time) a family of P systems (each one of a polynomial size) which will solve the instances of the problem in the following sense. In a well specified time, bounded by a given function, the system corresponding to the instances of a given size of the problem will sent to its environment a special object *yes* if and only if the instance of the problem introduced into the initial configuration of the system has a positive answer. During the computation, the system can grow exponentially (as the number of objects and/or the number of membranes) and can work in a nondeterministic manner; important is that it *always* halt. Standard **NP**-complete problems for illustrating this approach are SAT (satisfiability of propositional formulas in the conjunctive normal form) and HPP (the existence of an Hamiltonian path in a directed graph), but many other problems were also considered. Details can be found, e.g., in [25] and [27].

These two types of attractive results – computational universality and computational efficiency – as well as the versatility of the P systems explain the very rapid development of the membrane computing area.

There have also been several attempts to implement (actually, to simulate) such computing devices on the usual electronic computers. Of course, the bio-chemically inspired nice features of P systems (especially the nondeterminism and the parallelism) are lost, as they can only be simulated on the usual deterministic computers, but the simulators obtained simulators still can be useful for certain practical purposes (not to mention their didactic usefulness). Several applications of membrane computing were reported in the literature, in general, of the following type: one takes a piece of reality, most frequently from cell biology, but also from artificial life, abstract chemistry, biology of eco-systems, one constructs a P system modelling this piece of reality, then one writes a program which simulates this P system and one runs several experiments, carefully arranging the system parameters (especially the form of rules and their probabilities to be applied); statistics on the populations of objects in various compartments of the system are obtained, sometimes suggesting/supporting interesting conclusions. Typical examples can be found in [28–30] (including an approach to the famous Brusselator model, with conclusions which fit with the known ones, obtained by using continuous mathematics), [19] (an investigation of photosynthesis), and [6] (signalling pathways and T cell activation are addressed). Several other (preliminary) applications of P systems to cryptography, linguistics, and distributed computing can be found in [3, 4, 26].

2 P Systems with Symport/Antiport Rules

The chemicals do not always pass alone through membranes, but a *coupled transport* is often encountered, where two solutes pass together through a protein channel, either in the same direction or in the opposite direction. In the first case the process is called *symport*; in the latter case it is called *antiport*. Figure 2 illustrates these notions – for completeness, *uniport* refers to the case when a single molecule passes through a membrane.

More details on the biochemical mechanisms behind the coupled transport can be found, e.g., in [1] (page 620 ff.); see also [2]. An example of antiport is the exchange of Na^+ and K^+ ions through the *sodium-potassium pumps*: the potassium ion enters the cell at the same time as the sodium ion exits it. This kind of antiport is present in all cells and it is particularly active in nerve and muscle cells.

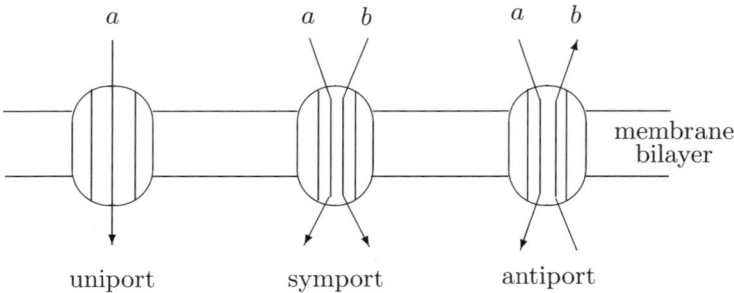

Fig. 2. Uniport and coupled transport

An example of symport appears in the way the bacterium *Escherichia coli* transports lactose from its environment, at the same time passing lactose and protons through the membrane: first, protons are pushed outside the cell by so-called proton pumps, creating a difference in concentration of H^+ between the inner and the outer regions; then the protons are coupled to lactose molecules with the help of certain proteins (lactose permease), and, because of the difference in the concentration of H^+, the lactose–proton pair is transported into the cell.

The idea of a coupled transport can be captured in membrane computing terms in a rather easy way: for the symport case, consider rules of the form (ab, in) or (ab, out), while for the antiport case write $(a, out; b, in)$, with the obvious meaning. Mathematically, we can generalize this idea and consider rules which move at the same time arbitrarily many objects through a membrane.

The use of such rules suggests a very interesting question (research topic): can we compute only by communication, i.e., only by transferring objects through membranes? This question leads to consideration of systems which contain only symport/antiport rules, which only change the places of objects, but not their "names". These types of systems are the main topic of our discussion.

We start by recalling the "standard" definition of a P system with symport/antiport rules, as introduced in [21] (hence used to generate natural numbers).

A few technical prerequisites are worth mentioning: the set of all strings over an alphabet V is denoted by V^*, the empty string is denoted by λ, and the length of $x \in V^*$ is denoted by $|x|$; the set of non-empty strings over V, that is, $V^* - \{\lambda\}$, is denoted by V^+; a multiset over a given alphabet V is represented by a string over V (the number of occurrences of each symbol in the string is the number of copies of that symbol in the corresponding multiset; thus, all permutations of a string represent the same multiset); a membrane structure with labelled membranes is represented by a string of matching labelled parentheses (for instance, the membrane structure from Fig.1 is represented by the string $[_1[_2 \ ]_2[_3 \ ]_3[_4[_5 \ ]_5[_6[_8 \ ]_8[_9 \ ]_9]_6[_7 \ ]_7]_4]_1$).

A *P system (of degree $m \geq 1$) with symport/antiport rules* is a construct Π of the form

$$\Pi = (V, T, \mu, w_1, \ldots, w_m, E, R_1, \ldots, R_m, i_o),$$

where:

1. V is an alphabet of *objects*;
2. $T \subseteq V$ is the alphabet of *terminal* objects;
3. μ is a membrane structure (with m membranes, labelled by the natural numbers $1, 2, \ldots, m$ in a one-to-one manner);
4. $w_1, \ldots, w_m$ are strings over V (representing multisets over V) associated with the regions $1, 2, \ldots, m$ of μ;
5. $E \subseteq V$ is the set of objects which are supposed to appear in an arbitrarily large number of copies in the environment;
6. $R_1, \ldots, R_m$ are finite sets of symport and antiport rules associated with the membranes $1, 2, \ldots, m$; a symport rule is of the form (x, in) or (x, out), where $x \in V^+$ (with the meaning that the objects specified by x enter, respectively exit, the membrane), and an antiport rule is of the form $(x, out; y, in)$, where $x, y \in V^+$ (the meaning is that the multiset x is sent out of the membrane and y is taken into the membrane region from the surrounding region);
7. i_o is the label of the *output membrane*, an elementary one in μ.

Starting from the *initial configuration*, which consists of μ and $w_1, \ldots, w_m, E$, the system passes from one configuration to another one by applying the rules from each R_i in a non-deterministic and maximally parallel way. (The environment is supposed to be inexhaustible; at each moment all objects from E are available in any number of copies we need. These objects are necessary, because otherwise we only handle a finite number of objects, those present in $w_1, \ldots, w_m$, hence we can compute only with numbers of a bounded value.) As usual, a sequence of transitions is called a *computation*, and a computation is *successful* if and only if it halts.

We associate a *result* with a successful computation, in the form of the number of objects from T present in membrane i_o in the halting configuration. The set of all such numbers computed by Π is denoted by $N(\Pi)$, and the family of all sets $N(\Pi)$, computed by P systems with at most m membranes, with symport rules $(x, in), (x, out)$ with $|x| \leq r$, and antiport rules

$(x, out; y, in)$ with $|x|, |y| \leq t$, is denoted by $NOP_m(sym_r, anti_t)$, $m \geq 1$ and $r, t \geq 0$. (One can say that r and t are the maximal *weight* of symport and antiport rules of Π, respectively.)

We illustrate the previous definition with the following **example**. Let

$$\Pi = (V, T, \mu, w_1, w_2, E, R_1, R_2, 2), \text{ with}$$
$$V = \{a, b, c, c', d, e, e', f, g, Z\},$$
$$T = \{a\},$$
$$\mu = [_1 [_2]_2]_1,$$
$$w_1 = ac,$$
$$w_2 = df,$$
$$E = \{a, b, c', e, e', g, Z\},$$
$$R_1 = \{(c, out; Z, in),\ (ca, out; cbb, in),\ (ca, out; c'bb, in),$$
$$(da, out; Z, in),\ (c'd, out; e, in),\ (eb, out; ea, in),$$
$$(eb, out; e'a, in),\ (fb, out; Z, in),\ (e'f, out; cdf, in),$$
$$(e'f, out; g, in)\},$$
$$R_2 = \{(d, out; c', in),\ (c', out),\ (f, out; e', in),\ (e', out),$$
$$(df, in),\ (Z, in),\ (Z, out),\ (ga, in),\ (g, out)\}.$$

For the reader's convenience, the initial configuration of the system, including the rules associated with each membrane and the objects present in the environment, is represented in Fig. 3.

Assume that at some moment we have in the skin region n copies of object a (initially, $n = 1$), and one copy of c, and in region 2 we have one copy of d and one copy of f (initially, this is the case). If we use the rule $(c, out; Z, in) \in R_1$, then Z gets into region 1 from the environment. In this case the computation will never stop, because of the rules $(Z, in), (Z, out)$ from R_2 which cause Z to oscillate between regions 1 and 2. Thus we have to begin by using the rule $(ca, out; cbb, in) \in R_1$ a number of times. When we replace all copies of a by b, then we have to use the rule $(c, out; Z, in) \in R_1$, and so the computation will not halt, hence we have to save one a and at some time to use the rule $(ca, out; c'bb, in) \in R_1$. The object c' must now go to region 2 and send from there the object d, by the rule $(d, out; c', in) \in R_2$. In the next step, if any copy of a is still present in region 1, then the rule $(da, out; Z, in) \in R_1$ must be used, and thus again the computation will never stop. If no copy of a is present in region 1, that is, the doubling of n was complete (the n copies of a were replaced by $2n$ copies of b), then d will wait one step in the skin region, as c' exits region 2, and then by rule $(c'd, out; e, in) \in R_1$ the two objects exit the system together, while e is brought in.

In the same way as c has changed all a to b, the object e will now change all copies of b into a. When this operation is completed (and only then, since otherwise Z will get into the system), e is replaced by e', and e' will send f out of region 2. The object f checks whether or not all copies of b were replaced by a, and only in the affirmative case it does exit the system together with e' and reintroduce the objects c, d, and f; d and f then enter region 2, and the whole process can be iterated. In this way, we can double the number of copies of a as many times as we want, but at least once. At any moment, instead of the rule $(e'f, out; cdf, in) \in R_1$ we can use the rule $(e'f, out; g, in) \in R_1$. The object g will carry each copy of a into membrane 2, the output one, and when this is completed g gets stuck in the skin region. Therefore, $N(\Pi) = \{2^n \mid n \geq 1\}$.

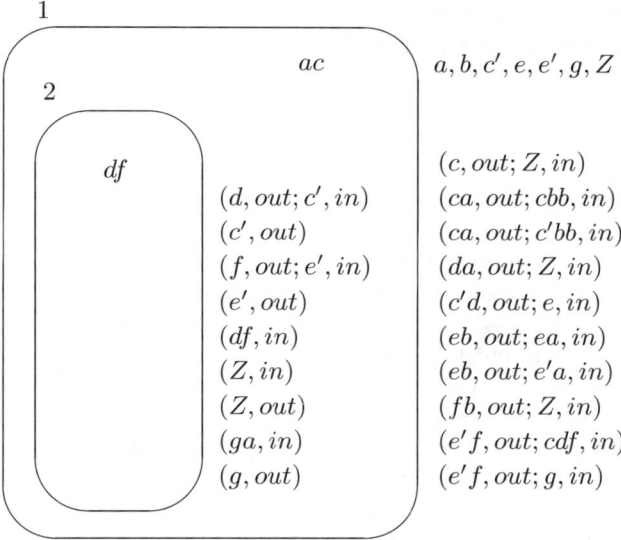

Fig. 3. An example of a symport/antiport P system

The previous system Π uses rather complex rules, but as we will see in the theorem given below systems with simpler rules can generate the same set, because P systems with symport/antiport rules are computationally universal (hence we can compute at the level of Turing machines by making use of communication only). The currently best results in this respect, in what concerns the number of membranes and the size of symport and antiport rules, are the following (NRE denotes the family of Turing computable sets of natural numbers; proofs can be found in [11, 12]); these results improve previous results from [21] and from several continuations of [21].

Theorem 1. $NRE = NOP_m(sym_r, anti_t)$, for $(m, r, t) \in \{(1, 1, 2), (3, 2, 0), (2, 3, 0)\}$.

The optimality of these results is not known. In particular, it is an *open problem* whether or not the families $NOP_m(sym_r, anti_t)$ with $(m, r, t) \in \{(2, 2, 0), (2, 2, 1)\}$ are also equal to NRE.

3 Following the Traces of Objects

We have mentioned above that it is a challenging task to work with numbers inside a system and to obtain a string as the result of a computation. Considering an external output in the case of P systems with symport/antiport rules does not look suitable, because the environment is a participant in the computation, it contains objects, and the objects which leave the system can come back during the computation. However, in the case of systems computing only by communication there is a natural way to find a string, namely by following the itineraries of certain objects through the membrane structure. In what follows we recall the definition from [14] – only the trace of a single distinguished object (called the *traveller*) is considered, also mapping this trace through a weak coding.

Specifically, we consider systems of the form

$$\Pi = (O, t, T, h, \mu, w_1, \ldots, w_m, E, R_1, \ldots, R_m),$$

where O is the alphabet of objects, $t \in O$ is a distinguished object (the traveller), T is an alphabet (no relation with O is assumed), $h : \{1, 2, \ldots, m\} \longrightarrow T \cup \{\lambda\}$ is a weak coding, μ is a membrane structure of degree m, with the membranes injectively labelled with $1, 2, \ldots, m$, $w_1, \ldots, w_m$ are strings representing the multisets of objects present in the m regions of μ, E is the set of objects present in the environment in an arbitrary number of copies, and $R_1, \ldots, R_m$ are the sets of symport and antiport rules associated with the m membranes. The traveller is present in exactly one copy in the system, in one of the multisets $w_1, \ldots, w_m$ (and not present in E).

For any configuration C of Π we define

$$C(t) = \begin{cases} i, & \text{if the traveller is in region } i \text{ of } C, \\ \lambda, & \text{if the traveller is outside the system.} \end{cases}$$

Then, *the trace* of t in a halting computation $\sigma = C_1 C_2 \ldots C_k, k \geq 1$, with respect to Π is

$$trace(t, \sigma) = C_1(t) C_2(t) \ldots C_k(t).$$

Note that the trace starts with the label of the membrane where t is placed in the initial configuration.

The computation σ is said to generate the string $h(trace(t, \sigma))$, hence the language generated by Π is $L(\Pi) = \{h(trace(t, \sigma)) \mid \sigma$ is a halting computation in $\Pi\}$. (By means of h we ignore some labels and we can rename the others.)

We denote by $LOP_m(sym_r, anti_t)$ the family of languages $L(\Pi)$ generated by P systems with at most m membranes, with symport rules of weight at most r and antiport rules of weight at most t, where $m \geq 1$ and $r, t \geq 0$.

Here is an **example**. Consider the system (of degree 5)

$$\Pi = (\{d,t\}, t, \{a,b,c\}, h, \mu, t, \emptyset, \emptyset, \emptyset, \emptyset, \{d\}, R_1, R_2, R_3, R_4, R_5),$$

with the weak coding $h(1) = a, h(3) = b, h(5) = c, h(2) = h(4) = \lambda$, the membrane structure $\mu = [_1[_2[_3[_4[_5\ ]_5]_4]_3]_2]_1$, and the following rules:

$$R_1 = R_3 = R_5 = \{(t, out), (td, in)\},$$
$$R_2 = R_4 = \{(t, in), (d, in)\}.$$

First, the traveller brings $n \geq 1$ copies of d from the environment (each of them immediately enters membrane 2), then t goes to membrane 2. From here, the traveller brings $m \leq n$ copies of d into membrane 3 (each of them immediately enters membrane 4), then t goes to membrane 4 (at that moment, it is possible that some copies of d remain in membrane 2). From membrane 4, all copies of d are carried into membrane 5; the computation stops with the traveller in membrane 4. Thus, for any computation σ of this type, we have

$$trace(t, \sigma) = 1^{n+1}(23)^m(45)^m4, \text{ for some } n \geq 1, m \leq n.$$

The traveller can also end up in membrane 2, after introducing all copies of d in membrane 3, and returning to membrane 2. In the case of such a computation σ we have

$$trace(t, \sigma) = 1^{n+1}(23)^n2, \text{ for some } n \geq 1.$$

Finally, we can also have the trivial computation where the traveller just enters membrane 2, without any copy of d present in the system, and this leads to $trace(t, \sigma) = 12$. Consequently,

$$L(\Pi) = \{a^{n+1}b^mc^m \mid n \geq 1, m \leq n\} \cup \{a^{n+1}b^n \mid n \geq 1\} \cup \{a\}.$$

This language is not context-free. Note that the system Π has only symport rules (of a weight at most two, hence as is the case in biology).

Clearly, a language over an alphabet with k letters cannot be generated in the previous way by a system with less than k membranes; thus, in a trivial manner we get the fact that the hierarchy on the number of membranes is infinite. This also suggests consideration of families of languages over alphabets of a given cardinality; that is why we denote by kRE the family of recursively enumerable languages over alphabets with at most k symbols; we also add k in front of the notation of families of trace languages, with the same signification.

The proof of the following results (they improve previous results from [14] and [25]) can be found in [12].

Theorem 2. $kRE = kLOP_{k+1}(sym_0, anti_2) = kLOP_{k+1}(sym_3, anti_0) = kLOP_{k+2}(sym_2, anti_0), \ k \geq 1$.

4 Automata-like Symport/Antiport P Systems

P systems with symport/antiport rules *generate* numbers in the basic variant and languages in the case of traces. A language *accepting* version of these systems was also considered, first in [7] and then in [8–10, 15].

The simplest variant is that from [10]. In short, one considers a usual symport/antiport system Π, without any output membrane, working in the standard way. During a computation, the system brings inside symbols from the environment. Then, a string w (over an alphabet T which can be a strict subset of the alphabet of all objects; the elements of T are called *terminal symbols/objects*) is recognized by the system Π if and only if there is a successful computation of Π such that the sequence of terminal symbols taken from the environment during the computation is exactly w. If more than one terminal symbol is taken from the environment in one step, then any permutation of these symbols constitutes a valid subword of the input string. The language of all strings $w \in T^*$ recognized by Π is denoted by $A(\Pi)$.

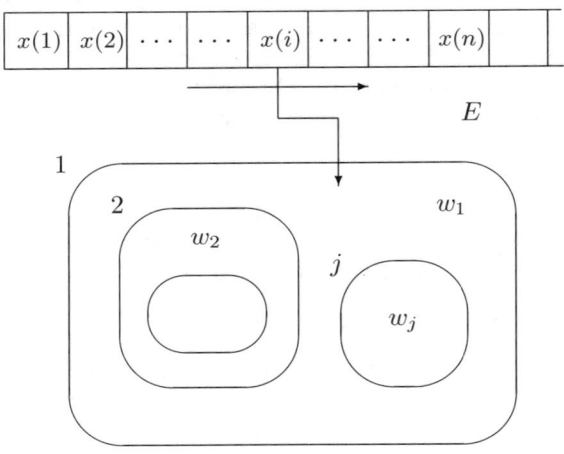

Fig. 4. An analysing P system in an automata-like representation

In this mode of analysing a string, if a string w is recognized by Π, then its symbols should exist in E; because each element of E appears in the environment in arbitrarily many copies, we cannot introduce a symbol of w into the system by using a symport rule, because an arbitrarily large number of copies would be introduced at the same time, hence the string will not be finite. Antiport rules are thus obligatory. In order to cope with this difficulty, and in order to recognize strings in a way closer to automata style, a restricted mode of accepting strings was introduced in [9], called *initial*: take a string $x = x(1)x(2) \ldots x(n)$, with $x(i) \in T, 1 \le i \le n$; in the steps $1, 2, \ldots, n$ of a computation we place one copy of $x(1), x(2), \ldots, x(n)$, respectively, in the environment (together with the symbols of E); in each step $1, 2, \ldots, n$ we request that the symbol $x(1), x(2), \ldots, x(n)$, respectively, is introduced into the system (by a symport or an antiport rule, alone or together with other symbols); after exhausting the string x, the computation may continue, possibly introducing further symbols into the system. If the computation eventually halts, then the string x is recognized. If the system halts before ending the string, or at some step i the symbol $x(i)$ is not taken from the environment, then the string is not accepted. The language of strings accepted by Π in the initial mode is denoted by $A_I(\Pi)$.

Figure 4 illustrates this mode of accepting strings, suggesting the automata-like behaviour of an analysing P system working in the initial mode.

It is important to note that in the initial mode of using a P system we can use, say, a rule of the form (a, in) for introducing a symbol of the analysed string into the system: a is not necessarily an element of E, it is only present on "the tape". Contrast this with the non-initial case, where such a rule will cause the computation to continue forever: if a is present in the environment, then it is present in arbitrarily many copies, thus the rule (a, in) can be used indefinitely, because the environment is inexhaustible.

The family of all languages $A_I(\Pi)$, accepted by systems Π with at most m membranes, with symport rules $(x, in), (x, out)$ with $|x| \leq r$, and antiport rules $(x, out; y, in)$ with $|x|, |y| \leq t$, is denoted by $A_I LP_m(sym_r, anti_t)$, $m \geq 1$, $r, t \geq 0$.

Let us examine an **example**. We consider the system

$$\Pi = (\{a, b, c, d\}, \{a, b, c, d\}, [_1 [_2 \ ]_2 [_3 \ ]_3]_1, \lambda, \lambda, \lambda, \emptyset, R_1, R_2, R_3),$$
$$R_1 = \{(a, in), \ (b, in), \ (c, in), \ (d, in)\},$$
$$R_2 = \{(ab, in), \ (cd, in)\},$$
$$R_3 = \{(\alpha, in), \ (\alpha, out) \mid \alpha \in \{a, b, c, d\}\}.$$

We examine the work of Π, in the initial mode, on strings of the form $a^{n_1} c^{n_2} b^{n_3} d^{n_4}$, $n_1, n_2, n_3, n_4 \geq 1$. In each time unit, the available symbol of the string (the only one present in the environment, because $E = \emptyset$) enters the system by the corresponding symport rule from R_1. Then, this symbol can pass back and forth through membrane 3, by the rules of R_3. In any moment when both a and b, or both c and d, are present in region 1, the pairs ab, or cd, respectively, can enter into membrane 2. Therefore, the computation stops only when $n_1 = n_3$ and $n_2 = n_4$. If n_1, n_2, n_3, n_4 are even numbers, then after introducing all symbols into the system, half of the occurrences of each a, b, c, d can be present in the skin membrane and half in membrane 3 (we assume that we always use the rules of R_3 and not those of R_2). In this way, in two consecutive steps, all symbols can be sent to membrane 2 – providing that $n_1 = n_3$ and $n_2 = n_4$. Therefore

$$A_I(\Pi) \cap (aa)^+ (cc)^+ (bb)^+ (dd)^+ = \{a^n c^m b^n d^m \mid n, m \geq 2 \text{ even numbers}\}.$$

This language is not context-free, hence also $A_I(\Pi)$ is not context-free.

Not all strings $a^n c^m b^n d^m$, $n, m \geq 1$, are recognized by the previous system Π. Consider, for instance, the case $n = 1, m = 2$. The first steps of a computation of Π are as follows:

step	environment	region 1	region 2	region 3
0	a	λ	λ	λ
1	c	a	λ	λ
2	c	c	λ	a
3	b	ac	λ	c
4	d	bc	λ	ac
5	d	acd	λ	bc
6	$-$	bcd	λ	acd
		or dbc	cd	a

The symbols a, b will never meet in region 1, hence the computation continues forever, and the string $accbdd$ is not accepted.

The above-mentioned analysing P systems are a less restrictive variant of the devices considered in [7], where one considers constructs of the form $\Pi = (V, T, \mu, w_1, \ldots, w_m, E, R_1, \ldots, R_m)$, with all components as above, but with the rules from the sets $R_i, 1 \leq i \leq m$, of the form $(y, in)|_x$, $x \in V^*$, $y \in V^+$. Such a rule is a symport rule (y, in) with a *promoter*, the multiset x: the objects from y are taken from the region surrounding membrane i only if the objects indicated by x are present in the region of membrane i. Several rules can have the same promoter multiset x or can share objects from their promoting multisets. Moreover, objects which appear in the promoter x of a rule associated with a membrane i may also appear in the "moving multiset" y of a rule associated with a membrane placed inside membrane i.

However, an important difference with the general case of P systems with symport/antiport rules is that in each step, in each region, at most one rule is used (the system works in a sequential manner in each region); if *at least* one rule can be used, then *one* of them is chosen and used.

As for the "standard" analysing P systems from [10], a string w over T^* is accepted if its symbols are taken from the environment during a halting computation, in the order they appear in w. We denote by $A(\Pi)$ the language of strings accepted by Π in the arbitrary mode, and by $A_I(\Pi)$ the language of strings accepted in the initial mode. We denote by $ALP_m(p_u sym_r, owts)$, $A_I LP_m(p_u sym_r, owts)$ the families of languages obtained; the meaning of all parameters is as usual, while p_u indicates that promoters of weight at most u are used; moreover, the indication $owts$ refers to the specific form (one-way communicatín, using only top-down symport rules) and working mode (sequential, one rule per region) of the systems we use. (There are some small differences between the above sketched definition of $A(\Pi)$ and the definition from [7], but we do not recall them here.)

With our notation, in [7] it is proved that $RE = ALP_7(p_4 sym_6, owts)$.

This result was improved in [8] from the point of view of all parameters (number of membranes, size of promoters, size of moved multisets):

Theorem 3. $RE = ALP_2(p_2 sym_2, owts)$.

The similar result for the initial mode is still an *open problem*, but for the one-letter alphabet we have:

Theorem 4. $1RE = 1ALP_2(p_2sym_2, owts) = 1A_ILP_2(p_2sym_2, owts)$.

A variant of the P automata of the form from [7] was introduced in [15], somewhat closer to the form of "standard" string automata, also involving *states*. We consider it here in a simplified version. Specifically, we deal with systems of the form $\Pi = (V, K, T, \mu, w_1, \ldots, w_m, E, R_1, \ldots, R_m)$, where all components are as usual, $K \subseteq V - T$ is the set of *states*, and the rules from sets $R_i, 1 \leq i \leq m$, are of the form $(qy, in)|_{px}$, where $p, q \in K$ and $x, y \in V^*$. The rules are used in the maximally parallel manner. Using a rule $(qy, in)|_{px} \in R_i$ is possible only if px is included in the multiset present in region i, and the effect is that the multiset y is introduced in region i from the surrounding region, while p is replaced by q (the objects from x are not affected by this rule, but can be simultaneously processed by other rules). Thus, in each region we can use at most as many rules as occurrences of state-objects existing in that region, and this is an efficient way to handle the parallelism. As usual, a string over T is accepted if it is the sequence of symbols from T taken into the system during a halting computation. The language of all strings accepted by Π is denoted by $A(\Pi)$ and the language of strings accepted in the initial mode is denoted by $A_I(\Pi)$. The family of all languages $A(\Pi)$, accepted by systems Π as above with at most m membranes, using rules $(qy, in)|_{px}$ with $|qy| \leq r, |px| \leq u$, and with at most t states, is denoted by $ALP_m(p_usym_r, state_t)$. If the strings are recognized in the initial mode, then we write A_ILP instead of ALP.

Clearly, the automata from [15] are extensions of those from [7]: we can consider a unique state, p, and instead of a rule $(y, in)|_x$ we consider the rule $(py, in)|_{px}$. Thus, the following results are consequences of Theorems 3 and 4:

Theorem 5. $RE = ALP_2(p_3sym_3, state_1)$.

Theorem 6. $1RE = 1ALP_2(p_3sym_3, state_1) = 1A_ILP_2(p_3sym_3, state_1)$.

The optimality of these results, as well as whether Theorem 5 holds also for the initial mode, remains an *open problem*.

Let us now examine accepting P systems with symport/antiport rules as those from [10]. In the arbitrary case they can recognize all recursively enumerable languages even when using only one membrane. More precisely, the following result was proven in [10]:

Theorem 7. *Each language $L \in RE$ can be recognized by an analysing P system with only one membrane using antiport rules $(x, out; y, in)$ with $(|x|, |y|) \in \{(1, 2), (2, 1)\}$ only (and no symport rule).*

Thus, there is nothing else to do with the general case; what remains is to examine the initial mode of accepting strings. This is largely a topic for further investigations, as only some preliminary results are known in this respect (see [9]).

Theorem 8. *Each regular language can be recognized by an analysing P system in the initial mode with only one membrane using antiport rules $(x, out; y, in)$ with $(|x|, |y|) \in \{(1, 2), (1, 1)\}$ only (and no symport rule). The converse assertion is not true: there are systems as above which recognize non-regular languages.*

A link between generating and accepting systems can be established:

Lemma 1. *Let $Q \in NOP_m(sym_r, anti_t)$ be a set of numbers generated by a P system of the type associated with the family $NOP_m(sym_r, anti_t)$, which counts in the halting configuration the symbols from a set T present in an elementary membrane i_o and such that the only rules associated with membrane i_o and involving elements of T are of the form $(a, in), a \in T$. Then $\{b^j \mid j \in Q\} \in 1A_I LP_{m+2}(sym_{r'}, anti_t)$, where $r' = \max(r, 2)$.*

As a consequence of this lemma and of Theorem 1 (as well as of the proofs from [12] and [11]) we get:

Theorem 9. *$1RE = 1A_I LP_n(sym_r, anti_t)$, for $(n, r, t) \in \{(3, 2, 2), (5, 2, 0), (4, 3, 0)\}$.*

However, when antiport rules with promoters are used, then the universality holds in general:

Theorem 10. *$RE = A_I LP_1(sym_0, p_1 anti_2)$.*

5 Concluding Remarks

The classes of P systems considered in the previous sections convincingly illustrate the versatility of membrane computing, as a framework for devising computing mechanisms directly inspired from biological facts and appealing from a mathematical point of view – simple definitions, large computational power, natural (sometimes challenging) research questions. Rather interesting is the fact that the universality can be reached by communication only: that we can compute at the level of Turing machines just by moving objects through membranes. The "explanation" lies in the fact that by changing the places of objects in a membrane structure we can count, compute, and check for zero, and this resembles the way the register machines work; thus, it is not incidental that most of the universality proofs mentioned in this chapter are obtained by simulating register machines (which are known from "old times" to be universal [18]) by means of various types of P systems with symport/antiport rules.

We conclude this short survey of symport/antiport P systems (several papers not cited above are mentioned in the bibliography, and further titles can be found on the web page mentioned in Introduction) by expressing our belief that the study of these devices deserves further efforts – possibly returning not only mathematical results, but also results with a significance for the biologist (as said repeatedly, one of the main tasks of biocomputing for the near future is to provide a model/simulation of a living cell as a whole, and to this end the basic features of P systems should be taken into consideration, such as the membrane structure, the multisets as the data structure to describe the solutions placed in the compartments of a cell, the parallelism – not necessarily the maximal one – the algorithmic approach, etc.)

References

1. Alberts, B., et al: *Molecular Biology of the Cell*, 3rd ed. (Garland, New York, 1994)
2. Ardelean, I.I.: The relevance of biomembranes for P systems. *Fundamenta Informaticae*, **49**(1–3), 35–43 (2002)
3. Calude, C.S., Dinneen, M.J., Păun, Gh. (eds.): *Pre-Proceedings of Workshop on Multiset Processing*, Curtea de Argeş, Romania. CDMTCS Technical Report 140, University of Auckland, 2000
4. Calude, C.S., Păun, Gh., Rozenberg, G., Salomaa, A. (eds.): *Multiset Processing. Mathematical, Computer Science, and Molecular Computing Points of View, Lecture Notes in Computer Science*, vol. 2235 (Springer, Berlin, Heidelberg, New York, 2001)
5. Cavaliere, M.: Evolution-communication P systems. In: [26], 134–145
6. Ciobanu, G., Dumitriu, D., Huzum, D., Moruz, G., Tanasă, B.: Client-server P systems in modeling molecular interaction. In: [26], 203–218
7. Csuhaj-Varju, E., Vaszil, G.: P automata. In: [26], 219–233
8. Freund, R., Martín-Vide, C., Obtulowicz, A., Păun, Gh.: On three classes of automata-like P systems. Submitted, 2003
9. Freund, R., Martín-Vide, C., Obtulowicz, A., Păun, Gh.: Initial automata-like P systems. Submitted, 2003
10. Freund, R., Oswald, M.: A short note on analysing P systems. *Bulletin of the EATCS*, **78**, 231–236 (October 2002)
11. Freund, R., Păun, A.: Membrane systems with symport/antiport: universality results In: [26], 270–287
12. Frisco, P., Hoogeboom, H.J.: Simulating counter automata by P systems with symport/antiport. In: [26], 288–301
13. Ionescu, M., Martín-Vide, C., Păun, A., Păun, Gh.: Membrane systems with symport/antiport: (Unexpected) universality results. *Proc. 8th Int. Meeting on DNA Based Computers* (ed. by Hagiya, M., Obuchi, A.), Sapporo, Japan, 2002, 151–160
14. Ionescu, M., Martín-Vide, C., Păun, Gh.: P systems with symport/antiport rules: The traces of objects. *Grammars*, **5**, 65–79 (2002)
15. Madhu, M., Krithivasan, K.: On a class of P automata. Submitted, 2002
16. Martín-Vide, C., Păun, A., Păun, Gh.: On the power of P systems with symport rules. *Journal of Universal Computer Science*, **8**(2), 317–331 (2002)
17. Martín-Vide, C., Păun, A., Păun, Gh., Rozenberg, G.: Membrane systems with coupled transport: Universality and normal forms. *Fundamenta Informaticae*, **49**(1–3), 1–15 (2002)

18. Minsky, M.: *Computation: Finite and Infinite Machines* (Prentice Hall, Englewood Cliffs, NJ, 1967)

19. Nishida, T.Y.: Simulations of photosynthesis by a K-subset transforming system with membranes. *Fundamenta Informaticae*, **49**(1–3), 249–259 (2002)

20. Papadimitriou, Ch.P.: *Computational Complexity* (Addison-Wesley, Reading, MA, 1994)

21. Păun, A., Păun, Gh.: The power of communication: P systems with symport/antiport. *New Generation Computing*, **20**(3), 295–306 (2002)

22. Păun, A., Păun, Gh., Rodríguez-Patón, A.: Further remarks on P systems with symport rules. *Annals of Al.I. Cuza University, Iaşi, Mathematics-Informatics Series*, **10**, 3–18 (2001)

23. Păun, A., Păun, Gh., Rozenberg, G.: Computing by communication in networks of membranes. *International Journal of Foundations of Computer Science*, **13**(6), 779–798 (2002)

24. Păun, Gh.: Computing with membranes. *Journal of Computer and System Sciences*, **61**(1), 108–143 (2000)

25. Păun, Gh.: *Computing with Membranes: An Introduction* (Springer, Berlin, Heidelberg, New York, 2002)

26. Păun, Gh., Rozenberg, G., Salomaa, A., Zandron, C. (eds.): *Membrane Computing, Lecture Notes in Computer Science* vol. 2597 (Springer, Berlin, Heidelberg, New York, 2003)

27. Pérez-Jiménez, M., Romero-Jiménez, A., Sancho-Caparrini, F.: *Teoría de la Complejidad en Modelos de Computatión Celular con Membranas* (Editorial Kronos, Sevilla, 2002)

28. Suzuki, Y., Fujiwara, Y., Tanaka, H., Takabayashi, J.: Artificial life applications of a class of P systems: Abstract rewriting systems on multisets. In: [4], 299–346

29. Suzuki, Y., Takabayashi, J., Tanaka, H.: Investigation of an ecological system by using an abstract rewriting system on multisets. In: *Recent Topics in Mathematical and Computational Linguistics* (ed. by Martín-Vide, C., Păun, Gh.) (The Publishing House of the Romanian Academy, Bucharest, 2000), 300–309

30. Suzuki,Y., Tanaka, H.: Artificial life and P systems. In: [3], 265–285

31. Zandron, C.: *A Model for Molecular Computing: Membrane Systems*. PhD Thesis, Universitá degli Studi di Milano, 2001

32. Zandron, C., Ferretti, C., Mauri, G.: Solving NP-complete problems using P systems with active membranes. In: *Unconventional Models of Computation* (ed. by Antoniou, I., Calude, C.S., Dinneen, M.J.) (Springer, Berlin, Heidelberg, New York, 2000), 289–301

Mathematical Modelling of the Immune System

Santo Motta[1] and Vladimir Brusic[2]

[1] Department of Mathematics and Computer Science, University of Catania, Italy
   motta@dmi.unict.it
[2] Institute for Infocomm Research, Singapore
   vladimir@i2r.a-star.edu.sg

Summary. The immune system is the natural defense of an organism. It comprises a network of cells, molecules, and organs whose primary tasks are to defend the organism from pathogens and maintain its integrity. The cooperation between components of the immune system network realizes effectively and efficiently the processes of pattern recognition, learning, and memory. Our knowledge of the immune system is still incomplete and mathematical modelling has been shown to help better understanding of its underlying principles and organization. In this chapter we provide a brief introduction to the biology of the immune system and describe several approaches used in mathematical modelling of the immune system.

1 Introduction

Mathematical modelling of the immune system is important from the perspective of both life sciences and mathematical modelling. Our knowledge of the immune system is incomplete and mathematical modelling helps better understanding of the underlying principles and organization of the immune system. This knowledge will ultimately help in finding new treatments and therapies for human disease. Modelling such complex systems requires applications of knowledge and methodology from other disciplines, such as applied mathematics, physics, and computer science. The immune system appears to be a distributed system which lacks central control, but which, nevertheless, performs its complex task in an extremely effective and efficient way. Lessons from modelling the immune system may help discover new strategies for modelling other complex phenomena. Both the modelling of the immune system and application of its paradigm in form of an artificial immune system have attracted interest of life scientists and mathematical modelers alike.

In this *review* chapter we will concentrate on the methodological aspects of the problem and describe several approaches used for modelling the immune system. In Sect. 2 we will briefly recall the major characteristics of the immune system. In Sect. 3 we introduce modelling purposes and describe different scales that one may adopt for describing complex phenomena. In Sect. 4 modelling strategies and techniques at the various scales will be described together with advantages and disadvantages of the individual strategies; we will also sketch some of the models proposed in the literature and describe in more detail the Celada–Seiden general model based on cellular automata. Finally in Sect. 5 we remark on the current trends for further investigations and modelling the immune system.

2 The Immune System

The immune system comprises a network of cells, molecules, and organs whose primary tasks are to defend an organism from potentially dangerous foreign agents (pathogens) and at the same time help maintain the integrity of the organism. Foreign agents include bacteria, fungi, parasites, viruses, various environmental and self-produced antigens, and toxins. The immune system is remarkably effective, most of the time. However, an immune system may be naive (when exposed to new pathogens) or compromised (does not function properly) and the host may catch a disease. In this section we will present the key immune system concepts of importance for mathematical modelling.

The most important function of the immune system is the self/nonself recognition that enables an organism to distinguish between the harmless self and the potentially dangerous nonself. The immune system displays a series of dual natures and its function can be divided into general/specific, adaptive/innate, cell-mediated/humoral, active/passive, and primary/secondary.

Parts of the immune system are **antigen-specific** (they recognize antigens specifically and act against them), **systemic** (not confined to the initial infection site, but work throughout the body), and have **memory** (recognize and mount effective responses to the same pathogen during subsequent exposures).

The mechanism of self/nonself recognition is mediated through major histocompatibility complex (MHC) molecules that bind short peptides intracellularly, transport them to the cell surface, and present for recognition by the T-cells of the immune system. These peptides act as markers; cells presenting self-peptides are tolerated while those presenting foreign peptides are subject to immune responses. The immune system is a learning system which can recognize even previously unseen foreign antigens. It also learns to recognize and tolerate new self-antigens that appear throughout the development and maturation of the host organism, as well as common non-pathogenic environmental antigens. In a healthy immune system the processes of immune learning and immune recognition are very effective.

Dysfunction of the immune system may have serious consequences. It may lead to disease susceptibility and chronic illness. Sometimes an immune system attacks healthy self-cells, leading to autoimmunity such as multiple sclerosis, some forms of arthritis, and autoimmune diabetes, among others. When the immune system mounts strong defensive responses to common environmental antigens that should be tolerated, we observe allergies. Agents that elicit allergic responses are called allergens. The immune system constantly monitors for pathological mutations and eliminates cells that undergo such changes. A failure of the immune system to recognize such changes may result in tumours and cancers.

There are two main fluid systems in the body: blood and lymph. The blood and lymph systems are intertwined throughout the body and they are both responsible for transporting the agents of the immune system across the organism.

Blood is composed of liquid plasma and cells. The *plasma* is mostly water and acts as a solvent for transporting cells and other materials . All blood *cells* originate from stem cells from bone marrow (*hematopoiesis*). These cells differentiate into precursors for all the different types of blood cells: *erythrocytes* (red blood cells), *leukocytes* (white blood cells), and *thrombocytes* (platelets). Leukocytes are further subdivided into granulocytes (containing large granules in the cytoplasm) and agranulocytes (without granules). The granulocytes consist of neutrophils, eosinophils, and basophils. The agranulocytes are lymphocytes (B and T cells) and monocytes (macrophages). Lymphocytes are the most important cells for *adaptive immunity*. They circulate in the blood and lymph systems, and make their home in lymphoid organs.

The Lymph System. Lymph is an alkaline (pH>7.0) fluid that is usually clear, transparent, and colorless. It flows in the lymphatic vessels and bathes tissues and organs in its protective covering. The lymphatic vessels, like veins, have one-way valves that prevent backflow. Additionally, along these vessels there are small bean-shaped **lymph nodes** that serve as filters of the lymphatic fluid. The lymph nodes are the usual places where antigens are presented to the cells of the immune system.

The human lymphoid system has primary and secondary organs. The primary organs are those where cells of the immune system are produced and differentiate into mature cells (the bone marrow and the thymus); the secondary organs are those where the immune system cells actually encounter antigens and fight the antigens. Thus they are located at or near possible portals of entry for pathogens. Secondary organs include adenoids, tonsils, lymph nodes, Peyer's patches in the gut, and the appendix.

The immune system function, also called immunity, has two features: innate immunity, and adaptive immunity. We will briefly describe innate immunity and then concentrate on adaptive immunity, which is more interesting for mathematical modelling of the immune system.

Innate Immunity

We are born with a functional innate immunity system which is nonspecific; all antigens are attacked pretty much equally. It is the first defense of the organism which eliminates the majority of potentially dangerous pathogens we encounter. This defense is achieved by many actions and components.

Surface Barriers and Mucosal Immunity. The first and most important barrier is the skin. The skin cannot be penetrated by most organisms unless it already has an opening, such as a nick, scratch, or cut. Pathogens can be expelled from our organism by many **mechanical actions**: ciliary action from the lungs; coughing and sneezing; the flushing action of tears, saliva, and urine. The sticky *mucus* of respiratory and gastrointestinal tracts traps many microorganisms. **Chemical barriers** prevent pathogen growth: the acid pH of skin; lactic acid and fatty acids in the sebum of hair; lysozyme in saliva, tears, nasal secretions, and perspiration; and others. The stomach is a formidable obstacle insofar as it secretes hydrochloric acid (very acidic) and protein-digesting enzymes that kill many pathogens. The stomach can even destroy drugs and other chemicals. **Normal flora** consists of microbes, mostly bacteria, that live inside and on the surface of the body and normally are harmless to the host. Normal flora fill almost all of the available ecological niches in the body and produce bacteriocidins, defensins, cationic proteins, and lactoferrin, all of which are harmful to other bacteria that compete for their niche in the body.

Phagocytes are cells that attract (by chemotaxis), adhere to, engulf, and ingest foreign bodies. *Promonocytes* are made in the bone marrow, after which they are released into the blood and called circulating monocytes, which eventually mature into **macrophages**. Once a macrophage phagocytizes a cell it places portions of its proteins, called T-cell epitopes, on the macrophage surface. These surface markers serve as an alarm to other immune cells which then infer the form of the invader. Such cells are called *antigen presenting cells* (**APCs**). The non-fixed or wandering macrophages roam the blood vessels and can even leave them to go to an infection site where they destroy dead tissue and pathogens (diapedesis or extravasation). **Natural killer cells** move in the blood and lymph to kill (cause to burst) cancer cells and virus-infected cells. They are large granular lymphocytes that attach to glycoproteins on the surfaces of infected cells. **Polymorphonuclear neutrophils** (or polys) provide a major defense against pyogenic (pus-forming) bacteria and are normally the first on the scene to fight an infection. Polys constitute 50-75% of all leukocytes. **The complement system** is a major enzyme-triggered innate immunity system in plasma. It coats microbes with molecules that make them susceptible to engulfment by phagocytes. They also encourage polys to adhere to the walls of capillaries (*margination*) from which they can squeeze through in a matter of minutes to arrive at a damaged area. **Eosinophils** are attracted to the cells coated with complement C3B, where they release major basic protein (MBP), cationic protein, perforins, and oxygen metabolites, all of which work together to burn holes in cells and parasites, such as helminths (worms). About 13% of the leukocytes are eosinophils. Their lifespan is about 8-12 days. Neutrophils, eosinophils, and macrophages are all subtypes of phagocytes. **Dendritic cells** are covered with a maze of membranous processes that look like nerve cell dendrites. Most dendritic cells are highly efficient antigen presenting cells. There are four basic types: Langerhans cells, interstitial dendritic cells, interdigitating dendritic cells, and circulating dendritic cells. *Langerhans cells* are found in the epidermis and mucous membranes, especially in the anal, vaginal, and oral cavities. Dendritic cells are efficient in processing and presentation of antigens to the T helper cells.

Each of the cells involved in the innate immune system bind antigens using pattern-recognition receptors. These receptors are encoded in the germ lines of each person. This immunity is passed from generation to generation. Over the course of human development these receptors for pathogen-associated molecular patterns have evolved via natural selection to be specific to certain characteristics of broad classes of infectious organisms. There are several hundred of these receptors and they recognize patterns of bacterial lipopolysaccharide, peptidoglycan, bacterial DNA, dsRNA, and other substances. These receptors target both Gram-negative and Gram-positive bacteria.

Adaptive Immunity

Adaptive immunity (or acquired immunity) is a function of the immune system underpinned by the dynamic processes and controls. Adaptive immunity is developed and modified throughout the life of the host organism. This immunity is specific; the immune system has to learn the shape of specific antigens before it can actually remove them and their source pathogens from the organism.

The most important components of adaptive immunity are the two major types of lymphocytes: B cells and T cells. Peripheral blood contains 20-50% of circulating lymphocytes; the rest move in the lymph system. Roughly 80% of them are T cells, 15% B cells, and the remainder are null or undifferentiated cells. Their total mass is about the same as that of the brain or liver. **B cells** are produced from the *stem cells* in bone marrow; B cells produce antibodies and oversee humoral immunity. **T cells** are nonantibody-producing lymphocytes which are also produced in the bone marrow but sensitized in the **thymus** and constitute the basis of cell-mediated immunity.

Parts of the immune system are changeable and can adapt to better attack the invading antigen. There are two fundamental adaptive mechanisms: cell-mediated immunity and humoral immunity. Cell-mediated immunity consists of the T lymphocytes or T cells. Each T cell has many identical antigen receptors ($\sim 10^5$/cell), which interact with antigens. Humoral immunity is mediated by serum antibodies, which are proteins secreted by the B cell compartment of the immune response.

Cell-Mediated Immunity

Macrophages engulf antigens, process them internally, and display parts of them on the cell surface together with MHC molecules. This mechanism sensitizes T cells to recognize these antigens. All cells are coated with various molecules and receptors. *CD* stands for *cluster of differentiation* and there are more than one hundred and sixty clusters, each of which is a different molecule that coats the surface. Every T and B cell has about 10^5 molecules on its surface. B cells are coated with CD21, CD35, CD40, and CD45 in addition to other non-CD molecules. T cells have CD2, CD3, CD4, CD28, CD45R, and other non-CD molecules on their surfaces.

The large numbers of molecules residing on the surfaces of lymphocytes produce huge receptor variability. They are produced with **random configurations** on their surfaces. There are some 10^{18} structurally different receptors. Essentially, an antigen is likely to find a near-perfect fit with a very small number of lymphocytes, perhaps as few as one.

T cells are primed in the thymus, where they undergo two selection processes. The first *positive selection process* weeds out only those T cells with the correct set of receptors that can recognize self-peptides presented by the MHC molecules. Then a *negative selection process* begins whereby T cells that can recognize MHC molecules complexed with foreign peptides are allowed to pass out of the thymus.

Cytotoxic or **killer T cells** (CD8+) do their work by releasing lymphotoxins, which cause cell lysis. **Helper T cells** (CD4+) serve as regulators that direct immune responses. They secrete chemicals called *lymphokines* that stimulate cytotoxic T cells and B cells to grow and divide, attract neutrophils, and enhance the ability of macrophages to engulf and destroy microbes. **Suppressor T cells** inhibit production of cytotoxic T cells providing a mechanism for prevention of self-damage that may be inflicted to healthy cells by the immune responses. **Memory T cells** are programmed to recognize and respond to a pathogen previously encountered by the organism.

Humoral Immunity

An immunocompetent but as yet immature B lymphocyte is stimulated to mature when an antigen binds to its surface receptors and there is a T helper cell nearby (to release a cytokine). This **sensitizes** or **primes the B cell** and it undergoes **clonal selection**, which means it reproduces asexually by mitosis. Most of the family of clones differentiate to plasma cells.

Plasma cells, after an initial lag, produce highly specific antibodies at a rate of as many as 2000 molecules per second for four to five days. Other B cells become long-lived **memory cells**.

Antibodies, immunoglobulins, or Igs, are soluble proteins secreted by the plasma offspring of primed B cells. The antibodies inactivate antigens by: *(a) complement fixation* (proteins attach to antigen surface and cause holes to form, i.e., cell lysis); *(b) neutralization* (binding to specific sites to prevent attachment; this is the same as taking their parking space); *(c) agglutination* (clumping); *(d) precipitation* (forcing insolubility and settling out of solution), and other more arcane methods. Constituents of gamma globulin are: IgG (76%), IgA (15%), IgM (8%), IgD (1%), and IgE (0.002%). IgE is responsible for autoimmune responses; IgG is the only antibody that can cross the placental barrier to the fetus and it is responsible for the three to six-month immune protection of newborns that is transferred from the mother; IgM is the dominant antibody produced in primary immune responses; IgG dominates in secondary immune responses.

The antibody molecule has many degrees of flexibility. This allows it to more easily conform to the nooks and crannies on an antigen. The antigen interaction part of an antibody molecule attaches to specific shapes on the surface of antigens, called **B cell epitopes**. B cells can produce as many as 10^{14} conformationally different forms of antibodies. All of these mechanisms hinge on the attachment of antigen and cell receptors. Since there are many receptor shapes available, white blood cells seek to optimize the degree of confluence between the two receptors. The number of these "best fit" receptors may be quite small, even as few as a single cell. This attests to the **specificity of the interaction**. Nevertheless, cells can bind to receptors whose fit is less than optimal when required. This property is referred to as **cross-reactivity**. Cross-reactivity has its limits. There are many receptors to which virions cannot possibly bind. Very few viruses can bind to skin cells.

This section is intended to present some basic concepts in a simplistic manner to non-immunologists interested in modelling the immune system. More complete descriptions of the immune system can be found in immunology textbooks (e.g., [1, 2]).

3 Modelling Purposes and Scales

Modelling Purposes

The history of natural sciences tell us that scientific endeavor begins with the observation of natural phenomena. It is usually followed by classification of observed phenomena mostly according to morphological aspects. Up to this stage one can have knowledge of the entities which take part in a phenomenon but no knowledge of the rules which regulate it. Then come the **hypotheses, heuristic, or qualitative theories** to suggest how the phenomenon can be described and explained. However, theories must be checked experimentally or by measurements. For this purpose they need a quantitative description of the phenomenon using a mathematical formulation.

The base of scientific method (*abduction*) is the constant comparison of theories with experimental data. A theory is accepted as long as it explains experiments and its predictions are verified by experiments. If a new experiment does not verify the theory then the theory must either be modified to satisfy the results from the new test, or just be abandoned. New theories usually originate from the observation of nature (simple observation of phenomena or experimental data) and the personal creativity of the scientist. Mathematical models are quantitative representations of phenomena built up in the framework of a theory using the language of mathematics (here we use the term *mathematics* in a broad sense which also includes all the quantitative approaches based on computer simulations). A mathematical model may be used only if a theory uses (and predicts) *measurable quantities* and gives relations between them. Qualitative explanations of a phenomenon are not sufficient to construct a model.

The basic constituents of a natural phenomenon are entities and relations between entities. Unfortunately we do not have the complete book of constituents and laws of nature and we face a problem of how to proceed with scientific research given this incomplete and imperfect knowledge. Both theories and models are always imperfect representations of reality. To keep models tractable we necessarily introduce simplifications and approximations when describing real phenomena. Furthermore, models need to be tested with experimental measurements. The model-experiment discrepancies often suggest modifications of the theory and underlying assumptions. The results from mathematical models may also be used to suggest new experiments to verify the theory.

The theory-model-experiment loop works clearly and efficiently when complex phenomena can be analysed in terms of simple rules and entities. An example of this type is the modelling of complex electric circuits using entities and rules of simple circuits. However, there are situations (if rules are not deterministic, if phenomena have chaotic behaviour, or if collective effects play new important roles, among others) where complex phenomena cannot be studied by reduction to simplistic ones. In the life sciences it is extremely difficult to isolate and study the behaviour of a single constituent. Even when this is possible many of the characteristics of living organism are due to collective effects (populations) and in most cases these are not deterministic. Moreover, living organisms are the result of evolution. New mechanisms have been built up by nature to solve new problems. The old mechanisms do not disappear from living bodies and can play a role in special situations. In this sense **redundancy** is a characteristic of living organisms, i.e., a specific function can be performed by different parts. This mechanism is often used in living organisms to recover from system errors and malfunctioning.

Living organisms are natural *complex systems* and modelling may play a crucial role since models can also be built with approximate and imperfect knowledge of the phenomenon. The model parameters (initial data, entities, and relations between entities) can be adjusted to fit modelling results to experimental measurements. These models can then be used to understand the general behaviour of the phenomenon in different situations; perform "model experiments" or "simulations" to understand the role of single constituents and relations; plan new experiments; test theoretical assumptions; and suggest theory modifications. Modelling can therefore stimulate scientific creativity and produce better theoretical descriptions of reality.

Modelling Scales

Theories and models always describe natural phenomena at a given scale. This reflects the fact that any natural phenomenon is the effect of a hierarchy of phenomena on different scales. Choosing a scale depends on which aspects of the phenomena, from *micro* to *macro*, one is interested in representing. In physics there is already a well-defined terminology which originates in different research areas, and the distinction between different scales is based upon the characteristic length of the phenomena.

In biology (and immunology) the definition of a scale is more ambiguous. A basic unit available for defining a scale is the *cell* with no regards to its physical dimension. So one can define three basic scales: the *subcellular scale*, the *cellular scale*, and the *macroscopic scale*.

Each phenomenon may be described only at a suitable level of scale. Allocations of phenomena inside a given scale are not well defined and may depend on an individual scientist's personal preference. Bellomo and Preziosi [3] proposed a classification for tumour evolution and its interaction with the immune system. Here we adapt their classification to our ends and propose the following classification:

- the **subcellular scale** refers to the main activities inside or at the surface of a cell, e.g., genetic changes, distortion in the cell cycle, loss of apoptosis; expression and transduction of signals between cells; etc.;
- the **cellular scale** refers to the main (interactive) activities of each cell, e.g., activation and proliferation of cells;
- the **macroscopic scale** refers to phenomena which are typical of continuous systems; for instance, cell migration, convection, diffusion of antibodies.

Phenomena identified at a certain scale can also be related to those at higher or lower scales. For instance, interactions developed at the cellular level are regulated by processes which are performed at the subcellular scale. The immune system shows interesting phenomena on each single scale and phenomena on different scales are related those at higher or lower scales.

Theories and models developed at the **microscopic** or **subcellular scale** deal with evolution of physical and biochemical states of a single cell. The evolution of a cell is regulated by genes contained in its nucleus. Receptors on the cell surface can receive signals which are transmitted to the cell nucleus where genes can be activated or suppressed. Particular signals can be responsible for cell reproduction, or they can induce programmed cell death (apoptosis). Modelling the overall activity of a single cell is a very hard problem as many biological details of this activity are still unclear or unknown. Biologists, mathematicians, physicists, computer scientists, and engineers have joined forces to develop and use mathematical and computer science techniques to model subcellular phenomena. Interested readers can find plenty of references in PubMed and specialized symposia (such as [4]).

At the **cellular scale** one is interested in describing the evolution of a system consisting of a large number of different cells. Cell interactions are regulated by signals emitted and received by cells through complex recognition processes. The connection with the subcellular scale is evident but at this level one may forget the details of single cell models and consider their outcome in the larger system. This is analogous to what is done in modelling complex circuits including electronic components where these elements are replaced with equivalent circuits. The overall system may be described with methods such as statistical mechanics, cellular automata, lattice gas, and others. In this framework observable quantities are obtained by suitable moments derived from their statistical distributions.

At the **macroscopic level** one is interested in describing the dynamical behaviour of observable quantities, in most cases the densities of various entities, using techniques from the framework of continuum phenomenological theories. This is analogous to population dynamics studies using Lotka-Volterra equations. Fitting model parameters to experimental data is always necessary to validate the model. The mathematical techniques used at this scale are mostly ordinary differential equations and partial differential equations. Nonlinearity, which is an intrinsic feature of all models, leads to very sophisticated mathematical problems.

4 Modelling in Immunology

The immune system has some important properties: uniqueness; distributed detection; imperfect detection; adaptability. *Uniqueness* means that the immune system of each individual is unique and therefore vulnerabilities differ from one system to the next. *Distributed detection* indicates that the small and efficient detectors used by the immune system are highly distributed, and are not subject to centralized control or coordination. *Imperfect detection* used by the immune system enables recognition which does not require absolute detection of every pathogen. This makes the immune system more flexible in allocating resources. *Anomaly detection* is the property of the immune system by which it can detect and react to pathogens that the body has never encountered before. *Adaptability* or *learning and memory* is the ability of the immune system to learn the structures of pathogens, and remember those structures, so that future responses to the pathogens can be targeted and much faster. These properties result in a system that is scalable, resilient to subversion, robust, very flexible, and that degrades gracefully.

The major problem that the immune system solves is distinguishing between self and nonself. Actually, the success of the immune system is dependent on its ability to distinguish between harmful nonself and everything else. This is a hard problem because: *(i)* there are so many patterns in nonself, $\sim 10^{16}$, that have to be distinguished from $\sim 10^6$ self patterns; *(ii)* the environment is highly distributed; *(iii)* the host organism must continue to function all the time; *(iv)* resources are scarce. The immune system solves this problem by using a multi-layered architecture of barriers: physical (the skin); physiological (e.g., the pH values); cells and molecules of the innate and acquired immune response (see Sect. 2).

The systemic models of immune responses have mainly been devoted to collective effects of actions of various immune system constituents. These models do not study single cells or single molecules, but focus on cell interactions and collective behaviour in initiation, control, and mounting of immune responses. Inside the scale framework these models focus on cellular and macroscopic levels. The panorama of the immune system models is quite broad. Nevertheless all these models are based on two biological theories underpinning our understanding of the immune system, namely the clonal selection theory and idiotypic network theory.

The **clonal selection theory** of the Nobel Prize laureate F. Burnet [5] was developed following the track first highlighted by P. Ehrlich at the beginning of the twentieth century. The theory of the clonal selection states that the immune response is the result of a selection of the "right" antibody by the antigen itself, much like the best adapted individual is selected by the environment in the theory of natural selection of Charles Darwin. The selected subsets of B cells (and T cells) grow and differentiate; they then turn off when the antigen concentration falls below some threshold. In the framework of this theory, memory B cells will be responsible for acquired immune tolerance.

The **idiotypic network theory** was formulated by the Nobel Prize laureate Niels K. Jerne in 1973 [6,7]. According to the idiotypic network theory the immune system is a regulated network of molecules and cells that recognize one another even in the absence of antigens. The idiotypic network hypothesis is based on the concept that lymphocytes are not isolated, but communicate with each other among different species of lymphocytes through interaction among antibodies. Accordingly the identification of antigens is not done by a single recognizing set but rather a system-level recognition of the sets connected by antigen-antibody reactions as a network. Jerne suggested that during an immune response antigens would directly elicit the production of a first set of antibodies, Ab_1. These antibodies would then act as antigens and elicit the production of a second set of "anti-idiotypic" (anti-id) antibodies Ab_2 which recognize idiotypes on Ab_1 antibodies, and so forth.

Nowadays immunologists consider these two independent theories as complementary and consistent with each other [8]. However, while clonal selection theory is believed to be the fundamental theory for understanding today's knowledge of the immune system, the idiotypic network theory is believed correct as far as the existence of anti-idiotypic reactions is concerned, but probably not relevant in the control of immune responses [9].

Most macroscopic-level models, also referred to as continuous models, have been formulated using the framework of both immunological theories [10,11]. The cellular-level models (also referred to as discrete models) are mostly based on Jerne's theory. The Celada–Seiden model [12] bases its foundation on the clonal selection theory, but may include both theories.

The main task of the immune system is to perform **pattern recognition** using cellular receptors to recognize target antigens. The binding mechanism, mostly unknown in detail, is based on different physical effects (short-range noncovalent interactions, hydrogen binding, van der Waals interactions, etc. [11]). A cellular receptor can recognize its target epitope if their matching surfaces have regions of extensive complementarity similar to the key and lock mechanism. Perelson and Oster [13] defined the constellation of features important in determining binding among molecules as the *generalized shape* of the molecules. Assuming this shape can be described by η parameters then a point in η-dimensional space (*shape space*) specifies the generalized shape of the receptor binding region. From this consideration Oster and Perelson estimated that completion of the receptor repertoire should satisfy the following conditions: *(i)* each receptor can recognize a set of related epitopes, each of which differs slightly in shape; *(ii)* the repertoire size is of the order of 10^6 or larger; *(iii)* at minimum a subset of the repertoire size is distributed randomly throughout the shape space [11]. Farmer, Packard, and Perelson [14] started using binary strings to represent shapes of receptors and epitopes, which enabled the use of numerous, readily available string-matching algorithms. A selection of string-matching algorithms are available for determining the degree of complementary between strings. Discrete models of the immune system widely use this representation to describe interactions between cell receptors and antigens.

Continuous models use *affinity functions* which globally represent interactions between populations of cells and those of antigens. The affinity function determines the behaviour of a model.

Continuous Models

The first attempts to model the immune system made use of continuous models. An excellent review of immune system models using continuous models can be found in Perelson and Weisbuch [11]. Here we will point out the main idea, the general mathematical formulation, and the advantages and disadvantages of these kinds of models.

Continuous models describe the time evolution of concentrations of different populations of cellular and molecular entities which play a significant part in the immune system. Each of these models is expressed as a continuous real function that defines the concentration (number of elements/volume) of elements. Some simplifying assumptions are inserted in the model: *(i)* concentrations do not depend on space variables; *(ii)* interactions between entities occur by uniformly random collisions. As a result the model will describe the change of mean values of population concentration defined by interactions described by theoretical approaches.

These models use systems of nonlinear ordinary differential equations, mainly representing conservation laws, in which unknowns represent the concentrations. When assumption (i) is released, then a description using partial differential equations is needed and the problem becomes mathematically more difficult.

The mathematical advantages of this approach are: (a) the mathematical methods used here have a well-known and theoretically solid background; (b) the behaviour of the solution can be found also by qualitative analysis and asymptotic analysis; (c) if a numerical solution is needed then well-known numerical methods are available; (d) from the biological point of view one can play with interaction strength to isolate systems defined by the strong interactions between constituents.

However, there are biological disadvantages: (a) approximations, needed to keep the equations tractable, may not be biologically evident so that the model may move away from real biology; (b) estimation of model parameters is not usually evident and requires fitting of parameters to experimental data; (c) inserting new biological details is not straightforward and may drastically change the mathematical structure of the model; (d) the model description fails if the concentration of some entities decreases drastically (a few individuals).

General Mathematical Description of Continuous Models

The nonlinear equations describing generic iteration networks have a general form

$$\partial_t x = G - L$$

where x is the vector describing concentrations, $x = \{x_1, \ldots, x_n\}$, and the vectors $G = G(x)$, $L = L(x)$ represent the gain and loss terms respectively.

The solution of these equations is represented by a curve in the state space (the n-dimensional state described by the x variables) describing the time evolution of the system (time is the curve parameter) starting from the initial condition $x^*(t = 0)$. By changing the initial conditions we get a family of curves (describing solution flux). In the simplest case, the behaviour of a system comprising two populations can be studied qualitatively in a plane.

Initially one needs to investigate if the system has stationary points (also called **fixed points**). If for some $x = x_1$ one has $G(x_1) = L(x_1)$ then $\partial_t \equiv 0$ and x_1 is a stationary point of the system. Then one studies the **local stability** to get information on the qualitative behaviour of the solution at fixed points. In a stable fixed point the trajectory of the solution tends to converge toward this point: this point is called the **attractor**. Different types of attractors can be distinguished: *limiting cycles* describe periodic behaviour while *strange attractors* describe chaotic behaviour. Next, the global stability of the system needs to be assessed for the global behaviour of the model. It may happen that the system is globally unstable. Such a model will not describe the real world as the system does not show a region where the solution lasts long enough to be observable. Globally unstable systems are not *"physically or biologically"* acceptable and the model has to be changed.

In these types of models it is crucial to properly describe the **affinity** between cells and molecules. The simplest affinity function is a bilinear form, $\propto x_i x_j$ (such as prey-predator systems). More detailed models need more complex nonlinear terms such as $\propto x_i f(h_i)$ where $f(h)$ is a bell-shaped function (see [11]).

As an example of this approach we sketch the model proposed by Behn, van Hemmen, and Sulzer [15]. The property they wanted to model was the memory of the immune system. This memory is remarkable in view of both its complexity and the finite lifetime of its constituents which is typically one or two orders of magnitude less than its memory span.

The model is based on the following theoretical ideas: (*a*) The memory is stored in a cyclical network consisting of elements (antibodies) that mutually stimulate and inhibit each other preserving a dynamic equilibrium. One of the antibodies in the cyclical network function as an internal image of the antigen. (*b*) The equilibrium is stabilized by memory (dormant) B cells, whose lifetimes exceed the lifetimes of their associated antibodies by about an order of magnitude (though still finite). (*c*) This whole structure is embedded in a network that is globally inhibitory (repressive) as advocated by Jerne [7] and hence globally stable. The simplest case of a symmetric two-cycle network interacting with an antigen shows explicitly the synergism of the mechanism of memory, i.e., a cycle and the associated memory B cells which stabilize it.

Consider two types of antibody x_1 and x_2 with mutual interaction $m_{12} = m_{21} = m$ and an antigen y which stimulate x_1 with matching m_0 in the presence of memory cells d_1 and d_2. The model equations are

$$\partial_t x_1 = x_1[(1-k)mx_2 - \gamma] + d_1 m x_2 + (d_1 + x_1)m_0 y$$
$$\partial_t x_2 = x_2[(1-k)mx_1 - \gamma] + d_2 m x_1$$
$$\partial_t y = (\alpha - m_0 x_1)$$

Defining the vector $\mathbf{z} = \{x_1, x_2, y\}$ the system has the form $\partial_t \mathbf{z} = F(\mathbf{z})$ and shows three fixed points

$$\mathbf{z}_1^s = (0, 0, 0)$$
$$\mathbf{z}_2^s = (\frac{\delta}{\beta_2}, \frac{\delta}{\beta_1}, 0)$$
$$\mathbf{z}_3^s = (a_1, a_2, b) = (\frac{\alpha}{m_0}, \frac{d_2}{\kappa - l + \frac{\gamma m_0}{\alpha m}}, \frac{\gamma(a_1 - a_2) + m(d_2 a_1 - d_1 a_2)}{(m_0 + \alpha)})$$

where $\delta = \gamma^2 - m^2 d_1 d_2$ and $\beta_i = m(1-\kappa)(\gamma + m d_i)$.

A fixed point is relevant if its components are nonnegative (coordinates represent concentrations). Thus there exist four regions where:

a) only the trivial $\mathbf{z}_1^s$ exists,

b) $\mathbf{z}_1^s$ and $\mathbf{z}_2^s$ exist,

c) $\mathbf{z}_1^s$ and $\mathbf{z}_3^s$ exist,

d) $\mathbf{z}_1^s$, $\mathbf{z}_2^s$ and $\mathbf{z}_3^s$ exist.

The solution behaviour in a flow diagram that enables elucidation of biological meaning can be obtained by linear stability analysis [15]. Multiple properties of the immune system have been studied by a variety of continuous models (see [16], [11], [17]).

Continuous models are very appealing but due to the mathematical approximations they can became non-realistic. Improvements are possible. We quote some of them here. (a) Time delay. The time delay for the finite interaction time between entities is not usually taken into account. This is usually of the same order as other characteristic times. The behaviour of delayed systems may be qualitatively different (e.g., the stability of fixed points can change). (b) Metadynamics. Usually new cells come into the system either from bone marrow or following hypermutations. This could be modelled using a stochastic source term in the equations. The overall results could be drastically modified. (c) Comparts models. In order to describe specialized body organs it would be preferable to build up special models for each of these organs.

Discrete Models

Cellular-scale models or discrete models consider that individuals (instead of populations) play a fundamental role and population dynamics arises in the simulation from collective effects.

These kinds of models use mathematical techniques, such as the generalized Boltzmann equation (GBE), cellular automata (CA), lattice gas (LG), and neural networks (NN), all widely employed to study *complex systems*. Most of these approaches are based heavily on computer simulations and have many advantages. Unlike continuous models, where the mathematical form of the model (such as the degree of nonlinearity) is important, the model form is less important in approaches using CA, LG, or NN. Moreover, models can be built using biological language and with biologically (not mathematically) based approximations. As an example, numerical values of the parameters are not crucial in defining the system even if they remain important for the simulation. Finally new biological details are easy to insert and do not change the mathematical structure of the model. However, there are also some disadvantages of discrete models: (a) Computer simulations are finite in time and space. (b) Qualitative and asymptotic analysis, and parameter analysis, are no longer possible. (c) Simulations can be computationally demanding.

In the following sections we will analyze some models based on cellular automata. We will give the general ideas of this modelling approach and we will discuss in some detail the Celada–Seiden model [12] which we believe is one of the most interesting biologically. The GBE approach is described in detail in Bellomo et al. [18, 19]. Recent reviews on discrete models can be found in [8, 11, 20]. Applications of the artificial immune system are described in the recent book by Dasgupta [21]. Models based on CA and LG have been developed over the last twenty years. These models are extensively used nowadays and they are known to produce interesting results [8]. Models based on *spin-glass* formalism have been proposed by Parisi [22, 23]. *Genetic algorithm*-based models have been mainly investigated at the Santa Fe Institute and University of New Mexico by Forrest and co-workers starting in the early 1990s [24, 25]. This is a *computer science*-driven approach which brings into the field the experience of the computer scientist. The guiding line of this approach is a deeper comprehension of the immune system in order to use the immune system *information processing algorithm* in applications (see the recent interesting review by Forrest and Hofmeyr [20]).

Automata-Based Models

We now consider automata-based models. In this section we will present some examples that represent the spectrum of this type of model.

The Model of Kaufman, Urbain, and Thomas

One of the first applications of discrete automata to investigate the logic of normal immune responses was introduced by Kaufman, Urbain, and Thomas (KUT) [26] in 1985. These authors were interested in the simplest way to describe the logic of interactions among a number of different cell types and their outcomes in terms of immune response. The original model considers five types of cells and molecules: helper cells (Th), suppressor cells (Ts), B lymphocytes (B), antibodies (Ab), and antigens or virus (Ag). Each entity is represented by a boolean variable denoting "spin up" (high concentration) and "spin down" (low concentration). The rules modelling the dynamic evolution of these variables are expressed by logical operations. Application of the rules is iterated over discrete time and the dynamics is observed. The discrete evolution rules are

$$Ab(t+1) = Ag(t) \text{ AND } B(t) \text{ AND } Th(t)$$
$$Th(t+1) = Th(t) \text{ OR } Ag(t) \text{ AND NOT } Ts(t)$$
$$Ts(t+1) = Th(t) \text{ OR } Ts(t)$$
$$B(t+1) = Th(t) \text{ AND } (Ag(t) \text{ OR } B(t))$$
$$Ag(t+1) = Ag(t) \text{ AND NOT } Ab(t)$$

where AND, OR, and NOT are the usual logical operators. There are five fixed points in the state space composed of $2^5 = 32$ points. Fixed points identify the global state of the immune system: naive, vaccinated, immune, paralyzed, paralyzed and sick.

The Model of Weisbuch and Atlan (WA)

This primitive model was followed by many other models. Weisbuch and Atlan [27] proposed a model based on Jerne's theory to study the special case of auto-immune diseases, like multiple sclerosis, in which the immune system attacks the cells of the nervous system of our own bodies. Similarly to the model of Kaufman et al., this model uses five binary variables representing: killer cells (S_1), activated killer cells (S_2), suppressor cells (S_3), helper cells (S_4), and suppressor cells produced by the helpers (S_5). The binary state of each threshold automaton represents the concentration of the corresponding cell type: 0 corresponds to small concentration, 1 to high concentration. The different types of cells influence each other with a strength which is 1, 0, or -1. The automaton evolves with the following rule: *at the next time step, the concentration of one cell is unity if the sum of the interactions with the various cell types is positive, otherwise the concentration is taken as zero.* Let $\mathbf{S} = \{S_1, S_2, S_3, S_4, S_5\}$ be the vector representing the five binary variables and

$$\mathbf{A} = \begin{bmatrix} 1 & 0 & -1 & 1 & 0 \\ 1 & 0 & -1 & 1 & 1 \\ 1 & 0 & 0 & 0 & 0 \\ 1 & 0 & 0 & 0 & 0 \\ 0 & 0 & 0 & 1 & 0 \end{bmatrix}$$

the matrix of the *synaptic connections*, then the evolution of WA network [27] can be described by

$$\mathbf{S}(t+1) = \mathrm{sgn}\,(\mathbf{A} \cdot \mathbf{S}(t))$$

where the function sgn(x) defined on the natural numbers set $\mathbf{N}$ is 1 for $x > 0$ and 0 otherwise.

This model shows the existence of only two basins of attractions over $2^5 = 32$ possible states: the empty state where all the concentrations are zero and the state 10111 where activated killers have a small concentration while the other four entities have high concentrations. This corresponds to a healthy carrier state, with killer cells only in the resting state, thus unable to harm the organism by developing an active autoimmune reaction.

These two models have been extensively studied [28–30]. Neumann [31] studied a different version of the KUT model using a boolean and threshold automata which include interactions with antigens. Dayan et al. [32] have extended the WA model using the same dynamics, but place the cells on a two-dimensional lattice to allow simulations using a statistical physics approach (Ising-like models). They use five variables on each lattice site corresponding to five boolean concentrations (0 or 1). The model can be classified as an integer lattice gas with $r = 5$ (five entities) and $K = 2$ (two states per entity). Each site influences itself and its nearest neighbours in the same way as in the WA model. For a square lattice of $L \times L$ sites there are $5 \times L^2$ spins. The main difference is that in this model the evolution of a single site includes both the site itself and its nearest neighbours. This lattice version of the WA model is found to have a simpler dynamics than the original model as the number of fixed points is smaller than in [27]. The description of other CA approaches to study different aspects of the immune system can be found in Perelson [10] and Atlan and Cohen [33].

Shape Space Model Approach

A series of models use the shape space approach [11]. Each point of the d-dimensional Euclidean space is associated with a different receptor generalized shape and each coordinate represents one of the main aspects involved in pattern recognition. Most of these models are represented on a 2-D cellular automata space.

The first of these models was proposed by Stewart and Varela [34]. They used a bit-string model in a 2-D shape space to analyze the metadynamics in which new clones are constantly generated in the bone marrow.

De Boer, Segel, and Perelson [35] proposed a model (BSP model) to describe the time evolution of the immune repertoire. The model is a discrete version of a set of population equations [11, 36, 37].

Each automaton in this model describes the time evolution of the concentration of a given clone. The updating of variables associated with the concentration of each clone is based on the activation windows. The updating is determined by the field function which depends on the concentration of the populations with complementary shapes of receptors. The BSP model considered only 1-D and 2-D lattices to obtain only stable behaviour.

Stauffer and Weisbuch [38] proposed a simpler higher dimensional version of the BSP model. They also obtained the "chaotic behaviours" which were also observed for larger lattices for probabilistic rules defined by Stauffer [39] and Dasgupta [40]. The latest version of the BSP model was further improved by Zorzenon dos Santos and Bernardes [41, 42].

CA Models for HIV

The dynamics involved in the response to HIV infection is the same as that involved in the response to any other virus. The main difference is in the much faster mutation rate of HIV. Many attempts have been made to simulate the immune system response to the HIV virus, the onset of AIDS, and the interaction between T cells and HIV.

Pandey and Stauffer further extended the KUT model using a probabilistic generalization of deterministic cellular automata. Their model focuses on a possible explanation of the time delay between HIV infection and the establishment of AIDS [30, 43]. They represent helper cells (H), cytoxic cells (S), virus (V), and interleukin (I). The interleukin molecules produced by helper cells induce the cytotoxic cells to kill the virus. The dynamics of the system interactions is described by the following rules:

$$V(t+1) = H(t) \text{ AND NOT } S(t)$$
$$H(t+1) = I(t) \text{ AND NOT } V(t)$$
$$I(t+1) = H(t)$$
$$S(t+1) = I(t)$$

showing an oscillatory behaviour ending in a fixed point where the immune system is completely destroyed, similarly to the real onset of AIDS.

Chowdhury et al. [44, 45] proposed a unified model of the immune system which recovers, as special cases, the KUT and WA models. The model describes the immune responses to HIV and reproduces some features of related experimental results. Extensions of this network approach for modelling HIV and cancer have been discussed by Chowdhury and Stauffer [46].

Recently Zorzenon dos Santos and Coutinho [47] proposed a new CA-based model. They considered a square lattice and to each site associated an automaton which can be found in four different states. Each state indicates that the automaton belongs to a different cell population: healthy, infected-A, infected-B, and dead cells. Healthy cells correspond to T helper cells and monocytes. Infected A cells represent the virus-producing cells at a given time which in a period τ (estimated as 4 weeks) generate new infected cells associated with new mutants. Infected-B cells represent the final stage of infected-A cells before their death, provoked by the immune response. The dead cell population simulates cell depletion due to the infection. The model is able to qualitatively reproduce the long time delay between HIV infection and the establishment of AIDS.

A different approach in modelling the HIV infection has been followed by Bernaschi and Castiglione [48]. These authors modelled the complex immune system response to HIV using a modified version of the Celada–Seiden model allowing the Th cells, along with the dentritic cells and macrophages, to be a possible target of antigen action. In this approach the CS model has been modified to take into account a rapidly mutating virus such as HIV-1. The helper T cells are infected by the virus, which in turn undergoes mutation at a high rate sufficient to escape the immune responses. This model *(i)* reproduces the two time scales in the course of the disease (fast acute phase and slow latent phase); *(ii)* indicates that the latent phase is a result of the selection of the slowly activating viral strain; and *(iii)* indicates that the immune deficiency is due more to the reduction in the lymphocyte T helper repertoire than just a decrease below critical values of the T helper count.

The Celada–Seiden Model

The Celada–Seiden model (CS model) was developed by Seiden and Celada in 1992 [12, 49]. This model is among the most outstanding attempts to cope with the quest for biological fidelity. The computational counterpart of this model is the **IMMSIM** (Immune Simulator) automaton. IMMSIM belongs to the class of immunological cellular automata, but its degree of sophistication sets it apart from simpler CA in the Ising-like class [26, 30].

The CS model explicitly encodes cellular and humoral immune responses in a comprehensive set of rules which apply to a variety of cellular and molecular entities. In particular, the following cells: lymphocyte B (**B**), lymphocyte T helper (**Th**), lymphocyte T killer (cytotoxic) (**Tk**), macrophage (**MA**), epithelial cell (as generic target cell) (**EP**) lymphocyte plasma B (**PLB**); and the molecules: antigen (bacteria or generic virus) (**Ag**), antibody (**Ab**), and immune complexes or Ab-Ag binding (**IC**). In addition, some intracellular signals are explicitly represented, for example, interferon-γ (**IFN**) and danger signal (**D**), while other cytokines are just "implicitly" taken into account in the interaction rules. The major difference among cellular and molecular entities is that cells may be classified on the basis of a *state* attribute. Computationally the state of a cell is an artificial label introduced by the logical representation of that cell's behaviour.

This model represents a portion of a lymph node of a vertebrate animal as a two-dimensional triangular lattice (six neighbour sites) $L \times L$, with periodic boundary conditions in both directions (up-down, left-right).

The CS model is based on the *theory of clonal selection*. Celada and Seiden have to look for a way to represent the lymphocytes' receptors shape space. They decided, as suggested in the early work of Farmer et al. [14], to use a bit-string to encode the information relative to the specificity of the antigens.

In the CS model a clonotypic set of cells is characterized by a bit-string-represented receptor. The bit-string length l is clearly a key parameter for determining both the time and space complexity of the algorithm. This algorithm simulates the behaviour of the whole set of entities as the number of potential receptors scales as 2^l (see [50,51]). The bonds among the entities are described in terms of *matching* binary strings with fixed directional reading frames. Bit-strings represent the generic "binding site" of both cells (read the receptor) and molecules (that is, peptides or epitopes). Every entity, including a receptor, is represented by a certain number of molecules. The repertoire is then defined as the cardinality of the set of possible instances of entities that differ in, at least, one bit of the whole set of binary strings used to represent its attributes. Indeed, the cells equipped with binding sites and the antibodies have a potential repertoire of $2^{N_e l}$, where N_e indicates the number of binary strings used to represent receptors, MHC-peptide complexes, epitopes, or other molecules of the entity e. Other entities do not need to be specified by binary strings so their repertoire is just one (i.e., $N_e = 0$). Examples include cytokine molecules such as interleukins or interferon-γ, and the danger signal.

In this model the two entities, each equipped with a receptor, interact with a probability calculated as a function of the *Hamming distance* between the binary strings representing the entities' binding site. This probability is called the *affinity potential*. For two strings s and s' such a probability is maximum (i.e., equal to 1) when all corresponding bits are complementary ($0 \leftrightarrow 1$), that is, when the Hamming distance between s and s' is equal to the bit-string length. A good and widely used analogy is the matching between a lock and its key. If l is the bit-string length and m is the Hamming distance between the two strings, the affinity potential is defined in the range $[\,0, \ldots, l\,]$ as

$$v(m) = \begin{cases} v_c^{(m-l)/(m_c-l)} & \text{for } m \geq m_c \\ 0 & \text{for } m < m_c \,, \end{cases}$$

where $v_c \in (0,1)$ is a free parameter which determines the slope of the function whereas m_c ($l/2 < m_c \leq l$) is a cut-off (or threshold) value below which no binding is allowed.

Equipped with their receptors, cells are free to diffuse on the lattice sites. At each time step, representing 8 hours of real time, cells and molecules residing on the same lattice site take the chance to interact among each other. The rules which implement these reactions are executed in a randomized order. Using immunological terms, they can be grouped as follows: phagocytosis, immune activation, opsonization, infection, and cytotoxicity [1,2]. Phagocytosis comprises the rules for the activity of antigen processing cells. Immune activation codes for the activity of helper T lymphocytes which recognize MHC-peptide complexes and release cytokines for activation of B cells for antibody production. Opsonization stands for the inactivation of the antigen by binding of antibodies. Infection is the action of the virus. Cytotoxicity collectively describes all the rules which account for the killing of the virus-infected cells by cytotoxic cells.

To complete this very concise description of the model we mention the mechanism of hematopoiesis. This takes into account the activity that generates pluripotent stem cells in the bone marrow. This production is realized by the periodic addition of newly randomly formed cells. The selective activity of the thymus accounts for positive and negative selection of T lymphocytes to avoid autoimmune reactions. A mean-reverting process is implemented to assure that in the absence of antigenic stimulus, the number of cells is in a steady state.

This model is modular and it allows easy addition or modification of cellular and molecular entities and integration of new sets of rules that determine the behaviour of model entities. The flexibility of the model has been exploited to construct a simulator of the HIV infection, one for the hypersensitive reactions, and another for the effect of the apoptosis mechanism on the immune responses. Moreover, as shown in [51,52], that model's results are independent of the computational platform used for the simulator.

The model, with such a degree of detail, has been used to address different questions revealing its versatility [12], namely: (i) What is the largest self-fraction of the total repertoire that is still compatible with reasonable alloreactivity? (ii) Why is the number of different MHC per individual so small (about four-six)? (iii) Why is the diversity of MHC phenotypes in a population so small compared to the theoretically possible diversity?

Later, the authors proposed slight variations of the original model to study autoimmunity and T lymphocyte selection in the thymus [53], the selection and hypermutation of the antibodies during an immune reaction [54], and the dynamics of various lymphocyte populations in the presence of different viruses characterized by infectivity, reproduction efficiency, and some other factors [55].

More recently, studies addressed additional questions. In [56] a simplified version of the model was used to construct an *artificial immune system* to study the pattern recognition properties of the immune system. In [57] a modification of the CS model has been used in a study of the apoptosis mechanism. This work was supported by *in vitro* experiments performed by Jamin et al. [58]. A subset of B cells was allowed to additionally present B-CD5 receptors and the behaviour of these receptors was modelled according to [58]. During the simulation this subset was stimulated, as in the experiment with two injections of anti-CD5 antibodies. This *naive* model showed good qualitative agreement with experimental results. In [48] the model has been modified to take into account a rapidly mutating virus such as the HIV-1. Results from this work have already been described in the text. Another modification of the CS model by Castiglione and Agur [59] has been used to study hypersensitivity to a drug during anticancer treatment. A major modification was the introduction of explicit representations for a number of cytokines (IL-4, IL-12, IL-2, IFN-γ), the subdivision of T helper cells in three classes (Th1, Th2, and Th0 or Th1/2 precursor), the subdivision of the antibodies in immunoglobulins IgE, IgG, and IgM, and the addition of mast cells. Mast cells are responsible for the release of histamine and other active mediators responsible for the symptoms of type I or IgE mediated allergy.

5 Conclusions

Observations of the immune system at the macroscopic level produce definite measurements, but are difficult to interpret at the molecular level. The idiotypic network theory, for example, can only be translated into speculative interpretations at the molecular level. Mathematical modelling implemented as computational programs can easily translate speculative hypotheses into quantitative descriptions. The parameters of the mathematical models can then be tuned to represent the real behaviour of the immune system. These models can then be used to determine the framework for studying the kinetics of immune responses, and for practical applications such as the prediction of immune interventions. Mathematical models of the immune system can model interactions of a large number of elements thereby approaching the complexity of the human immune system. Systemic-level mathematical models provide a framework for understanding the immune system as a whole. We foresee the convergence of mathematical models at the systemic and molecular levels in the future. Huge experimental data sets produced by genomics, proteomics, and molecular biology efforts will ultimately be integrated with mathematical models of the immune system at the organism level to produce models of the whole organism.

Acknowledgements

S.M. acknowledges partial support by MIUR Cofin and University of Catania grants for Research and Scientific Upgrading. This work started when S.M. was Visiting Professor at the Institute for Infocomm Research, Singapore.

References

1. Benjamini, E., Coico, R., Sunshine, G.: *Immunology: A Short Course* (Wiley, New York Singapore Toronto 2000)
2. Roitt, I., Brostoff, J., Male, D.: *Immunology*, 6th edn. (Harcourt, Edinburgh London New York Sydney Toronto 2001)
3. Bellomo, N., Preziosi, L.: Modeling and mathematical problems related to tumor evolution and its interaction with the immune system. Math. Comput. Model. **32**, 413-452 (2000)
4. International Symposium on Computational Cell Biology (2001). http://www.nrcam.uchc.edu/conference/.
5. Burnet, F.: *The Clonal Selection Theory of Acquired Immunity.* (Vanderbilt University, Nashville, 1959)
6. Jerne, N.K.: The immune system. Sci. Am. **229**(1), 52-60 (1973)
7. Jerne, N.K.: Towards a Network Theory of the Immune System. Ann. Immunol. (Inst. Pasteur) **125C**, 373-389 (1974)
8. Zorzenon Dos Santos, R.M.: Immune Responses: Getting Close to Experimental Results with Cellular Automata Models. In Stauffer, D. (ed.): *Annual Reviews of Computational Physics*, Vol.**V** (World Scientific, Singapore 1999), 159-202
9. Lollini, P.L.: private communication (2002)
10. Perelson, A.S. (ed.): *Theoretical Immunology, Part One & Two*, SFI Studies in the Sciences of Complexity (Addison-Wesley, Boston 1988)
11. Perelson, A.S., Weisbuch, G.: Immunology for physicists. Rev. Mod. Phys. **69**, 1219-1267 (1997)
12. Celada, F., Seiden, P.E.: A computer model of cellular interactions in the immune system. Immunol. Today **13**(2), 56-62 (1992)
13. Perelson, A.S., Oster, G.F.: Theoretical studies on clonal selection: Minimal antibody repertoire size and reliability of self-nonself discrimination. J. Theor. Biol. **81**, 645-670 (1979)
14. Farmer, J.D., Packard, N., Perelson, A.S.: The immune system, adaptation and machine learning. Physica D **22**, 187-204 (1986)

15. Behn, U., Leo van Hemmen, J., Sulzer, B.: Memory to Antigenic Challenge of the Immune System: Synergy of Idiotypic Interactions and Memory B-Cells. J. Theor. Biol. **165**, 1-25 (1993)
16. Lippert, K., Behn, U.: Modeling the Immune System: Architecture and dynamics of idiotypic networks. In Stauffer, D. (ed.): *Annual Reviews of Computational Physics*, Vol.**IV** (World Scientific, Singapore 1997), 287-311
17. Behn, U., Celada, F., Seiden, P.E.: Computer modeling in immunology. In Lanzavecchia, A., Malissen, B., Sitia, R. (eds.): *Frontiers of Life*, Vol.**II** (Academic Press, London 2001), 611-630.
18. Bellomo, N., Pulvirenti, M. (eds): *Modeling in Applied Sciences: A Kinetic Theory Approach* (Birkhäuser, Boston 1996)
19. Bellomo, N., Lo Schiavo, M.: *Lecture Notes on the Generalized Boltzmann Equation* (World Scientific, London Singapore 2000)
20. Forrest, S., Hofmeyr, S.A.: Immunology as information processing. In Segel, L.A., Cohen, I. (eds.): *SFI Studies in the Sciences of Complexity: Design Principles for the Immune System and Other Distributed Autonomous Systems* (Oxford University Press, New York 2001)
21. Dasgupta, D. (ed.): *Artificial Immune Systems and Their Applications* (Springer, Berlin Heidelberg New York 1999)
22. Parisi, G.: Immunological memory in a network perspective. In Livi, R., Ruffo, S., Ciliberto, S., Buiatti, M. (eds.): *Chaos and Complexity* (World Scientific, Singapore 1988), 394-401
23. Parisi, G.: A simple model for the immune network. Proc. Natl. Acad. Sci. USA **87**, 429-433 (1990)
24. Smith, R.E., Forrest, S., Perelson, A.S.: Searching for diverse, cooperative populations with genetic algorithms. Evol. Comput. **1**(2), 127-149 (1993)
25. Forrest, S., Javornik, B., Smith, R.E., Perelson, A.S.: Using genetic algorithms to explore pattern recognition in the immune system. Evol. Comput. **1**(3), 191-211 (1993)
26. Kaufman, M., Urbain, J., Thomas, R.: Towards a logical analysis of the immune response. J. Theor. Biol. **114**, 527 (1985)
27. Weisbuch, G., Atlan, H.: Control of the Immune Response. J. Phys. A **21**, 189-192 (1988)
28. Stauffer, D.: In Pires, A., Landau, D.P., Herrmann, H.J. (eds.): *Computational Physics and Cellular Automata* (World Scientific, Singapore, 1989)
29. Cohen I.R., Atlan, H.: Network regulation of autoimmunity: an automation model. J. Autoimmun. **2**(5), 613-625 (1989)
30. Pandey, R., Stauffer, D.: Metastability with probabilistic cellular automata in an HIV infection. J. Stat. Phys. **61**, 235 (1990)
31. Neumann, A.U.: Control of the immune response by a threshold automata model on a lattice. Physica A **162**, 1-19 (1989)
32. Dayan, I., Havlin, S., Stauffer, D.: Cellular automata generalization of the Weisbuch-Atlan model for immune response. J. Phys. A **21**(3), 2473-2476 (1988)
33. Atlan, H., Cohen, I.R. (eds.): *Theories of Immune Networks* (Springer, Berlin Heidelberg New York 1989)
34. Stewart, J., Varela, F.J.: Morphogenesis in shape-space. Elementary metadynamics in a model of the immune network. J. Theor. Biol. **153**, 477-498 (1991)

35. De Boer, R.J., Segel, L.A., Perelson, A.S.: Pattern formation in one- and two-dimensional shape-space models of the immune system. J. Theor. Biol. **155**(3), 295-333 (1992)
36. Segel, L.A., Perelson, A.S.: A paradoxical instability caused by relatively short range inhibition. J. Appl. Math. **50**, 91-107 (1990)
37. Weisbuch, G., De Boer, R.J., Perelson, A.S.: Localized memories in idiotypic networks. J. Theor. Biol. **146**(4), 483-99 (1990)
38. Stauffer, D., Weisbuch, G.: High dimensional simulation of shape space model for immune system. Physica A **180**, 42-52 (1992)
39. Stauffer, D.: Monte-Carlo simulation of Ising-like immunological shape space. Int. J. Mod. Phys. C **5**(3), 513-518 (1994)
40. Dasgupta, S.: Monte Carlo simulation of the shape space model of immunology. Physica A **189**, 403-419 (1992)
41. Bernardes, A.T., Zorzenon dos Santos, R.M.: Immune network at the edge of chaos. J. Theor. Biol. **186**(2), 173-187 (1997)
42. Zorzenon dos Santos, R.M., Bernardes, A.T.: Immunization and Aging: A Learning Process in the Immune Network. Phys. Rev. Lett. **81**, 3034-3037 (1998)
43. Pandey, R., Stauffer, D.: Immune response via interacting three dimensional network of cellular automata. J. de Physique **50**, 1 (1989)
44. Chowdhury, D., Stauffer, D., Choudary, P.V.: A unified discrete model of immune response. J. Theor. Biol. **145**(2), 207-215 (1990)
45. Chowdhury, D.: Immune Network: An Example of Complex Adaptive Systems. In Dasgupta, D. (ed.): *Artificial Immune Systems and Their Applications* (Springer, Berlin Heidelberg New York 1999)
46. Chowdhury, D., Stauffer, D.: Statistical Physics of Immune Networks. Physica A **186**, 61-81 (1992)
47. Zorzenon dos Santos, R.M., Coutinho, S.C.: The dynamics of the HIV infection: a cellular automata approach. Phys. Rev. Lett. **87**, 168102-14 (2001)
48. Bernaschi, M., Castiglione, F.: Selection of escape mutants from immune recognition during HIV infection. Immunol. Cell Biol. **80**, 307-313 (2002)
49. Seiden, P.E., Celada, F.: A model for simulating cognate recognition and response in the immune system. J. Theor. Biol. **158**, 329-357 (1992)
50. Castiglione, F., Bernaschi, M., Succi, S.: Simulating the immune response on a distributed parallel computer. Int. J. Mod. Phys. C **8**, 527-545 (1997)
51. Castiglione, F., Mannella, G., Motta, S., Nicosia, G.: A network of cellular automata for the simulation of the immune system. Int. J. Mod. Phys. C **10**, 677-686 (1999)
52. Motta, S., Nicosia, G.: A plain cellular automata for the simulation of the immune system. In Heemink, A.W., Dekker, L., Arons, H.d.S., Smit, I., v. Stijn, T.L. (eds.): *EUROSIM 2001: Shaping Future with Simulation*, TU Delft, The Netherlands, 2001
53. Morpurgo, D., Serenthá, R., Seiden, P.E., Celada, F.: Modeling thymic functions in a cellular automaton. Int. Immunol. **7**, 505-516 (1995)
54. Celada, F., Seiden, P.E.: Affinity maturation and hypermutation in a simulation of the humoral immune response. Eur. J. Immunol. **26**(6), 1350-1358 (1996)
55. Kohler, B., Puzone, R., Seiden, P.E., Celada, F.: A systematic approach to vaccine complexity using an automaton model of the cellular and humoral immune system. Vaccine **19**, 862-876 (1999)

56. Castiglione, F., Motta, F., Nicosia, G.: Pattern Recognition by Primary and Secondary Response of an Artificial Immune System. Theory Biosci. **120**(2), 93-106 (2001)

57. Castiglione, F., Motta, S., Nicosia, G., Zammataro, L.: The effects of an apoptosis mechanism on the immune response. In Heemink, A.W., Dekker, L., Arons, H.d.S., Smit, I., v. Stijn, T.L. (eds.): *EUROSIM 2001: Shaping Future with Simulation*, TU Delft, The Netherlands, 2001

58. Jamin, C., Le Corre, R., Lydyard, P.M., Youinou, P.: Anti-CD5 extends the proliferative response of humane CD5+ B cells activated with anti-IgM and interleukine-2. Eur. J. Immunol. **26**, 57-62 (1996)

59. Castiglione, F., Agur, Z.: The Effect of Dose and Inter-Dosing Interval on the Patient's Hypersensitivity to the Drug: Analyzing a Cellular Automata Model of the Immune System. In Preziosi, L. (ed.): *Cancer Modeling and Simulation* (Chapman & Hall/CRC Press 2003)

The π-calculus as an Abstraction for Biomolecular Systems

Aviv Regev[1] and Ehud Shapiro[2]

[1] Bauer Center for Genomics Research, Harvard University, Cambridge, USA
    aregev@cgr.harvard.edu
[2] Dept. of Computer Science, Weizmann Institute, Rehovot, Israel
    Ehud.Shapiro@weizmann.ac.il

1 Introduction

Biochemical processes, carried out by networks of proteins, underlies the major functions of living cells ([8,60]). Although such systems are the focus of intensive experimental research, the mountains of knowledge about the function, activity, and interaction of molecular systems in cells remain fragmented. While computational methods are key to addressing this challenge ([8,60]), they require the adoption of a meaningful mathematical abstraction [50]. The research of biomolecular systems has yet to identify and adopt such a unifying abstraction.

An abstraction (that is, a mapping from a real-world domain to a mathematical domain) highlights some essential properties while ignoring other, complicating, ones. A good scientific abstraction has four properties: it is *relevant*, capturing an essential property of the phenomenon; *computable*, bringing to bear computational knowledge about the mathematical representation; *understandable*, offering a conceptual framework for thinking about the scientific domain; and *extensible*, allowing the capture of additional real properties in the same mathematical framework.

For a good abstraction for biomolecular processes to be *relevant*, it should capture two essential properties of these systems in one unifying view: their molecular organization and their dynamic behaviour. A *computable* abstraction will then allow both the simulation of this dynamic behaviour and the qualitative and quantitative reasoning on these systems' properties. An *understandable* abstraction will correspond well to the informal concepts and ideas of molecular biology, while opening up new computational possibilities for understanding molecular systems, by, for example, suggesting formal ways to ascribe biomolecular function to a biological system, or by suggesting objective measures for behavioural similarities between systems. Finally, the desired abstraction should be *extensible*, scaling to higher levels of organization, in which molecular systems play a key, albeit not exclusive, role.

In this work we aim to adopt the much-needed abstraction for biomolecular systems from computer science. We investigate the "molecule-as-computation" abstraction, in which a system of interacting molecular entities is described and modelled by a system of interacting computational entities. Based on this abstraction, we adapt process algebras for the representation and simulation of biomolecular systems, including regulatory, metabolic, and signalling pathways, as well as multicellular processes.

1.1 Previous Work

Biochemical systems are traditionally studied using "dynamical systems theory". In recent years, additional approaches from computer science have been adapted for the representation of pathways. The various approaches can be roughly divided into four groups.

- **Chemical kinetic models.** Dynamical systems theory views biomolecular systems based on the "cell-as-collection-of-molecular-species" abstraction. Such models describe different molecular systems from a pure biochemical perspective (e.g., [55], [38], [37], [7], [5], [34], [3], [70], [56]). This well-developed approach has several advantages: it is clear, well understood, and extremely powerful, with an extensive theoretical background, and a variety of methodologies and tools. However, it has been convincingly argued that this approach, adopted from the physical sciences, is lacking in *relevance* and *understandability*, when handling biological systems [17]. Its main limitation lies in its indirect treatment of biomolecular *objects*. Biological systems are composed of molecular objects, which maintain their overall identity while changing in specific attributes, such as their chemical modification, activation state, or location. Chemical models, however, handle the cell as a monolithic entity and are mostly insusceptible to more structured descriptions that are typical of biological thinking [17]. Thus, each modified state of a molecule, a molecular complex, machine, or compartment, must be handled as a distinct variable rather than as one entity with multiple states.
- **Generalized models of regulation.** Boolean network models use a "molecule-as-logical-expression" abstraction to describe and simulate gene regulatory circuits. These models were introduced by Kaufmann and subsequently applied to various biological systems (e.g., [54], [64], [53], [39], [63], [62], [1], [2], [14]). These approaches have proved highly useful when studying general properties of large networks, and in handling systems for which only qualitative knowledge is available. However, as a general model of biomolecular systems (rather than a specific model of regulation) they are limited in *relevance* (as they apply only to the regulation aspect of molecular systems), *computability* (as they have only limited predictive power), and *extensibility* (as, apparently, a Boolean abstraction of activation and inhibition suffices to handle only specific types of molecular systems).

- **Functional object-oriented databases.** Pathway databases are based on the "molecule-as-object" abstraction. They store information on molecular interactions (e.g., the DIP [69], BIND [6], and INTERACT [15] databases) and complete pathways (e.g., the MetaCyc [30], WIT [57], KEGG [43], CSNDB [27], aMAZE [65], and TransPath [68] databases). Each such database has a sophisticated object-oriented schema that provides a biologically appealing hierarchical view of molecular entities. Most databases are equipped with querying tools of variable levels of sophistication, from simple queries to pathway reconstruction. Functional databases provide an excellent solution for organizing, manipulating, and (sometimes) visualizing pathway data. However, in most cases, they provide little if any dynamic capabilities (e.g., simulation) and their querying tools are thus seriously limited. Note that recently developed **exchange languages** (e.g., [31], [42], and [16]) attempt to address these limitations by providing means to integrate models and tools from various sources. Most are XML-based markup languages (e.g., CellML and SBML). However, the utility of the language depends on its underlying pathway model, which is still primarily a kinetic one.

- **Abstract process languages applied to biomolecular systems.**
  Approaches based on abstract process languages use the "molecule-as-computation" abstraction, and have gained increasing importance during the past few years. They were first proposed by Fontana and Buss [17], who employed the λ-calculus and linear logic as alternatives to dynamics systems theory for studying (bio)chemistry. Notable examples use existing formalisms for concurrent computation (often with a strong graphical component) to model real biomolecular systems. The most comprehensive works used Petri Nets (e.g., [21], [20], [35], [36], [33], and [25]) for representation, simulation, and analysis of metabolic pathways. Petri nets considerably improve the traditional kinetic model, due to their clarity and graphical convenience, and augment it further with specific analytic tools. However, Petri Nets essentially suffer from the same *relevance* problem as "dynamical systems theory": each molecule (and state or modification) is represented by an individual place, and there is no immediate use of a more structured representation of complex molecular entities. Indeed, Petri nets are mostly (and successfully) used for the study of metabolic pathways, handled from a largely chemical perspective. Recent advances [36] try to overcome these limitations by combining Petri nets with object-oriented representation.

  Recent studies employed statecharts ([26], [28], [29]) to build qualitative graphical models primarily for cellular systems with a molecular component. Unlike typical graph-based visualization of pathways, statecharts provide a rich and expressive process language with clear semantics. Statecharts are highly expressive, and are proving successful in qualitatively representing complex biological systems, from a molecular to an organismic scale. Unfortunately, statecharts are currently a primarily qualitative framework, and do not handle quantitative aspects in a direct and accessible manner. Statecharts are, however, extremely useful in describing complex multicellular scenarios, as indeed has been their primary biological application.

These recent attempts highlight the promise in using abstract process languages for pathway informatics. Our approach belongs to this last category, based on the "molecule-as-computation" abstraction, and using the π-calculus and its extensions as a computational framework, as suggested by us and others ([17], [10], [51], [11]). We develop ways to represent and simulate biomolecular processes, describing the set of steps taken to adapt, extend, and implement this core language to conform to the unique requirements of biochemical systems, first on a semi-quantitative and then on a fully quantitative, stochastic scale. We present the language intuitively[3] and provide a multitude of simple and complex examples to which it is applied, highlighting its capabilities to represent and study a variety of molecular systems, including compositional and modular ones.

2 Abstracting Biomolecular Systems as Concurrent Computation: an Overview

An abstraction is a mapping from a real-world domain to a mathematical one, which preserves some essential properties of the real-world domain while ignoring other properties, considered superfluous. Building the abstraction has three components. First, we informally organize our knowledge on the real-world domain, identifying essential entities in this domain, their properties, and behaviour. Second, we select a mathematical domain which we believe can be useful for the abstraction of the real-world one, motivated by some informal correspondence between those properties or behaviours we deem essential in the real-world domain and those of the mathematical one. Third, we design and perform the abstraction, starting by laying down the principles for the mapping, and then using these *pragmatic* guidelines to build specific representations of actual real-world entities.

The Real-World Domain: Essential Properties of Biomolecular Systems

We identify several essential aspects of biomolecular systems (also illustrated by a toy example in Fig. 1).

---

[3] Due to space limitations we do not formally present the π-calculus, but rather use an informal presentation of the language, focused on biological examples. We assume that biological readers will benefit the most from this decision. Readers from a formal background may skip the informal introduction (Sect. 2) if familiar with the calculus, or are referred to [41] for details.

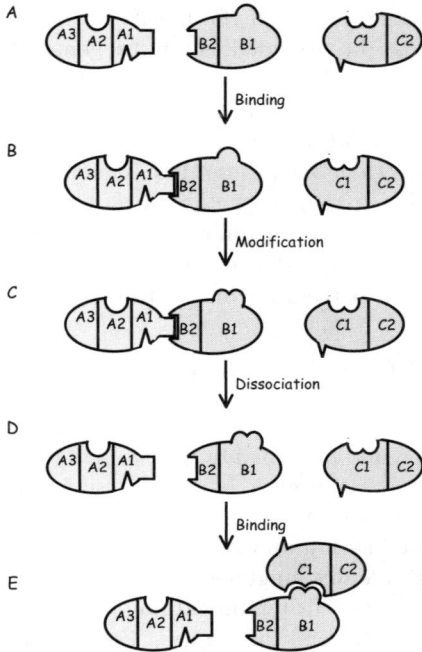

Fig. 1. A toy biomolecular system. A. Three molecular species: A, B, and C, each with several sub-domains (dividing lines) and motifs (protrusions and depressions). B. Domains A1 and B2 interact via complementary motifs. C. A motif on domain B1 is modified. D. The proteins dissociate. E. The modified motif on B1 is used for interaction with domain C1.

- **Molecular species and populations.** Biomolecular systems are composed of a population of molecules. The molecular population may include multiple molecular species (e.g., Fig. 1A), with multiple copies of each molecule type.
- **Molecular composition.** Each protein molecule may be composed of several domains – independent structural and functional parts (Fig. 1A). Each domain may have one or more structural-chemical motifs (Fig. 1A).
- **Complementarity and specific interaction.** Different domains specifically interact through complementary motifs. Different motifs in domains allow for different specific interactions (Fig. 1A,B).
- **Interaction outcomes: reconstitution, modification, and change in molecular state.** One or more things can happen to molecules following interaction. First, the molecule (e.g., an enzyme) may be reconstituted without any change. In other cases, the molecule may be changed follow-

ing interaction (Fig. 1C,D). The change may be attributed to a specific chemical modification of one of the motifs in the molecule by its interaction partner, or may be regarded as a more general change in the state of the molecule, such as a change from an inactive to an active state. In each state the molecule may potentially participate in different interactions.

- **Alternative interactions.** In many cases a molecule may be able to participate in more than one interaction. The alternative interactions may be independent. For example, protein A may interact with protein B on one domain and with protein C on another, such that one interaction does not preclude the other. In other cases, the same domain may have several potential interaction partners, which compete with each other.

The Mathematical Domain: Concurrent Computation

In the "molecule-as-computation" abstraction, a system of interacting molecular entities is described and modelled by a system of interacting computational entities. Several abstract computer languages, such as Petri nets [52], statecharts [22], concurrent logic programming [58], and the π-calculus [41], were developed for the specification and study of such systems.

We focus on the π-calculus language. π-calculus programs specify networks of *communicating* processes. Communication occurs on complementary *channels* that are identified by specific names (Fig. 2). The programs describe the potential behaviour of processes: what communications the processes can participate in on such channels.

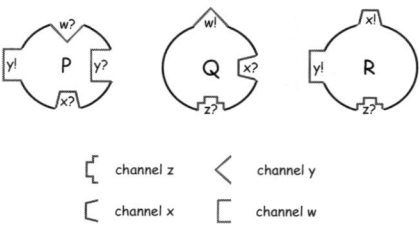

Fig. 2. π-calculus processes and channels: an intuitive view. Three processes, P, Q, R (ovals), with four communication channels (complementary shapes of protrusions and depressions).

There are two different kinds of communications in the π-calculus. In the first type (also present in simpler languages, such as CCS [41]), the processes only send *alerts* (or empty *"nil"* messages) to one another. In this case (depicted in Fig. 3A), a sender process may communicate with a receiver process if they share the same communication channel. Following communication, the action is removed, and each process may either iterate or change its state, becoming another process with different channels and behaviour.

The second type of communication is more complex and unique to the π-calculus. In this case (Fig. 3B), the sender process sends a *message* to the receiving process. The content of this message is one or more channel names. These channels can be used by the receiving process to communicate with other processes. As before, following communication the processes may either iterate or change state. However, passing channels as messages allows the process (and its continuations) to acquire communication capabilities dynamically that were not specified *a priori* in its explicit program. Such a change in future communication capabilities as a result of passing messages is termed *mobility*.

As shown in Fig. 3 the two communication capabilities may serve to describe similar systems. However, in the former mechanism the communication capabilities of a process are all strictly pre-defined (and thus may change only as a result of a pre-specified state change), while in the latter messages may change the communication capabilities of a process dynamically. Thus, messages allow us to leave certain components under-specified, to be determined only following interaction with different processes which may send different messages along the same channel.

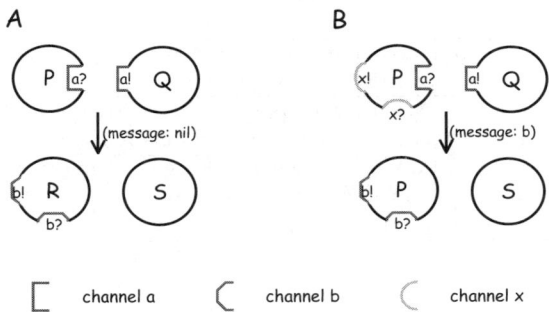

Fig. 3. π-calculus communication: a schematic view. A. Communication with state change. B. Communication with mobility. P,Q,S,R (ovals): processes; a,b: communication channels.

The Molecule-as-Computation Abstraction

We find an intuitive correspondence between the world of computational processes and of biomolecular systems, based on the representation of molecules as the processes in which they may participate. Note that this view is essentially a biological one. Biologists typically characterize molecules by *what they can do*. For example, enzymes are named by the reaction that they can catalyze, and binding domains in proteins by the entities which they bind.

Table 1 gives a general intuition of the guidelines underlying this abstraction. In the next one we build up the abstraction step by step with a real biological system.

Table 1. Guidelines for the abstraction of biomolecular systems to the π-calculus.

Biomolecular entity	π-calculus entity
Molecular species	Process species
Molecular population	System of concurrent processes
Complementary motif types	Complementary input and output occurrences on the same channel name
Motif occurrence in molecule or domain	Communication offers on channels in process
Biomolecular event	**π-calculus event**
Specific interaction on complementary motif	Specific communication on complementary channels
Outcome following interaction	Communication *prefix* preceding process creation or other communication
Reconstitution following interaction	Communication *prefix* preceding process *recreation*
Changed molecular state following interaction	Communication *prefix* preceding creation of new process
Modification of motifs during interaction	Non-mobile approach: recreation of a different type of process with a different channel set
	Mobile approach: message (a tuple of channel names) sent in communication to replace channels in the receiving process

Note that we may represent modification either as a special case of molecular state change or by employing the mobility mechanism of the π-calculus, abstracting state change and modification as a *message* sent from one of the processes to the other one. Each of these two abstractions for molecular interactions has benefits and drawbacks. In the non-mobile approach, the abstracted system is fully specified, and thus often simpler to write and follow. The mobile approach is more complicated, but also closer to biological reasoning: the same process continues to abstract molecules throughout modification, and message passing corresponds directly and specifically to the modification event.

3 Biomolecular Systems in the π-calculus

To fully describe the abstraction of biomolecular process in the π-calculus, we now turn to a series of biological examples, which we will abstract step by step using the various constructs and rules of the π-calculus. A formal presentation of the calculus is given in the next section. Our examples are derived from a model of a canonical signal transduction pathway starting from a receptor tyrosine kinase (RTK) on a cell's membrane, and ending with transcriptional activation of immediate early genes.

3.1 Molecules and Domains as Concurrent Processes

We abstract each biomolecular system as a process, denoted by a capitalized name, e.g., P. For example, the RTK–MAPK signal transduction pathway is a system, and will be denoted as

$$RTK\_MAPK\_pathway$$

Each constituent molecule in the pathway is a process, too, as are its domains. For example, the growth factor unbound ligand, receptor tyrosine kinase, and Ras molecules will be denoted as $Free\_Ligand$, RTK, and Ras, respectively. Similarly, the extracellular, intra-cellular, and transmembranal domains of the receptor molecule will be denoted by $Extra$, $Transmem$, and $Intra$ processes.

We next define a system process as a collection of molecule processes, occurring in parallel. This concurrency is denoted by the PAR operator, $|$, inserted between the process names. For example,

$$RTK\_MAPK\_pathway ::=$$
$$Free\_ligand \mid RTK \mid RTK \mid RTK \mid Ras \mid Ras \mid Ras \mid Ras$$

denotes that the RTK–MAPK pathway is composed of several ligand, RTK, and Ras molecules.

Analogously, we define a protein molecule as several domain processes composed in parallel. For example, we represent a homo-dimer ligand composed of two identical binding domains, and a receptor composed of intra-cellular, transmembranal, and extracellular domains, by

$$Free\_ligand ::= Free\_binding\_domain \mid Free\_binding\_domain$$
$$RTK ::= Extra \mid Transmem \mid Intra$$

3.2 Molecular Complementarity as Communication Channels

Two molecules (or domains) interact with each other based on their structural and chemical complementarity [46]. We abstract molecular complementarity as complementary offers on communication channels. For example, if the ligand's free binding domain and the receptor's extracellular domain have complementary binding motifs, we denote this motif pair as a *ligand_binding* channel, which appears as an output *ligand_binding* ! [ ] offer in the ligand's *Free_binding_domain* process and as an input *ligand_binding* ? [ ] offer in the receptor's *Free_extracellular_domain* process:

$$Free\_binding\_domain ::= ligand\_binding \ ! \ [\ ] \cdots$$
$$Free\_extracellular\_domain ::= ligand\_binding \ ? \ [\ ] \cdots$$

Molecules (processes) with complementary motifs (communication offers) are graphically depicted in Fig. 4. The motifs on molecules may often vary based

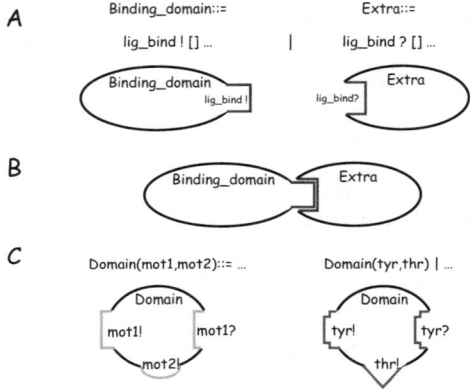

Fig. 4. Molecular complementarity as input and output channels. A. The *Free_binding_domain* and the *Extra* processes (ovals) have complementary output (protrusion) and input (depression) offers on the *ligand_binding* channel. B. Process (molecular) interaction is enabled by channel (motif) complementarity. C. We may define a process with channel parameters (left), to be instantiated with actual channels (right) only when the process is *created*.

on different biochemical modifications. We often abstract such motifs as *channel parameters* of the process. For example, the intra-cellular domain of the receptor may have three potential binding motifs, which we will represent as parameters when defining the process:

$$Free\_intracellular\_domain(motif1, motif2, motif3) ::= \cdots$$

Then, when we *create* the *Free_intracellular_domain* process as part of a particular *RTK*, we replace these parameters with actual channels, representing the specific state for those motifs, e.g.,

$$RTK ::= \ Free\_intracellular\_domain(p\_tyr1, tyr2, sh2) \ | \cdots$$

Molecule processes with motif channel parameters are graphically depicted in Fig. 4C.

3.3 Sequential and Mutually Exclusive Events

Biochemical events may occur in sequence, in parallel with other independent events, or in a mutually exclusive, competitive fashion. We abstract a sequence of interactions as a sequence of input and output offers, separated by a prefix operator: "," (the comma sign). For example, if the extracellular domain can first bind a ligand, and then bind to another domain of another molecule, we define

$$Extra ::= \quad ligand\_binding \; ? \; [\;], \; rtk\_binding \; ? \; [\;], \; \cdots$$

In other cases, there may be several alternative interactions in which a molecule may participate: once one occurs, the other options are no longer possible. Such mutually exclusive offers are *summed* together, using the "+" or ";" *choice* operator.[4] For example, if the receptor is able to bind either the (agonist) ligand or an antagonist but not both, we write

$$Extra ::= ligand\_binding \; ?[\;], \; rtk\_binding \; ?[\;], \; \cdots +$$
$$antagonist\_binding \; ?[\;], \; \cdots$$

Finally, several interactions may occur in parallel, without directly affecting each other. Such cases will be handled by composing different offers in parallel, as shown above.

3.4 Compartmentalization as Private Channels

A pathway is not merely a bag of molecules and their domains. Rather, it is composed of defined compartments, from individual molecules, through molecular complexes form, to cellular compartments, often bounded by membranes. In all cases molecules which share a common compartment may interact with each other, while molecules excluded from the compartment may not [24]. We abstract only the limits that compartments impose on communication by introducing "private" channels with restricted communication scopes ([49] discusses an extension that handles compartments directly).

A private channel x is created using the "new x" operator. For example, if we wish to represent the fact that the three domains of a receptor are linked by a common backbone and belong to a single molecule, we introduce a new *backbone* channel that is shared by the three processes, and is distinct from all other (external) channels:

---

[4] We will use both notations interchangeably. For their syntactic distinction, see the implementation section.

$$RTK ::= (\text{new } backbone)(Extra|Transmem|Intra)$$

The private *backbone* channel can be used for communication only between the three sub-processes of one *RTK*, and thus represents the limits that a shared backbone poses on interaction.

3.5 Interaction and Biochemical Modification as Communication

Each biochemical interaction in a system may affect subsequent interactions in several ways. First, interaction is often accompanied by modification of one molecule by the other, such that the modified molecule may now be complementary to (and interact with) other motifs, which it was not "suitable for" before (and vice versa). Second, a particular interaction may lead to the exclusion of other possibilities. Finally, an interaction may lead to a general "state change" in both interacting molecules, enabling events that were disabled before. We will represent these three results of interaction (and modification) by the communication rules of the π calculus, using either a *non-mobile* or a *mobile* (message passing) abstraction.

Non-mobile Communication

In the abstraction of interaction as *non-mobile* communication we represent modification and state change *together* by the introduction of a newly created (or recreated) process into the system. The different channel set owned by the created process will represent the modified residues in the corresponding molecule. For example, consider an active protein kinase (*Active_kinase* process) and a target binding domain (*Mod_Bind_domain* process). The binding domain has a motif (*phosph_site* ?) that may be identified and bound by a complementary motif (*phosph_site* !) in the kinase. The kinase may then phosphorylate a tyrosine residue in another motif in the binding site. This modification is represented by the creation of a new instance of a different process type (*Phospho_Bind_domain*) that uses the *phosphotyrosine* channel rather than the *tyrosine* one:

$$Active\_kinase ::= phosph\_site \;! \;[\;] \;, \; Kinase$$
$$Mod\_Bind\_domain ::= phosph\_site \;? \;[\;] \;, \; Phospho\_Bind\_domain$$
$$Phospho\_Bind\_domain ::= phosphotyrosine \;! \;[\;] \;, \; \cdots$$

If an interaction takes place between *Active_kinase* and *Mod_Bind_domain*, two things will happen simultaneously. First, both the input and output events on the *phosph_site* channel will be removed. Second, the remainder of the two processes will be allowed to continue, creating a *Kinase* process and a *Phospho_Bind_domain* process. The latter one uses the *phosphotyrosine* channel, representing the modification. Overall, the non-mobile communication process can be summarized in the following reduction:

$$phosph\_site \ ! \ [\ ] \ , \ Kinase \quad | \quad phosph\_site \ ? \ [\ ] \ , \ Phospho\_Bind\_domain$$

$$\longrightarrow$$

$$Kinase \quad | \quad Phospho\_Bind\_domain$$

The newly created *Phospho_Bind_domain* can use its *phosphotyrosine* channel for further communications with other processes which harbor this motif, such as the *SH2* process:

$$SH2 ::= phosphotyrosine \ ? \ [\ ] \ , \ \cdots$$

The full sequence of events is summarized in Fig. 5.

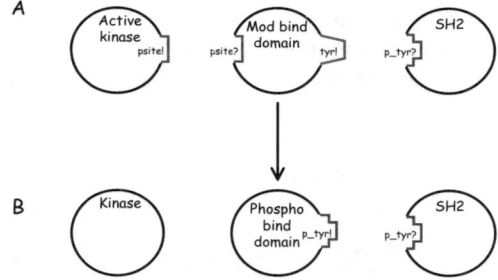

Fig. 5. Interaction and modification as communication and state change.
A. Pre-communication. Three processes (*Active_kinase*, *Mod_Bind_domain*, *SH2*) and their respective channels. *Active_kinase* and *Mod_Bind_domain* may interact on *phosph_site*. B. Post-communication. A *Phospho_Bind_domain* process is introduced into the system instead of the *Mod_Bind_domain* process. It may communicate with *SH2* on *p_tyr*.

Mobile Communication

In the mobile abstraction we represent *modification* as sending and receiving *messages*. Each message is composed of one or more channel names, representing the modified motif(s). These names are received into "placeholders" in the receiving process which they substitute, representing the modification. Then, they may be used by the receiving process in subsequent communication, representing the effect of modification on subsequent interactions. For example, consider again the active protein kinase (*Active_kinase* process) and a target binding domain (*Mod_Bind_domain* process), where the kinase may phosphorylate a tyrosine residue in another motif in the binding site. In the mobile approach, this modification is represented by the sending of a p_tyr message, to be received by the tyr "placeholder":

$$Active\_kinase ::= phosph\_site \;!\{p\_tyr\} \;,\; Kinase$$

$$Mod\_Bind\_domain ::= phosph\_site \;?\{tyr\} \;,\; tyr!\{\cdots\} \;,\; \cdots$$

If an interaction takes place between $Active\_kinase$ and Mod_Bind_domain, the received p_tyr channel will replace the tyr channel in the receiving Mod_Bind_domain. Overall, the communication-with-message process can be summarized in the following reduction:

$$phosph\_site \;!\{p\_tyr\} \;,\; \cdots \;\;\mid\;\; phosph\_site \;?\{tyr\} \;,\; tyr!\{\cdots\} \;,\; \cdots$$

$$\rightarrow$$

$$Kinase \;\mid\; p\_tyr!\{\cdots\} \;,\; \cdots$$

representing the interaction, modification, and release events involved in the chemical reaction in the system. Most importantly the p_tyr received by the Mod_Bind_domain process may now be used for further communications with other processes which harbor this motif, such as the $SH2$ process:

$$SH2 ::= p\_tyr?\{\cdots\} \;,\; \cdots$$

The full sequence of events is summarized in Fig. 6.

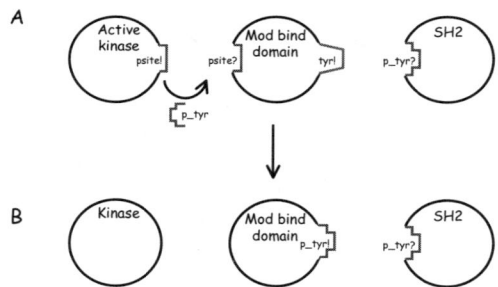

Fig. 6. Interaction and modification as communication and mobility. A. Pre-communication. Three processes ($Active\_kinase$, Mod_Bind_domain, $SH2$) and their respective channels. $Active\_kinase$ and Mod_Bind_domain may interact on $phosph\_site$. B. Post-communication. The tyr channel in Mod_Bind_domain is replaced with p_tyr, which may be used for communication with $SH2$.

3.6 Compartment Change as Extrusion of a Private Channel's Scope

We also use the *mobility* mechanism to abstract compartment changes, such as complex formation. For example, consider a three-molecule complex that forms between a dimeric ligand and the extracellular domain of two RTK receptors. As a result of complex formation, the two RTK domains may interact "privately", without the participation of any other RTK molecule. To represent this, the private *backbone* channel is sent from the *Ligand*'s *Binding_domain* sub-processes to the *RTKs*' *Extra* processes:

$$Ligand ::= \ (\mathsf{new}\ backbone)(Binding\_domain \mid Binding\_domain)$$

$$Binding\_domain ::= \ ligand\_binding\ !\ \{backbone\},\ Bound\_domain$$

$$Extra ::= \ ligand\_binding?\{cross\_backbone\}, Bound\_Extra(cross\_backbone)$$

As a result of two communication events (of each of the *Binding_domain*s with a different *RTK Extra* sub-process), we end up with two *Extra* processes which both "own" the same private *backbone* channel, representing their indirect link via a commonly bound ligand molecule:

$$(\mathsf{new}\ backbone)$$
$$(\ ligand\_binding\ !\ \{backbone\},\ Bound\_domain\ \mid$$
$$ligand\_binding\ !\ \{backbone\},\ Bound\_domain\ )\ \mid$$
$$ligand\_binding\ ?\ \{cross\_backbone\}\ ,\ Bound\_Extra(cross\_backbone)\ \mid$$
$$ligand\_binding\ ?\ \{cross\_backbone\}\ ,\ Bound\_Extra(cross\_backbone)$$

$$\longrightarrow$$

$$(\mathsf{new}\ backbone)$$
$$(\ Bound\_domain \mid Bound\_domain \mid$$
$$Bound\_Extra(backbone) \mid Bound\_Extra(backbone)\ )$$

The *Bound_Extra* processes will subsequently use this private *backbone* channel to interact with each other, without external interruptions.

3.7 Molecular Objects as Parametric Processes

The use of parametric process definitions allows us to abstract the constant identity of a molecule (process) as its structure (public channels) and compartment (private channels) change with interaction. For example, we may define the *Mod_Bind_domain* described above (which can either interact with a modifying kinase, or bind to an SH2 domain) as

$$Mod\_bind\_domain(kinase\_site, sh2\_site) ::=$$
$$kinase\_site?\{mod\_sh2\_site\}, Mod\_bind\_domain(kinase\_site, mod\_sh2\_site) +$$
$$sh2\_site\ ?\ [\ ],\ \cdots$$

3.8 Competition as Choice

A molecule may often participate in one of several mutually exclusive interactions. We abstract this by the *choice* operator (denoted ";" or "+"). For example, consider a *System* in which the *Extra* process may interact either with an agonist *Ligand*'s *Binding_domain* (on the *ligand_binding* channel), or with an *Antagonist* (on the *antagonist_binding* channel):

$$
\begin{aligned}
System ::=&\quad Extra \mid Ligand \mid Antagonist \\
Extra ::=&\quad ligand\_binding\ !\ [\ ]\ ,\ Extra\_Bound\_Agonist\ + \\
&\quad antagonist\_binding\ !\ [\ ]\ ,\ Extra\_Bound\_Atagonist \\
Ligand ::=&\quad ligand\_binding\ ?\ [\ ]\ ,\ Bound\_ligand \\
Antagonist ::=&\quad antagonist\_binding\ ?\ [\ ]\ ,\ Bound\_antagonist
\end{aligned}
$$

If *ligand_binding* is chosen, then the *antagonist_binding* option is discarded and the resulting system is

$$Extra\_Bound\_Agonist \mid Ligand\_bound \mid Antagonist$$

Vice versa, if *antagonist_binding* is chosen, then the *ligand_binding* option is discarded and the resulting system is

$$Extra\_Bound\_Antagonist \mid Ligand \mid Bound\_antagonist$$

Choice is resolved non-deterministically: all enabled communications (where both input and output are available) are equi-potent. A more biologically realistic model would assign different probabilities to different interactions, based on reaction rates. We extend the language to handle such a model in the next section.

4 Simple Examples

The general principles outlined in Sect.2 and Sect.3 allow us to formally represent detailed information on complex pathways, molecules, and biochemical events. In this section we illustrate these capabilities with several small programs representing different aspects of molecular systems.

4.1 Molecular Complexes

Our first two examples handle the formation and breakage of a bimolecular complex. To represent the formation and breakage of a complex between two molecules, Molecule1 and Molecule2, we use both public and private channels (Fig. 7). Each of the molecules is represented by a process (*Molecule1* and *Molecule2*, Fig. 8A). The two processes share a public *bind* channel, on which one process (*Molecule1*) is offering to send a message, and the other (*Molecule2*) is offering to receive (Fig. 8A). These complementary communication offers represent the molecular complementarity of the two molecules, and the communication event represents binding. The private

```
System::= Molecule1 | Molecule2 .
Molecule1::= (new backbone) . bind ! {backbone} ,
                          Bound_Molecule1(backbone) .
Bound_Molecule1(bb)::= bb ! [ ] , Molecule1 .
Molecule2::= bind ? {cross_backbone} ,
                  Bound_Molecule2(cross_backbone) .
Bound_Molecule2(cbb)::= cbb ? [ ] , Molecule2 .
```

Fig. 7. π calculus code for a heterodimer complex

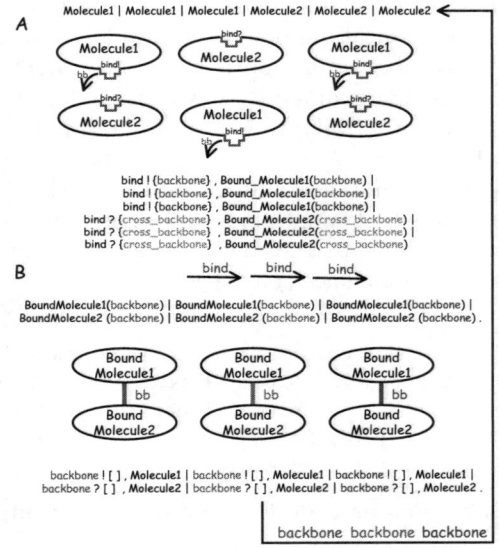

Fig. 8. Formation and breakage of a heterodimer complex. A. Separate molecules. B. Three complexes, represented by distinct private *backbone* channels.

backbone channel sent from *Molecule*1 to *Molecule*2 represents the formed complex, and the two processes change to a "bound" state (*Bound_Molecule*1 and *Bound_Molecule*2, Fig. 8B). A communication between the two "bound" processes on the shared private *backbone* channel represents complex breakage. As a result, the two processes return to the initial "free" state (*Molecule*1 and *Molecule*2), completing a full cycle. Note that in a system with many copies of these processes, any two particular copies of *Molecule*1 and *Molecule*2 may communicate (i.e., "bind") on the *bind* channel. However, the two resulting "bound" processes share a private channel, which is distinct from all other channels, and may allow only this particular pair to communicate ("unbind") with each other. Three such complexes and their specific private channels (distinguished in red, green, and purple) are shown in Fig. 8B.

Special Case: Homodimerization

The π-calculus program presented above for a heterodimer must be modified to represent homodimerization (Fig. 9). The reason is that a homodimer forms between two identical molecules that are represented by two copies of the same process. Since it is impossible *a priori* to break down the symmetry between the two components of the homodimer, both input and output must be offered simultaneously by the same process, analogously to the way molecules should have a "zipper"-like structure to allow homodimerization. The asymmetry is broken by using *non-deterministic choice*: while both processes offer both a send and a receive communication, one will (non-deterministically) send and the other receive. Since once an option is chosen the other is automatically withdrawn, a process cannot communicate with itself (in the same sense that a molecule cannot bind onto itself). In Fig. 9), each *Molecule* process offers a choice between sending and receiving a message on the *bind* channel. Once one is selected, the other is withdrawn, and the *Molecule* changes to a "bound" state (*Bound_Mol* process). The message, as in heterodimerization, is a private *backbone* channel, to be used for an "unbinding" communication. This, too, is done by a symmetric choice construct.

```
System::= Molecule | Molecule | Molecule | Molecule | Molecule .
Molecule::= (new backbone) . bind ! {backbone} , Bound_Mol(backbone);
                            bind ? {cbb}      , Bound_Mol(cbb) .
Bound_Mol(bb)::= bb ! [ ] , Molecule ;
                 bb ? [ ] , Molecule .
```

Fig. 9. π calculus code for a homodimer complex

4.2 Enzymes

Our next use-case tackles "classical", single-substrate, enzymatic reactions typical of metabolic pathways. The Michaelis–Menten mechanism provides a detailed account of an enzymatic reaction [61], breaking it down to its elementary steps. A single-substrate reversible enzymatic reaction includes four elementary reactions: binding of an enzyme to the substrate or to the product and formation of an EX complex, and release of either a product or a substrate from this complex. In this case (coded in Fig. 10, and schematically depicted in Fig. 11), we have five process types, representing the free enzyme (*Enzyme*), bound enzyme (*EX*), substrate (*Substrate*), product (*Product*), and intermediate (*X*). *Enzyme* includes a *choice* between an interaction with *Substrate* (on *bind_s*) and an interaction with *Product* (on *bind_p*). In both cases, two private channels, *rel_p* and *rel_s*, are sent from *Enzyme* to its counterpart (Fig. 11A). Following communication, *Enzyme* changes to *EX*

```
System::= Enzyme | Substrate .
Enzyme::= (new rel_s, rel_p) .
                 bind_s ! {rel_s,rel_p} , EX(rel_s,rel_p) ;
                 bind_p ! {rel_s,rel_p} , EX(rel_s,rel_p) .
EX(release_s,release_p)::= release_s ! [ ] , Enzyme ;
                          release_p ! [ ] , Enzyme .
Substrate::= bind_s ? {erel_s , erel_p} , X(erel_s , erel_p) .
Product::= bind_p ? {erel_s , erel_p} , X(erel_s , erel_p) .
X(rel_es,rel_ep)::= rel_es ? [ ] , Substrate ;
                   rel_ep ? [ ] , Product .
```

Fig. 10. π calculus code for single-substrate enzymatic reactions. An elementary (Michaelis–Menten) reaction model for a single-substrate reversible reaction with one product.

("bound enzyme"), and its counterpart (be it *Substrate* or *Product*) changes to X (reaction intermediate) (Fig. 11B). The two private channels shared between EX and X represent the formed complex , and are used to finish the reaction, resulting in *Enzyme* reconstitution and either *Product* or *Substrate* release (Fig. 11C). Note that as a result of the non-determinism of the *choice* construct, *Substrate* binding to *Enzyme* may end up either as a *Product* (by reaction of X and EX on *rel_p*) or be released as an intact *Substrate* (reaction of X and EX on *rel_s*). The same is true for *Product*.

4.3 Enzymes in Signal Transduction

Enzymatic reactions play a critical role in signal transduction (ST) pathways. However, unlike metabolic pathways where enzymes (proteins) and substrates (metabolites) are distinct kinds of biochemical entities, in ST pathways most substrates are proteins, serving as binding partners, transcription factors, or enzymes. We represent such "modification of modifiers" by extensively using the *mobility* mechanism of the π-calculus.[5] Previously, we distinguished the substrate from the product by using distinct process names. In representing the modification of ST molecules, we maintain process identity throughout modification.

Phosphorylation and Dephosphorylation by Protein Kinases and Phosphatases

We consider a toy example involving a binding protein (*Binding_Protein* process), a protein tyrosine kinase (*Kinase* process), and a protein tyrosine phosphatase (*Phosphatase* process). In this system, the protein kinase phosphorylates a tyrosine residue, and the phosphatase de-phosphorylates it. For

---

[5] An alternative non-mobile representation of such events was discussed above, and will not be presented in further detail.

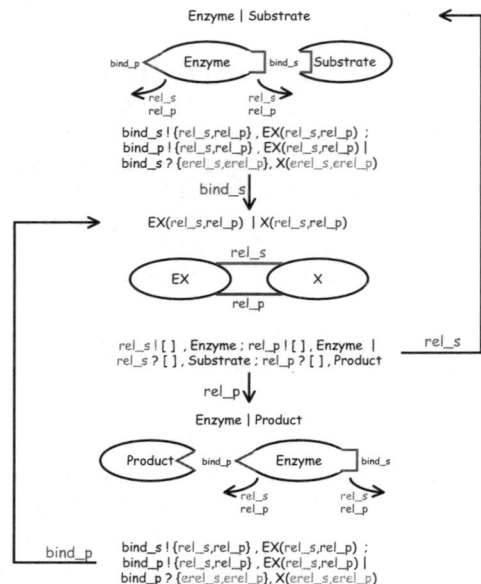

Fig. 11. Reversible, single-substrate enzymatic reaction with a single product: elementary reaction model. A. *Enzyme* may interact with either *Substrate* (on *bind_s*) or *Product* (on *bind_p*). B. Following communication, *Enzyme* becomes *EX*, while *Substrate* (or *Product*) becomes "intermediate" *X*. C. *X* (intermediate) and *EX* (bound enzyme) may interact either on *rel_p* (releasing *Product*) or *rel_s* (releasing *Substrate*).

simplicity, we handle the enzymatic reaction as a "single-step" reaction, rather than the elaborate model discussed above (Fig. 12 and Fig. 13).

```
System::= Binding_Protein(tyr) | Kinase | Phosphatase .
Binding_Protein(res)::= res ? {mod_res} , Binding_Protein(mod_res) .
Kinase::= tyr ! {p_tyr} , Kinase .
Phosphatase::= p_tyr ! {tyr} , Phosphatase .
```

Fig. 12. π calculus code for phosphorylation and de-phosphorylation of a binding domain. Model is based on a simplified (single-step) enzymatic reaction.

We represent the modifiable residue as two channels: *tyr* for the non-phosphorylated residue, and *p_tyr* for the phosphorylated one. *Kinase* sends *p_tyr* as a message on the *tyr* channel, representing the fact that the tyrosine kinase identified non-phosphorylated tyrosines and modifies them to the phosphorylated form. Conversely, *Phosphatase* sends a *tyr* message on the *p_tyr* channel.

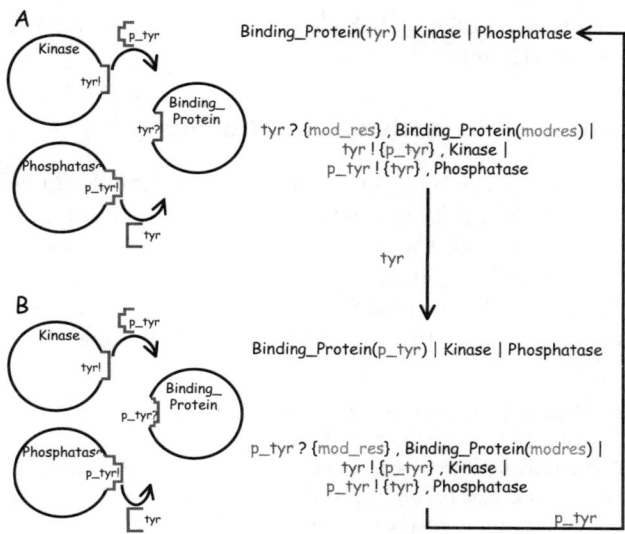

Fig. 13. Modification of a protein residue by kinases and phosphatases.
A: A system composed of a *Kinase*, a *Phosphatase*, and a *Binding_Protein* with
a *tyr* channel. B: Following communication between *Kinase* and *Binding_Protein*
on *tyr*, *Kinase* is recreated, and *Binding_Protein* now has a *p_tyr* channel, rather
than a *tyr* one, allowing interaction with *Phosphatase*. Following communication
between *Phosphatase* and *Binding_Protein* on *p_tyr*, *Phosphatase* is recreated,
and *Binding_Protein* now has a *tyr* channel, rather than a *p_tyr* one, returning to
the original state (A).

The *Binding_Protein* has a channel parameter, representing the phosphorylation state of its tyrosine residue. When the system is initialized we
assume that it is non-phosphorylated, so we create a *Binding_Protein(tyr)*.
This process offers to receive on *tyr*, and may thus interact with *Kinase*,
but not with *Phosphatase*. Upon communication, the *Binding_Protein* receives *p_tyr*, and is recreated, but now with the newly received channel (i.e.,
Binding_Protein(p_tyr)). This process offers to receive on *p_tyr*, and may
thus interact with *Phosphatase*, but not with *Kinase*. As before, upon communication, *Binding_Protein* is recreated with the received *tyr* channel, returning to the initial state.

In this way, the iteration of *Binding_Protein* between *tyr* and *p_tyr* parameters represents the cycling of the protein molecule between a phosphorylated and a non-phosphorylated state. These channel parameters affect the
communication capabilities of *Binding_Protein*, reflecting the effect of the
residue's modification state on the protein's ability to be identified by kinases and phosphatases. This principle can be extended and incorporated
into more detailed models. For example, we can devise a model that com-

bines the elementary reaction (Michaelis–Menten) scheme with the *tyr/p_tyr* channel parameter scheme.

Modifiable Residues in Activity and Binding

The *channel parameter* approach can be extended to handle the effect of modified residues on the binding and activity of the proteins that harbor them. For example, assume that the phosphorylated tyrosine (but not the non-phosphorylated one) of the binding protein discussed above can be bound by an SH2 domain in an adaptor protein. To model this situation we now extend the previous program (Fig. 14). Rather than using a single channel parame-

```
System::= Binding_Domain(tyr_mod,tyr_bind) |
          Kinase | Phosphatase | Adaptor .
Binding_Domain(res_mod,res_bind)::=
    res_mod ? {res_mod1,res_bind1} ,
            Binding_Domain(res_mod1,res_bind1) ;
    res_bind ? {unbind} ,
            Bound_Domain(res_mod,res_bind,unbind) .
Bound_Domain(res_mod,res_bind,ub)::=
    ub ? [ ] , Binding_Domain(res_mod,res_bind) .
Kinase::= tyr_mod ! {p_tyr_mod, p_tyr_bind} , Kinase .
Phosphatase::= p_tyr_mod ! {tyr_mod,tyr_bind} , Phosphatase .
SH2_Adaptor::= (new unbind) . p_tyr_bind ! {unbind} ,
                            Bound_Adaptor(unbind) .
Bound_Adaptor(ub)::= ub ! [ ] , SH2_Adaptor .
```

Fig. 14. π **calculus code for phosphorylation and de-phosphorylation of a binding domain**

ter in *Binding_Protein* (either *tyr* or *p_tyr*), we now introduce two channel parameters. The first parameter, *res_mod* (either *tyr_mod* or *p_tyr_mod*), is used for communication with *Kinase* and *Phosphatase*. The second parameter, *res_bind* (either *tyr_bind* or *p_tyr_bind*), represents the ability (or lack thereof) to bind to the adaptor protein. *Kinase* and *Phosphatase* send a tuple of two channels to *Binding_Protein*, representing the fact that residue modification affects the ability both to interact with the modifying enzyme and to bind to the adaptor protein.

5 The Biochemical Stochastic π-calculus: a Quantitative Extension

The original framework of the π-calculus is semi-quantitative by definition: all *individual* communications are equally likely to occur. Thus, the π-calculus abstraction of the molecular realm reflects the number of molecules, but not the rates of the reactions in which these molecules participate. This results in two inter-dependent inaccuracies. First, reactions do not occur at the correct rate. Second, all time steps are equal and do not represent the time evolution of the real system.

While previous studies with qualitative (e.g., [39]) and semi-quantitative (e.g., [23] and our work [51]) modelling of biomolecular systems show that even such coarse abstractions have some merit, quantitative aspects are key to the function of many biomolecular systems, a fact that is recently gaining growing recognition (e.g., [32]), and must be appropriately addressed when devising a relevant abstraction of biomolecular systems.

In the π-calculus abstraction, we have associated our representation so far with a particular non-deterministic and discrete semantics. While the discrete aspect of the calculus is fundamental, the non-deterministic semantics is not, and will now be replaced with a stochastic one, in which different communications have different rates, and communications are selected based on probabilistic conditions. The programming language mechanism is realized using a previously proved Monte Carlo algorithm for the stochastic simulation of systems of coupled chemical reactions, known as the Gillespie algorithm [18].

To provide the *pi*-calculus with a stochastic extension we introduce several changes into the π-calculus domain and the π-calculus based abstraction:

- **Channels with base rates as elementary reactions with mesoscopic rate constants.** Each channel is now associated with a *base rate*. The channel's base rate is identical to the mesoscopic rate constant of the corresponding elementary reaction.
- **Channels' actual rates as reactions' actual rates.** At each state in the π-calculus system we determine the actual rate of a channel based on that channel's base rate, and the number of input and output offers on the channel at that state (which represent the number of reactant molecules in the corresponding reaction). The actual channel rate is identical in calculation and value to the actual rate of the corresponding reaction.
- **π-calculus time steps as the time evolution of a chemical system.** The discrete even time steps implicit to the unfolding of a semi-quantitative simulation[6] are replaced by an explicit clock. The clock is advanced in uneven steps, according to the Gillespie algorithm, based on the actual channel rates. The resulting time evolution of the abstract π-calculus system corresponds to the time evolution of a single (representative) trajectory of the chemical system.

---

[6] The semantics of the π-calculus and similar languages is originally purely concurrent, where events can happen in parallel. In practice, however, concurrency is implemented by an *interleaving semantics*, where systems unfold in a step-by-step manner, with a single event chosen at each step. Here we refer to this standard, well-accepted, approach.

- **Stochastic selection of communication according to the probability of a reaction.** The non-deterministic way in which *enabled* communication actions are chosen at each step in the unfolding of a π-calculus system is replaced by a stochastic selection process, based on the actual channel rates at each state of the system. The selection process is identical to the one specified in the Gillespie algorithm. The resulting state evolution of the π-calculus system corresponds to the state evolution of a (statistically representative) trajectory of the chemical system.

The actual rates of chemical reactions are dependent on their rate constants (our "base rate") and reactant quantities. However, we distinguish several types of elementary reactions, each of which uses a different kind of rate calculation rule and different abstraction guidelines (summarized in Table 2). Briefly, each reaction type will be represented by a different channel type, and

Table 2. Bimolecular, unimolecular, and instantaneous reactions in the stochastic extension of the π-calculus.

Reaction type	Channel type	Explanation
Asymmetric bimolecular reaction	Regular	One reactant is represented by a process offering to send on the reaction channel and the other by a process offering to receive on the reaction channel. **Rate**: Base rate $\times$ #senders $\times$ #receivers
Symmetric bimolecular reaction	Symmetric	The reactants are represented by two instances of the same process, each offering a *choice* between send and receive on the reaction channel. **Rate**: Base rate $\times \frac{\text{\#senders} \times (\text{\#receivers}-1)}{2}$ (The number of senders and receivers is equal)
Unimolecular reaction	Regular (public)	The single reactant is represented by a process offering to receive on the public reaction channel. A single *Timer* process offers the complementary communication. **Rate**: Base rate $\times 1 \times$ #receivers
	Regular (private)	The internally interacting parts of a single reactant are represented as two parallel processes communicating on a shared private reaction channel. Each potential "copy" of the reaction is represented by a separate private channel. **Rate**: Base rate $\times 1 \times 1$ (An actual rate is calculated for each individual private reaction channel)
Instantaneous reaction	Instantaneous	Same as for bimolecular reaction **Rate**: Infinite: executed immediately when enabled, prior to any reaction on channels with finite rates, and without advancing the clock

the channel type will determine which rate calculation rule to apply. Asymmetric bimolecular reactions involving two reactants from two different species are represented by two different processes using a regular type channel. Symmetric bimolecular reactions involving two reactants from the same species are represented by two identical processes using a symmetric type channel (each offering a choice between a send and a receive action). Unimolecular reactions involve only a single reactant. As the π-calculus allows only pair-wise interaction, we represent these either as an asymmetric communication (on a regular channel) with an additional, single-copy, *Timer* process, or by a symmetric communication (on a *private* symmetric channel, one per molecule) between two sub-processes of the *Reactant* process. Finally, we also allow instantaneous reactions, occurring immediately when enabled, on an instantaneous type channel. These are very useful for various encoding needs, as well as when we would like to focus on slower events.

The different ways in which the actual rate of channels is calculated for different types of reactions, and the different handling of communications on instantaneous channels, mean that a given channel name may only be used in one way throughout the unfolding of a stochastic π-calculus system. Thus, once a channel name is used as one type, it may not be used as another.

6 BioSpi: Simulating π-calculus Programs

Once a detailed π-calculus model of a particular pathway is built, one would like to utilize it in different ways. The most natural use is to treat the model as an executable computer program, and simulate the behaviour of the pathway by running the program.

To this end, we have developed a computer application, called BioSpi. BioSpi is based on the Logix system [59], which implements Flat Concurrent Prolog (FCP [58]). Two unique features of FCP made it suitable for our purposes. First, the ability to pass logical variables in messages was used to implement the mobility ("sending channels as messages") mechanism of the π-calculus. Second, its support of guarded atomic unification allowed the implementation of synchronized interaction with both input and output guards. Note that previous implementations of the π-calculus or of related formalisms (e.g., [41] and references therein) do not provide such full guarded synchronous communication.

BioSpi receives as input π-calculus code and executes it. An appropriate surface syntax was devised for the π-calculus syntax, allowing also for a simple but useful arithmetic extension[7]. The ASCII-based[8] syntax was devised in such a way that the original π-calculus primitives and operations (including

---

[7] The arithmetic extension allows us to employ variables and arithmetically manipulate their values as well as set conditions on these values.

[8] In order to maintain pure ASCII representation and comply with some limitations of the Logix system, some of the original π-calculus notation was replaced in

match/mismatch constructs) are clearly separated from the arithmetic ones. Furthermore, the syntax is insulated from general Logix procedures, and a pure π-calculus representation is maintained in spite of the use of an FCP-based platform.

In the simulation, each instance of a π-calculus process is realized as a running computational one. Processes run concurrently and interact using channel objects, following the reaction rules of the calculus. The simulation follows, step by step, the evolution of the system. At each step, a single communication (reaction) step is realized in one atomic operation: a pair of complementary communication actions (input and output on the same channel in two concurrent processes) is chosen, a message is transmitted between the processes, the communication actions and alternative *choices* are eliminated, the received channel(s) are substituted (*instantiated*) for the appropriate place holder, and the processes are allowed to continue.

BioSpi has several versions, consistent with our biologically-motivated extensions of the *pi*-calculus. In BioSpi 1.0, communication selection is nondeterministic, and all enabled communication actions are equally likely to be selected. As a result, the simulation process is *semi-quantitative*: communication (i.e., reaction) rates are effectively identical, therefore process (i.e., molecule) numbers determine which reactions are more probable than others, and higher-copy-number processes are more likely to participate in communication than lower-copy ones.

Several tools are available for tracing the execution of BioSpi 1.0 programs. Thus, not only the net outcome of a computation can be studied, but also the specific scenario that has led to this outcome. First, the simulation may be executed in a step-by-step fashion, with the ability to set specific break points. This approach is useful to follow small detailed systems. Second, an ordered trace of all the communications in the system and the processes that participated in them is maintained. The trace can be viewed in form of a tree, ordered chronologically, or according to process evolution. Third, the simulation can be suspended at any desired moment, and the contents (so-called *resolvent*) of the system's current state of computation are examined. The level of detail in which a system is traced (process, channels, messages, senders, etc.) can be determined dynamically throughout a session. We applied the BioSpi 1.0 system to study a complex full model of the RTK–MAPK pathway, gleaning meaningful biological insights even in this limited, semi-quantitative framework (see [51] for details).

BioSpi 2.0 implements the Gillespie algorithm and serves as a stochastic simulation platform. In BioSpi 2.0, each channel object is also associated with a base rate and type (instantaneous, symmetric, or asymmetric). The channel base rate is supplied when the channel is defined. An "infinite" base rate serves

---

the BioSpi syntax. Only BioSpi notation is used throughout this work with the exception of the (new x) construct.

to define an instantaneous channel type. Asymmetric and symmetric channel types are identified upon their first use.

The central monitor in BioSpi 2.0 maintains the simulation clock and is responsible for the correct execution of the Gillespie algorithm: the selection of the next communication and time step. The monitor operates in the following way. Requests to instantaneous channels are satisfied as soon as possible. Requests to channels which have a finite rate (> 0) are queued. Each time that a new event is required the central monitor and all channel objects with a finite, non-zero rate jointly determine a communication event. In each iteration, each channel object determines its actual rate, according to its type, its basal rate, and the numbers of send and receive offers. The monitor then uses these actual rates to select the next reaction channel and time step, exactly according to the selection procedure of the Gillespie algorithm [18]. The monitor then advances a "clock" counter according to the selected time step and allows the chosen channel to complete one transmission (send/receive pair), relaying the sent message to the receiver.[9] As in BioSpi 1.0, the completion of the send and receive requests is synchronized by the channel. In addition, other messages offered on this and other channels by the same two processes whose requests were completed are withdrawn. The withdrawals are not synchronized, but they do precede continuation of their respective processes, and the next step of the clock.

In addition to the tracing and debugging tools available for BioSpi 1.0 simulations, BioSpi 2.0 allows us to maintain a full *record* of the time evolution of the system. The record specifies each communication in the system, the time at which it occurred, the communicating processes, the channel on which the communication occurred, and the resulting processes. The record file can be post-processed to generate the time evolution of the system's state (i.e., number of each process species at each time point) at any desired level of resolution. In the next section, we will see the utility of the record and trace tools in studying the stochastic behaviour of specific systems.

Importantly, the changes in the stochastic extension of the π-calculus abstraction are mostly confined to the *semantic* interpretation of π-calculus programs, and to the *pragmatic* use of the abstraction (e.g., the representation of unimolecular reactions). As a result, the new guidelines for stochastic abstraction can be seamlessly composed on top of the existing semi-quantitative ones. Practically, this means that with the addition of appropriate channel rates, any semi-quantitative π-calculus abstraction (BioSpi 1.0-compatible) can become stochastic (and BioSpi 2.0-compatible).

---

[9] To be totally accurate, the selection of the send/receive pair should be done in a fully random way. In the current implementation, the pair is selected in a "top-of-the-queue" fashion, which is non-random. In most cases, there are only two process species communicating on a given channel, and this does not raise a problem, as all senders (or receivers) are equivalent. In some cases, however, this may be inaccurate. We are currently developing a fully random selection of a communication pair to be available in the next release.

7 Studying Biomolecular Systems with BioSpi 2.0: a Compositional Abstraction of Glycogen Biosynthesis

The incorporation of a stochastic semantics allows us to build more accurate abstractions of biomolecular systems. In this section we study glycogen biosynthesis as one example of these benefits. In this example, we use the distinction between channels with finite rates and instantaneous channels to build a detailed model of glycogen biosynthesis. We also extensively use an arithmetic extension of the calculus and its ability to abstract the *composition* of entities of growing complexity from simple building blocks.

7.1 Glycogen Biosynthesis

Polymerization is a key event in the creation of various biomolecules including DNA, RNA, proteins, various protein filaments (e.g., actin, microtubuli) and polysaccharides. In contrast to proteins and nucleic acids, polysaccharides can form both branched and linear polymers.

Glycogen is a branched polysaccharide composed of glucose monomers [67]. Glycogen is biosynthesized in a combination of two alternative enzymatic reactions. In the first *elongation* reaction, catalyzed by the glycogen synthase enzyme, an $\alpha(1 \rightarrow 4)$ glycosidic bond is formed between the C_4 of the polymerized glucose at the end of a growing polymer and the C_1 of an "incoming" UDP-glucose monomer.[10] In the second *branching* reaction, catalyzed by the glycogen branching enzyme, an $\alpha(1 \rightarrow 6)$ glycosidic bond is formed between the C_6 of a polymerized glucose within the glycogen polymer and the C_1 of an "incoming" glycogen oligomer. The oligomer is branched "off" an existing polymer chain. In addition, the glycogen polymer is *initiated* by a third enzyme, called glycogenin. This enzyme forms an initial 7-residue primer, which we term a *seed*.

The biosynthesis of glycogen follows specific "rules". First, the specificity of glycogen synthase ensures that elongation is allowed only from growing $4'$ ends. After branching, there may be more than one such end, and elongation may proceed independently at each of these ends. The branching "rules" are more complex. A branch is specifically created by transferring a 7-residue segment from the end of one chain to the C_6-OH group of a polymerized glucose residue on the same or another glycogen chain. Each transferred segment must come from a chain of at least 11 residues, and the new branch point must be at least 4 residues away from any other branch point.

The balance between elongation and branching determines the architecture of the glycogen molecule: chain length and extent of branching. This architecture determines the compactness of glucose storage and its availability and is thus critical for the functional role of glycogen as an energy reserve.

---

[10] The glucose $\rightarrow$ UDP-glucose reaction is catalyzed by a separate enzyme which we will not discuss here. We will use the terms glucose and UDP-glucose interchangeably from here on.

7.2 Building the Abstraction in the π-calculus

The first step in building an abstraction is identifying and defining the basic entities in the real-world domain. To this end, we now break down the above description into several components.

First, each glycogen strand is directional (Fig. 15). We term one end, with a C_1 carbon that cannot grow, as the *root side*. We term the other end, with a C_4 carbon that can grow, as the *leaf side*. Second, we distinguish three

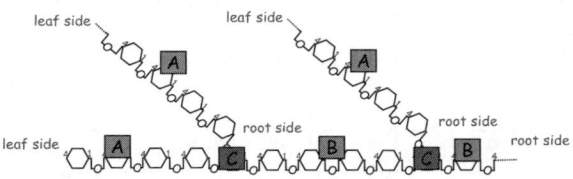

Fig. 15. The architecture of a glycogen molecule: definitions. The C_1 and C_4 ends of a glycogen strand defined as "root" and "leaf" respectively. A polymerized residue may assume one of three potential positions: on a straight segment (A), on a branched segment (B), or at a branch point (C).

potential positions for a polymerized residue (Fig. 15). First, it may reside on a *straight segment*, where no branch points exist to the leaf side. Second, it may be part of a *branched segment* that has a branch point to the leaf side. Third, the residue may be the *branch point* itself.

The exact position of the residue determines the types of reactions in which it may participate. A *polymerization enabled* residue may participate in an elongation reaction (as the C_4 donor). A *cleavage enabled* residue may be cleaved in a branching reaction and linked into another glycogen branch. A *branch enabled* residue may serve as a branch point onto which a new glycogen oligomer is added. The various types of residues and their reaction capabilities are listed in Table 3.

We are now ready to abstract the system in the stochastic π-calculus. First, each glucose residue is abstracted as a process. Different process states represent the different positions of a glucose residue (Table 3, rightmost column). Processes also represent the two types of enzyme molecules (*Glycogen_Synthase* and *Branching_Enzyme*) and the seed (*Seed_Glucose* for the "free" seed and *Root_Glucose* for the polymerized seed). The formation of bonds is abstracted as private channels, with one private channel shared between each pair of immediate neighbours. Thus, each process representing a fully polymerized residue has two private channels: one linking it with its root-side neighbour and the other with its leaf-side neighbour. The processes representing the root and leaf processes share only one private channel with a neighbour (Fig. 16).

Table 3. Different types of residues defined by their reaction capabilities and abstracted as π-calculus processes.

General position	Residue capabilities	Detailed position	Process name
Non-polymerized	Polymerization enabled	Free UDP-glucose ($1'$ end)	$UDP\_Glucose$
Straight segment	Polymerization enabled	Leaf ($4'$ end)	$Leaf\_Glucose$
	Cleavage and branch enabled	At distance 7 exactly from leaf and at distance 4 at least from closest branch point or root	$BCE\_Glucose$
	Branch but not cleavage enabled	At distance other than 7 from leaf and at distance 4 at least from closest branch point or root	$BNCE\_Glucose$
	Disabled	Not leaf and distance less than 4 from closest branch point or root	$Disabled\_Glucose$
Branched segment	Branch but not cleavage enabled	At distance at least 4 from both flanking branch points/root	$BNCE\_Glucose$
	Disabled	At distance less than 4 from at least one of its flanking branch points/root	$Disabled\_Branched\_Glucose$
Branch point	Disabled	At the branch point (C_6 donor)	$Branch\_Point$

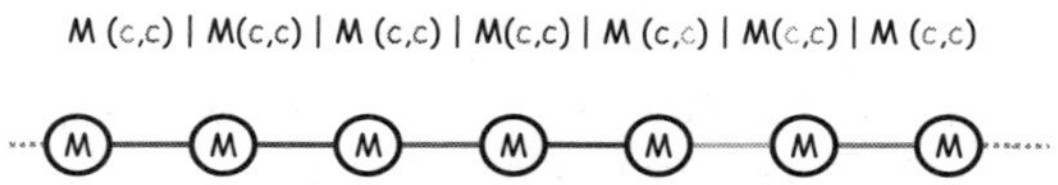

$$\mathsf{M}\,(c,c)\mid \mathsf{M}(c,c)\mid \mathsf{M}\,(c,c)\mid \mathsf{M}(c,c)\mid \mathsf{M}\,(c,c)\mid \mathsf{M}(c,c)\mid \mathsf{M}\,(c,c)$$

Fig. 16. A polymer abstracted as a chain of processes linked by private channels. Circles represent processes. Connecting lines represent individual private channels.

The exact position of a residue is critical to determine its behaviour. Similarly, the exact "position" of the corresponding process is critical to determine its state. We abstract a residue's position in the glycogen polymer by the *values* of three *position variables* that are associated with each individual residue process (Fig. 17):

- **Leaf counter (LC)** measures the distance of the residue from the leaf. It is initialized to zero, incremented ($+1$) upon extension, and decreased (-7) upon cleavage in the remaining segment residues.
- **Root counter (RC)** measures the distance of the residue from the root or from the flanking branch point on the root side. It is initialized to the value of the RC variable of the neighbour residue on the root side $+ 1$, updated upon cleavage in the cleaved segment residues to the RC of the

new root side neighbour + 1, and updated upon insertion in the segment on the leaf side of the new branch point to RC − (RC of new branch point).

- **Leaf-branch counter (LBC)** measures the distance of the residue from the flanking branch point on the leaf side. It is initialized to zero and updated upon insertion in the segment on the root side of the new branch point to the LBC of the leaf side neighbour + 1.

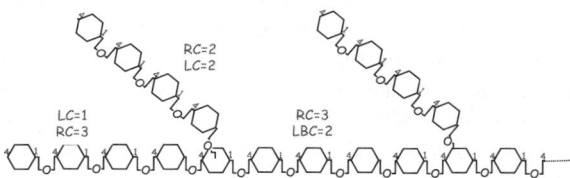

Fig. 17. The leaf (LC), root (RC), and leaf-branch (LBC) counters: an example. The values of the LC, RC, and LBC counters for several residues are shown.

As counters abstract the position of each residue in the polymer, and as the residue's position determines its state, we use the counters to define the corresponding process state. We use a *choice* construct, in which the counter values are checked according to the different rules:

$\{LBC = 0\}$,   %Straight segment

$(\{LC = 0\}$, $Leaf\_Glucose$;

$\{LC = 7 , RC >= 4\}$, $BCE\_Glucose$;

$\{LC > 0 , LC = / = 7 , RC >= 4\}$, $BNCE\_Glucose$;

$\{LC > 0 , RC < 4\}$, $Disabled\_Glucose$) ;

$\{LBC > 0\}$,   %Branched segment

$(\{RC >= 4 , LBC >= 4\}, BNCE_{G}lucose$;

$\{RC < 4\}$, $Disabled\_Branched\_Glucose$;

$\{LBC < 4\}$, $Disabled\_Branched\_Glucose$) .

The choice construct is examined each time that the counter value changes.

The position of a residue may change *directly*, when the residue participates in an interaction (polymerization or cleave/branch) or *indirectly*, when the residue is part of a segment which participates in those reactions (e.g., upon branching). In order for the abstract counters to correctly represent the position of the corresponding residues in the "real-world" polymer, they must be constantly updated. Since only two processes participate in the immediate interaction, we incorporate an *update propagate mechanism* into our abstract model.

Counter updating will be performed by communication on the private channels linking the process residues. Importantly, all such private channels will be instantaneous, while all the communications representing *actual* reactions (elongation, cleavage, and branching) will be carried out on channels with a real finite rate. As a result, counter updating will be done in zero simulation time, and will not interfere with the kinetics of biochemical reactions, while allowing us to maintain a detailed view of the glucose monomer's position within the glycogen polymer.

The full abstraction of the glycogen biosynthetic system in the π-calculus is given in Fig. 18. The enzymatic elongation reaction is represented by the *glycogen* and *udp_glucose* channels, and the cleave and branch reaction by the *branch* and *cleave* channels. Each $UDP\_Glucose$ process is defined with two private channels (*to_root* and *to_leaf*) that will be subsequently used as specific links to residues in the resulting polymer. For simplicity, all the "reaction" channels are given an equal base rate of 1 (we will revisit this decision below), while all private channels are instantaneous with an infinite base rate.

Abstraction of Initiation

In the initiation step (after a seed oligomer is already produced by glycogenin), glycogen synthase catalyzes the formation of a glycosidic bond between

```
global(glycogen(1), udp_glucose(1), dummy(1), branch(1), cleave(1)).
baserate(infinite).

Seed_Glucose(RC,LC,LBC)::= glycogen ? {to_leaf},to_leaf ! {RC,LBC},
                                Root_Glucose(to_leaf,RC,LC,LBC).
Root_Glucose(to_leaf,RC,LC,LBC)::=
    to_leaf ? {LC,LBC} , {LC++} , Root_Glucose(to_leaf,RC,LC,LBC) .
UDP_Glucose(LC,LBC)+(to_root,to_leaf)::=
    udp_glucose ! {to_root} , to_root ? {RC,LBC} , {RC++} ,
        to_root ! {LC,LBC} , Glucose(to_root,to_leaf,RC,LC,LBC) .
Glucose(to_root, to_leaf, RC, LC, LBC)::=
{LC>=0},
    << {LBC = 0} , << {LC = 0} ,  Leaf_Glucose ;
                    {LC = 7 , RC >= 4} , BCE_Glucose ;
                    {LC > 0 , LC =\= 7 , RC >= 4} , BNCE_Glucose ;
                    {LC > 0 , RC < 4} , Disabled_Glucose >> ;
        {LBC > 0} , << {RC >= 4 , LBC >=4} , BNCE_Glucose ;
                    {RC < 4} , Disabled_Branched_Glucose ;
                    {LBC < 4} , Disabled_Branched_Glucose >> .
```

Fig. 18. π calculus code for the glycogen system. The entire system is shown with the exception of initialization of position counters. The code is continued on the next page.

```
Leaf_Glucose::=
 glycogen ? {to_leaf} , to_leaf ! {RC,LBC} , to_leaf ? {LC,LBC} ,
   {LC++} , to_root ! {LC,LBC} , Glucose(to_root,to_leaf,RC,LC,LBC);
 to_root ? {RC,_} , <<  {RC >=0} , {RC++} , Glucose ;
                               {RC < 0} , Disabled_Leaf_Glucose >> .
Disabled_Leaf_Glucose::=
 to_root ? {RC,_} , {RC++} , Glucose .
BNCE_Glucose::=
 to_leaf ? {LC,LBC} , {LC++} , << {LBC = 0} , to_root ! {LC,LBC} ,
   Glucose ; {LBC > 0} , {LBC++} , to_root ! {LC,LBC} , Glucose >> ;
 to_root ? {RC,_} , << {RC >=0} , {RC++} , to_leaf ! {RC,LBC} ,
   Glucose ; {RC < 0} , to_leaf ! {RC, LBC} , Disabled_Glucose >> ;
 branch ? {to_branch} , Branch_Synch1(to_branch,RC,LC,LBC) .
Branch_Synch1(to_branch,RC,LC,LBC)+(RC1,LBC1)::=
 {RC1=0} | {LBC1=1} |
   << to_branch ! {RC1,LBC} ,
       << to_leaf ! {RC1,LBC} , to_root ! {LC,LBC1} ,
         Branch_Point(to_root,to_branch,to_leaf) >> >>  .
Disabled_Glucose::=
 to_leaf ? {LC,LBC} , {LC++} ,
   << {LBC = 0} , to_root ! {LC,LBC} , Glucose ;
       {LBC > 0} , {LBC++} , to_root ! {LC,LBC} , Glucose >> ;
 to_root ? {RC,_} , {RC++} , to_leaf ! {RC,LBC} , Glucose .
BCE_Glucose+(new_to_root,RC1,LC1,LBC1)::=
 << to_leaf ? {LC,LBC} , {LC++} ,
   << {LBC = 0} , to_root ! {LC,LBC} , Glucose ;
       {LBC > 0} , {LBC++} , to_root ! {LC,LBC} , Glucose >> ;
  to_root ? {RC,_} , << {RC >=0} , {RC++} , to_leaf ! {RC,LBC} ,
   Glucose ; {RC < 0} , to_leaf ! {RC,LBC} , Disabled_Glucose >> ;
 branch ? {to_branch} , Branch_Synch(to_branch,RC,LC,LBC) ;
 cleave ! {new_to_root} , {LC1 = -1} | {RC1 = -1} |
   Cleave_Synch(to_leaf) .
Cleave_Synch(to_leaf)::=
 to_root ! {LC1,LBC} , to_leaf ! {RC1, LBC} , new_to_root ? {RC,_} ,
   {RC++} , to_leaf ! {RC,LBC} ,
     Glucose(new_to_root,to_leaf, RC, LC, LBC) >> .
Branch_Synch(to_branch,RC,LC,LBC)+(RC1,LBC1)::=
 {RC1=0} | {LBC1=1} |
   << to_branch ! {RC1,LBC} ,
       << to_leaf ! {RC1,LBC} ,
 to_root ! {LC,LBC1} , Branch_Point(to_root,to_branch,to_leaf) >> >> .
Disabled_Branched_Glucose::=
 to_leaf ? {LC,LBC} , {LC++} ,
   << {LBC = 0} , to_root ! {LC,LBC} , Glucose ;
       {LBC > 0} , {LBC++} , to_root ! {LC,LBC} , Glucose >> ;
 to_root ? {RC,_} , {RC++} , to_leaf ! {RC,LBC} , Glucose >> .
Branch_Point(to_root,to_branch,to_leaf)::=
  to_root ? {_,_} , self ;
  to_branch ? {_,_} , self ;
  to_leaf ? {_,_} , self .
```

```
Glycogen_Synthase::=
    udp_glucose ? {to_root} , glycogen ! {to_root} , Glycogen_Synthase   .
Branching_Enzyme::=
    cleave ? {to_branch} , branch ! {to_branch} , Branching_Enzyme  .
```

a UDP-glucose and the 4′ leaf end of the seed oligomer. In the abstract π-calculus model we represent the initiation as two consecutive communication steps. In the first step, the *to_root* private channel of *UDP_Glucose* is sent to *Glycogen_Synthase* on the *udp_glucose* channel. In the second step, this private channel name is further relayed in a communication from *Glycogen_Synthase* to *Seed_Glucose* on *glycogen* (Fig. 19A and B). Following

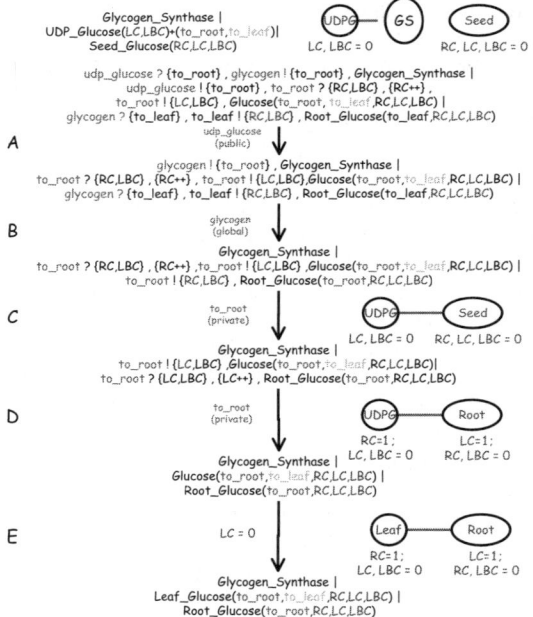

Fig. 19. Initiation as a multi-step communication between *UDP_Glucose*, *Glycogen_Synthase*, **and** *Seed_Glucose*. A. Communication between *UDP_Glucose* and *Glycogen_Synthase*. B. Communication between *Glycogen_Synthase* and *Seed_Glucose*. C,D. Counter update using a shared private channel. E. *UDP_Glucose* changed to *Leaf_Glucose* based on counter values.

the two communications, *UDP_Glucose* and *Seed_Glucose* now share a private (instantaneous) channel. In the next two steps, the position counters of each of the processes are updated to reflect their relative position using this shared channel (Fig. 19C and D). First, the *RC* counter is sent on the private channel from *Seed_Glucose* to *UDP_Glucose*, and is used to update *UDP_Glucose*'s *RC* value. This represents the change in the position of this now polymerized residue relative to its newly acquired root. Then, the *LC* and *LBC* counters are sent on the private channel from *UDP_Glucose* to the now *Root_Glucose*, and are used to update the respective counters in *Root_Glucose*. This represents the concomitant change in the position of the root oligomer relative to its newly acquired leaf. Finally, the new counter values are evaluated in *UDP_Glucose*, changing it to its new *Leaf_Glucose* state (Fig. 19E).

Abstraction of Elongation

Elongation is similar to initiation and involves the same processes and communications, with the exception of *Seed_Glucose* which is "replaced" by *Leaf_Glucose*. Two consecutive communications (first on *udp_glucose* and then on *glycogen*) serve to relay a private channel from *UDP_Glucose* to *Leaf_Glucose* via *Glycogen_Synthase*. This private channel is then used to initiate an update of the positional counters of *all* the processes along the growing chain (each time using the relevant private channel). Finally, the updated counters determine the new state of each of the linked processes.

Abstraction of Branching

Consider a relatively simple branching scenario in which a 12-residue chain is cleaved at the fifth residue. The 4-residue "rooted" part is then elongated by two more residues, followed by linking of the 8-residue "clipped" part on its fifth residue (Fig. 20).

A process chain of size 12 is shown in Fig. 20, where process state is determined according to its relative position. Each pair of processes is linked by a unique private channel, established during the elongation step, as described above. Cleavage is abstracted as a communication on *cleave* between a *Branching_Enzyme* and a *BCE_Glucose*. Due to the short chain length in this example, there is only a single *BCE_Glucose*. In the communication, a private channel (*new_to_root*) is sent from *BCE_Glucose* to *Branching_Enzyme*. Following the communication, the *LC* and *RC* counters of *BCE_Glucose* are set to special values (−1), followed by a series of communications on the chains of private channels to update the relevant positional information. First, the "cleave event" is propagated to the root side of the cleave point. As a result, the immediate root-side neighbour of the cleaved residue process will become a *Leaf_Glucose*. The other root-side processes will be updated accordingly with the communication propagated up to the *Root_Glucose*.

In parallel, the cleavage event is propagated to the leaf side of the cleavage point. Since cleavage and branching may be separated by additional steps (e.g., in the example scenario we have additional elongation of the "rooted

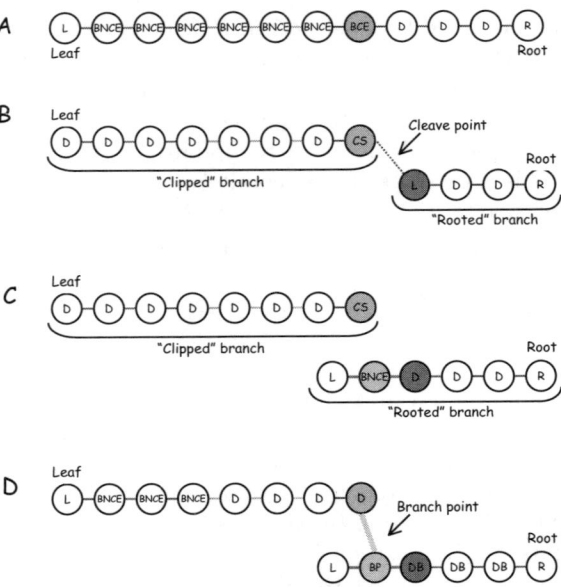

Fig. 20. A simple branch and cleave scenario. A. A chain of length 12. B. Following cleavage: a rooted chain (length 4) and a clipped chain (length 8). C. The rooted chain is elongated by two more residues. D. Linking the clipped chain to residue 5 of the rooted chain. Corresponding process names are shown (R: *Root_Glucose*, L: *Leaf_Glucose*, D: *Disabled_Glucose*, BP: *Branch_Point*, DB: *Disabled_Branched_Glucose*, CS: *Cleave_Synch*).

branch"), the "clipped" branch is "frozen" until it is linked to a new position. The "freeze" is based on propagating the special *RC* counter value (-1) throughout the clipped chain. The special value serves to set each process to a *Disabled_Glucose*, and will change only upon updating the positional counters (following branch linking, see below). On the other hand, the "rooted" chain may participate like any other chain in elongation (e.g., the two elongation steps in our example scenario).

We now examine how to abstract the linking of a cleaved ("frozen") branch to form a branch point. Assuming two additional elongation steps, we start with a system with two process chains, one representing a "rooted" branch of size 6 and the other representing a cleaved branch of size 8. Branching is represented by a communication between *Branching_Enzyme* and *BNCE_Glucose* on *branch*, in which the private *new_to_root* channel, originating in the "cleaved" *BCE_Glucose*, is relayed to *BNCE_Glucose*. This results in a private link between the two *Glucose* processes. A series of communications on private channels propagate the change as an update in positional counters from the cleave point (to the leaf direction, resulting in "unfreezing" of the cleaved branch) and from the branch point (to both the root and leaf directions).

7.3 Studying Glycogen Metabolism with BioSpi 2.0

The detailed glycogen biosynthesis model, based on a "monomer-as-process" abstraction, allows us to monitor the exact architecture of glycogen molecules through the architecture of the process "polymer". The tracing tools of BioSpi 2.0 provide us with the information (process names, counter values, and private channel identity) necessary to reconstruct this architecture.

A simple example, based on a system initialized with a single *Seed_Glucose*, 15 *UDP_Glucose* processes, a single *Glycogen_Synthase* process, and a single *Branching_Enzyme*, is shown in Fig. 21. The processes available in the system after it is run until suspension (no additional enabled communications) are shown in the so-called resolvent in Fig. 21A, together with the channels they use and the values for the RC, LC, and LBC counters. This information is sufficient to specify the molecular architecture of the corresponding glycogen molecule (Fig. 21B).

BioSpi 2.0 also allows us to follow the time evolution of various quantitative properties in the abstracted population of glycogen molecules, such as the number of branch points, leaves, and seeds, giving us an overall measure of molecular architecture. For example, Fig. 22 shows the number of various processes representing different properties of the polymers (leaves, polymerized residues, etc.) under different relative polymerization and branching rates.

8 Perspectives: Molecules as Computations

The behaviour of biomolecular systems is traditionally studied with *dynamical systems theory* (described via differential equations) [17], which abstracts the cell and its molecular constituents to their quantifiable properties (e.g., concentration, position) and their couplings. While this approach is powerful, clear, and well understood, the abstraction it makes is problematic when describing the behaviour of complex biomolecular systems [17]. It treats the cell as a monolithic unstructured entity, and does not identify molecular *objects* as they are modified. Biological systems are composed of dynamic molecular *objects*, but dynamical systems theory is not compositional. Thus, it only indirectly captures the identity and fate of a single molecule, molecular complex, or machine with several states (and/or sub-components).

8.1 The Molecule-as-Computation Abstraction

Computer science offers an alternative starting point in the search for good abstractions of biological systems composed of dynamically changing objects.

. . .

A
```
. Root_Glucose. comm( . UDP_Glucose. to_root! )
Disabled_Branched_Glucose. comm( . UDP_Glucose. to_root! , . UDP_Glucose. to_root! , 1, 4,
4, global . branch(1)! , global . cleave(1)! , global . glycogen(1)! )
Disabled_Branched_Glucose. comm( . UDP_Glucose. to_root! , . UDP_Glucose. to_root! , 2, 3,
3, global . branch(1)! , global . cleave(1)! , global . glycogen(1)! )
Disabled_Branched_Glucose. comm( . UDP_Glucose. to_root! , . UDP_Glucose. to_root! , 3, 2,
2, global . branch(1)! , global . cleave(1)! , global . glycogen(1)! )
. Branch_Point. comm( . UDP_Glucose. to_root! , BCE_Glucose. new_to_root! ,
. UDP_Glucose. to_root! )
Disabled_Glucose. comm( BCE_Glucose. new_to_root! , . UDP_Glucose. to_root! , 1, 8, 0,
global . branch(1)! , global . cleave(1)! , global . glycogen(1)! )
Disabled_Glucose. comm( . UDP_Glucose. to_root! , . UDP_Glucose. to_root! , 2, 7, 0,
global . branch(1)! , global . cleave(1)! , global . glycogen(1)! )
Disabled_Glucose. comm( . UDP_Glucose. to_root! , . UDP_Glucose. to_root! , 3, 6, 0,
global . branch(1)! , global . cleave(1)! , global . glycogen(1)! )
BNCE_Glucose. comm( . UDP_Glucose. to_root! , . UDP_Glucose. to_root! , 4, 5, 0,
global . branch(1)! , global . cleave(1)! , global . glycogen(1)! )
BNCE_Glucose. comm( . UDP_Glucose. to_root! , . UDP_Glucose. to_root! , 5, 4, 0,
global . branch(1)! , global . cleave(1)! , global . glycogen(1)! )
BNCE_Glucose. comm( . UDP_Glucose. to_root! , . UDP_Glucose. to_root! , 6, 3, 0,
global . branch(1)! , global . cleave(1)! , global . glycogen(1)! )
BNCE_Glucose. comm( . UDP_Glucose. to_root! , . UDP_Glucose. to_root! , 7, 2, 0,
global . branch(1)! , global . cleave(1)! , global . glycogen(1)! )
BNCE_Glucose. comm( . UDP_Glucose. to_root! , . UDP_Glucose. to_root! , 8, 1, 0,
global . branch(1)! , global . cleave(1)! , global . glycogen(1)! )
Disabled_Glucose. comm( . UDP_Glucose. to_root! , . UDP_Glucose. to_root! , 1, 1,
0, global . branch(1)! , global . cleave(1)! , global . glycogen(1)! )
Leaf_Glucose. comm( . UDP_Glucose. to_root! , . UDP_Glucose. to_leaf , 2, 0, 0,
global . branch(1)! , global . cleave(1)! , global . glycogen(1)! )
Leaf_Glucose. comm( . UDP_Glucose. to_root! , . UDP_Glucose. to_leaf , 9, 0, 0,
global . branch(1)! , global . cleave(1)! , global . glycogen(1)! )
. Glycogen_Synthase. comm( global . glycogen(1)! , global . udp_glucose(1)! )
. Branching_Enzyme. comm( global . branch(1)! , global . cleave(1)! )
```

B

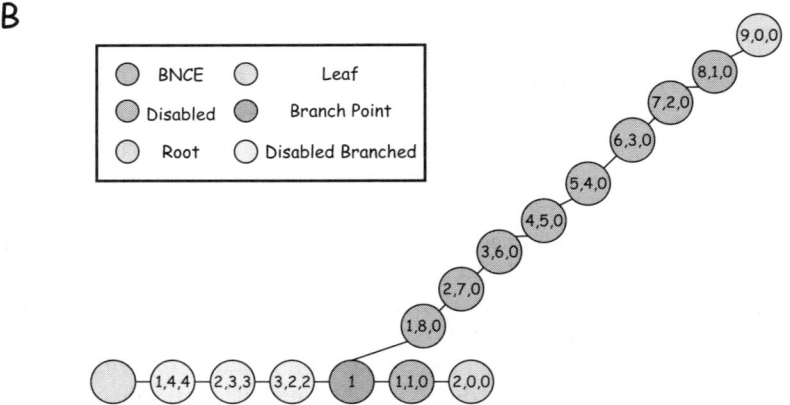

Fig. 21. A BioSpi 2.0 resolvent allows reconstruction of the molecular architecture of a glycogen molecule. A. A BioSpi resolvent following a complete run of a system initialized with a single *Seed_Glucose*, 15 *UDP_Glucose*, a single *Glycogen_Synthase*, and a single *Branching_Enzyme*. B. The molecular architecture reconstructed from the resolvent. Counter values for *RC*, *LC*, and *LBC* shown inside circles.

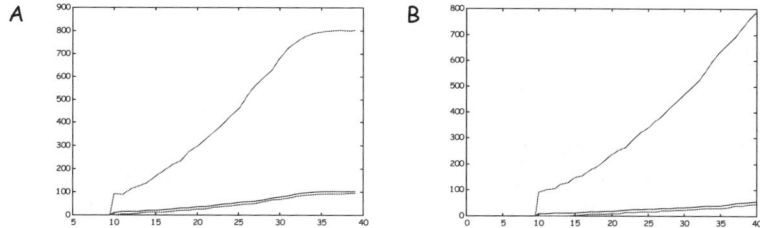

Fig. 22. BioSpi 2.0 simulation of glycogen synthesis under different conditions. The number of *Branch_point* (bottom curve), *Leaf_Glucoose* (middle curve), and "polymerized" (sum of *BCE_Glucose*, *BNCE_Glucose*, *Disabled_Glucose*, and *Disabled_Branched_Glucose*) processes (top curve) is shown as a function of time either when the base rates of the polymerization and branching communications are equal (A) or when the polymerization rate is 10 times faster (B).

The view of computational systems as composed of interacting processes renders them a tempting source of novel abstractions for the behaviour of systems of interacting biological entities, such as molecules and cells. In this work, we began the search for this abstraction with the π-calculus, a process algebra for the representation and study of *concurrent mobile systems*.

As a first step we identified the basic entities of biomolecular systems and the events in which they participate. We then developed general guidelines for the abstraction of these entities to the mathematical domain of the π-calculus, which we then employed for the modelling and study of real complex biomolecular systems, such as the receptor tyrosine kinase RTK–MAPK signalling pathway [51]. While the function of certain biomolecular systems may be abstracted in a qualitative or semi-quantitative way, the functionality of others critically depends on quantitative aspects. We therefore further extended the original abstraction with a stochastic component [47], embedding the well-accepted Gillespie algorithm [18] within the π-calculus framework. Although the original stochastic abstraction is limited to elementary reactions, BioSpi 2.0 allows us, in principle, to define alternative rate calculation rules per (asymmetric) channel and extend the channel type scheme accordingly. In this case, however, the correctness of the simulation of the resulting system is no longer ensured.

8.2 Benefits and Limitations

How *relevant*, *computable*, *understandable*, and *extensible* is the "molecule-as-computation" abstraction we devised?

Relevance

A *relevant* abstraction of biomolecular systems should capture two essential properties of these systems: their molecular organization and their dynamic behaviour. We believe that the π-calculus abstraction is indeed highly relevant. On the one hand, essential biomolecular entities are directly abstracted to computational ones and the abstraction handles a variety of biomolecular events. On the other hand, the dynamic behaviour of the derived computational system closely mimics that of the molecular one.

The two alternative (non-mobile and mobile) approaches we present for abstracting molecular interaction offer a tradeoff between relevance and utility. The non-mobile approach is often simpler, but suffers from a *relevance* problem, similar to that of dynamical systems theory, as each state and modification are represented as an individual entity. The mobile approach combines mobility with channel parameters to represent chemical modification as a change in the process's channel set. Thus, processes are under-defined in such a way that allows them to assume different states based on the channels they received. While we believe this approach is more *relevant*, we also appreciate the greater difficultly to specify and understand mobile systems. In practice, when building individual abstractions we use a mixture of the two approaches, guided by pragmatic considerations.

A key benefit of our proposed abstraction is in its ability to handle complex entities, composed of lower-level entities in a dynamic fashion. This is demonstrated by our glycogen biosynthesis example, representing the challenge posed by *open ended* biological entities such as polymers whose boundaries and structure change dynamically and are difficult to pre-define, but are composed of (relatively) simple monomers, whose capabilities are well-known. The *compositionality* of the "molecule-as-computation" abstraction, i.e., the ability to compose complex entities from constituent parts according to pre-defined rules, offers a unique way to handle such entities, as shown with the glycogen biosynthesis model. Note that the compositional approach has drawbacks as well. First, it requires us to maintain each monomer as an independent entity (process) in the abstract model (and simulation). Second, in a compositional representation all properties are "distributed" among the individual components. In the glycogen example this required constant status updates of a series of counters throughout the affected area. This can both pose a computational burden and affect the readability of the abstraction. One solution to this problem is to build a hybrid abstraction in which sections of the polymer are abstracted as a single process.

Computability

A *computable* abstraction should allow both the simulation of dynamic behaviour and qualitative and quantitative reasoning on systems' properties. We have seen that with the aid of BioSpi we can simulate the behaviour of biomolecular systems by executing their computational π-calculus abstraction. However, simulations of molecular systems as computational ones can also suffer from performance and scaling problems. The main difficulty lies in the abstraction of each molecule as a separate *instance* of a computational process. One solution is to handle only *molecular species* explicitly, while the specific molecules are represented by counter variables. While this approach significantly improves performance and scalability, it loses some of the immediate transparency of the "molecule-as-process" abstraction.

A π-calculus abstraction also supports qualitative and quantitative reasoning on the modelled system's properties. An extensive behavioural theory has been developed for the mathematical domain of the π-calculus (and similar languages), prior to and regardless of its proposed use for biological systems. This provides methods and tools to *formally verify* certain properties of systems described in the π-calculus (e.g., [44], [66]). In principle, we can verify certain qualitative assertions on biomolecular systems (e.g., "will a signal reach a particular molecule?") by verifying a corresponding assertion on the π-calculus abstraction of the system (e.g., "will a particular communication be realized in a particular process?"). Furthermore, methods exist to *compare* two π-calculus programs to determine the degree of mutual similarity of their behaviour, termed *bi-simulation* ([41], [44], [66]). Different levels of similarity, of weakening strength, have been defined [41].

While such methods have been generally developed they have not been applied in the past to computational systems at the scale and complexity of the ones we are building as abstractions of biomolecular systems. Furthermore, the types of queries and analysis methods developed for computational purposes may not be the same ones necessary to elucidate biological questions. In particular, stochastic and quantitative issues (such as the ones explored in the next chapter) were hardly handled in the purely computational setting.

Thus, while we deem these *computable* aspects of the "molecule-as-computation" abstraction as critical components of its success, they are beyond the scope of this work, and will require further extensive research. Nevertheless, the potential computational possibilities opened up by the use of the π-calculus serve as additional support for its importance. Indeed, in recent work several researchers have started to adapt analysis methods from process algebra to the study of biomolecular systems (e.g., [12], [13], [9], [45], [4]).

Understandability

An *understandable* abstraction offers a conceptual framework for thinking about the scientific domain: it corresponds well to the informal concepts and ideas of science, while opening up new computational possibilities for understanding it. The properties which lend relevance and computability to the "molecule-as-computation" abstraction also render it *understandable*. On the one hand, the abstraction attempts to translate our informal knowledge of the biological realm to formal concepts in a transparent and visible way.

On the other hand, the *computational* theory for the study and analysis of concurrent computational systems described above opens up new possibilities for understanding molecular systems. For example, once biological behaviour is abstracted as computational behaviour, we can distinguish between implementations (related to a real biological system, e.g., the MAP kinase cascade) and its corresponding specification (related to its biological function, e.g., an amplifier or a switch). Then, the theory and methodology of semantic equivalence can be applied in this new context to formally ascribe a biological function to a biomolecular system. Similarly, two molecular-level abstractions of similar systems in different cells or organisms can be compared for their behavioural similarity, to determine their level of *homology* (if evolutionary related) or *analogy* (if evolutionary distinct).

The abstraction we presented in this work, however, is still *obscure* in certain respects. The use of private channels as an abstraction of biomolecular compartments raises several problems including cumbersome, multi-step encoding of the formation of multi-molecular complexes, and a difficulty to express simultaneous conditions on proximity and biochemical complementarity. Another limitation lies in the need for each communication to involve exactly two processes. While intra-molecular events can still be modelled by using subprocesses, interactions between more than two molecules cannot be abstracted by a single event in the computational realm. Although this limitation is biochemically correct (elementary molecular interactions are extremely rare), it may raise problems in abstracting systems for which knowledge is limited and cannot be broken down to elementary processes. Another current limitation to the transparency and utility of the abstraction is its purely textual nature and the lack of a graphical component for the specification and visualization of the abstracted system and its unfolding.

Extensibility

Such limitations in the *understandability* of the abstraction can often be overcome by appropriate extensions to the abstraction or to the mathematical domain. Since we do not expect a single, concise, computational language to be an immediate all-inclusive solution, the desired abstraction (or the mathematical domain) should be easily *extensible* to capture additional real-world properties without introducing major changes to the core abstraction. As we have shown in the stochastic case, the well-defined and concise domain of the π-calculus is relatively amenable to extensions. Indeed, we have recently extended the calculus and the abstraction to directly handle compartments, as will be reported elsewhere [49].

A graphical extension or translation of the abstraction is beyond the scope of this work. Note, however, that graphical variants of the π-calculus have been previously proposed (e.g., [40]) and can serve as a starting point for the development of a graphical extension. Alternatively, the abstraction may be translatable into other well-developed graphical formalisms, such as the unified modelling language (UML) [48]).

Comparison to Other Representations

The two prominent existing computational approaches to biomolecular systems are "dynamical systems theory", based on the "cell as a collection of molecular species" abstraction, and the functional pathway databases that employ the "molecule as object" abstraction. In order to render the former abstractions *understandable* another layer of representation (graphics, object scheme, etc.) is typically required; while in order to render the latter *computable* an additional level of semantics and implementation is required that adds dynamics and behavioural properties to an essentially static scheme.

Abstract computer languages, such as the π-calculus, offer a synthesis of the benefits of both representations. On the one hand, the molecular world is clearly abstracted into the computational one, in a way which is as *relevant* and *understandable* as possible. On the other hand, the generated abstract model is *computable* allowing both simulation and analysis of the modelled system's behaviour.

How does the π-calculus abstraction compare to that offered by other, related languages, such as Petri nets [20], statecharts [29], and linear logic [17], that are now being used to represent various biomolecular systems? While the abstraction to linear logic has only been rudimentarily developed [17], both Petri Nets and statecharts are actively employed, and must be considered as alternatives when evaluating our π-calculus based approach.

Petri nets are one of the first uses of the "molecule-as-computation" abstraction, and have been primarily successful as an abstraction of metabolic systems. The *computability* of this abstraction is well-developed, in terms of both simulation tools (e.g., [20]) and verification ones [19]. Furthermore, it is accompanied by a pleasing graphical representation which greatly enhances its utility (and which the π-calculus lacks). However, the Petri net based abstraction is close to a "cell as a collection of molecular species", and suffers from the same *relevance* problems: it does not allow for the structured representation of molecular objects (which the π-calculus does).

Statecharts [29] are a highly *relevant* and *understandable* model of biomolecular and multicellular systems, providing a structured and intuitive framework for the abstraction of biomolecular systems. The graphical nature of the language is much more developed than the Petri net one and very useful, and a variety of computational tools exist for qualitative simulation and verification. Despite these major advantages, the abstraction lacks in *understandability* as, unlike the π-calculus, it is a purely qualitative framework, and does not currently accommodate process quantities or reaction rates.

Clearly, each of these approaches has a different balance of benefits and limitations, and none represents a final solution to the challenge posed by biological systems. A full understanding of these differences requires further dedicated research. We believe such research is important as a first step towards distilling and synthesizing an optimal abstraction of molecules as computations.

References

1. T. Akutsu, S. Miyano, and S. Kuhara. Algorithms for identifying boolean networks and related biological networks based on matrix multiplication and fingerprint function. *Journal of Computational Biology*, 7(3):331–343, 2000.
2. T. Akutsu, S. Miyano, and S. Kuhara. Algorithms for inferring qualitative models of biological networks. In R.B. Altman, A.K. Dunker, L. Hunter, and T.E. Klein, editors, *Pacific Symposium on Biocomputing*, volume 5, pages 293–304, World Scientific Press, 2000.
3. R. Alves and M.A. Savageau. Extending the method of mathematically controlled comparison to include numerical comparisons. *Bioinformatics*, 16(9):786–798, 2000.
4. M. Antoniatti, B. Mishra, C. Piazza, A. Policriti, and M. Simeoni. Modeling cellular behavior with hybrid automata: bisimulation and collapsing. In *Proceedings of the first international workshop on Computational Methods in Systems Biology*, Lecture Notes in Computer Science. Springer-Verlag, 2003.
5. A. Arkin, J. Ross, and H.H. McAdams. Stochastic kinetic analysis of developmental pathway bifurcation in phage lambda-infected Escherichia coli cells. *Genetics*, 149:1633–1648, 1998.
6. G.D. Bader, I. Donaldson, C. Wolting, B.F. Ouellette, T. Pawson, and C.W. Hogue. BIND - the biomolecular interaction network database. *Nucleic Acids Research*, 29(1):242–245, 2001.
7. U.S. Bhalla and R. Iyengar. Emergent properties of networks of biological signaling pathways. *Science*, 283(5400):381–387, 1999.
8. D. Bray. Reductionism for biochemists: how to survive the protein jungle. *Trends in Biochemical Sciences*, 22(9):325–326, 1997.
9. N. Chabrier and F. Fages. Symbolic model checking of biochemical networks. In *Proceedings of the first international workshop on Computational Methods in Systems Biology*, Lecture Notes in Computer Science. Springer-Verlag, 2003.
10. G. Ciobanu. Formal description of the molecular processes. In C.Martin-Vide and Gh. Paun, editors, *Recent Topics in Mathematical and Computational Linguistics*, pages 82–96. Publishing House of the Romanian Academy, 2000.
11. G. Ciobanu and M. Rotaru. Molecular interaction. *Journal of Theoretical Computer Science*, 289(1):801–827, 2002.
12. G. Ciobanu, V.Ciubotariu, and B.Tanasa. A π-calculus model of the Na pump. In *Genome Informatics*, pages 469–472, Tokyo, 2002. Universal Academy Press.
13. M. Curti, P. Degano, and C.T. Baldari. Casual calculus for biochemical modeling. In *Proceedings of the first international workshop on Computational Methods in Systems Biology*, Lecture Notes in Computer Science. Springer-Verlag, 2003.
14. P. Dhaeseleer, S. Liang, and R. Somogyi. Genetic network inference: from co-expression clustering to reverse engineering. *Bioinformatics*, 16(8):707–726, 2000.
15. K. Eilbeck, A. Brass, N. Paton, and C. Hodgman. Interact: an object oriented protein-protein interaction database. In *Intelligent Systems for Molecular Biology*, volume 7, pages 87–94, AAAI Press, 1999.
16. A. Finney, H. Sauro, M. Hucka, and H. Bolouri. An XML-based model description language for systems biology simulations. Technical report, California Institute of Technology, September 2000.

17. W. Fontana and L.W. Buss. The barrier of objects: from dynamical systems to bounded organizations. In J. Casti and A. Karlqvist, editors, *Boundaries and Barriers*, pages 56–116. Addison-Wesley, 1996.

18. D.T. Gillespie. Exact stochastic simulation of coupled chemical reactions. *Journal of Physical Chemistry*, 81(25):2340–2361, 1977.

19. C. Girault and R. Valk. *Petri Nets for Systems Engineering: A Guide to Modeling, Verification, and Applications.* Springer-Verlag, 2002.

20. P.J.E. Goss and J. Peccoud. Quantitative modeling of stochastic systems in molecular biology by using stochastic Petri nets. *Proceedings of the National Academy of Sciences USA*, 95(12):6750–6755, 1998.

21. P.J.E. Goss and J. Peccoud. Analysis of the stabilizing effect of ROM on the genetic network controlling cole1 plasmid replication. In R. B. Altman, A. K. Dunker, L. Hunter, and T. E. Klein, editors, *Pacific Symposium on Biocomputing*, volume 4, pages 65–76, World Scientific Press, 1999.

22. D. Harel and E. Gery. Executable object modeling with statecharts. *IEEE Computer*, 30(7):31–42, 1997.

23. K.R. Heidtke and S. Schulze-Kremer. Deriving simulation models from a molecular biology knowledge base. In *Proceedings of the 4th Workshop on Engineering Problems for Qualitative Reasoning of the 16th International Joint Conference on Artificial Intelligence*, 1999.

24. C.-H. Heldin and M. Purton, editors. *Signal Transduction*. Modular Texts in Molecular and Cell Biology 1. Chapman and Hall, London, 1996.

25. R. Hofestadt and S. Thelen. Quantitative modeling of biochemical networks. *In Silico Biology*, 1(1):39–53, 1998.

26. M. Holcombe and A. Bell. Computational models of immunological pathways. In M. Holcombe and R. Paton, editors, *Information Processing in Cells and Tissues: Proceedings of IPCAT '97*, pages 213–226, Plenum Press, New York, 1998.

27. T. Igarashi and T. Kaminuma. Development of a cell signalling networks database. In R.B. Altman, A.K. Dunker, L. Hunter, and T.E. Klein, editors, *Proceedings of the Pacific Symposium of Biocomputing '97*, pages 187–197, World Scientific Press, Singapore, 1997.

28. N. Kam, I.R. Cohen, and D. Harel. The immune system as a reactive system: Modeling T cell activation with statecharts. *Bulletin of Mathematical Biology*, 2002. An extended abstract of this paper appeared in the Proceedings of the Symposium on Human-Centric Computing Languages and Environments, Stresa, Italy, pages 15-22, 2001.

29. N. Kam, D. Harel, and I.R. Cohen. Modeling biological reactivity: statecharts vs. boolean logic. In *Proceedings of the Second International Conference on Systems Biology*, Pasadena, CA, 2001.

30. P.D. Karp, M. Krummenacker, S. Paley, and J. Wagg. Integrated pathway/genome databases and their role in drug discovery. *Trends in Biotechnology*, 17(7):275–281, 1999.

31. T. Kazic. Semiotes: a semantics for sharing. *Bioinformatics*, 16(12):1129–1144, 2000.

32. H. Kitano. Computational systems biology. *Nature*, 420:206–210, 2002.

33. R. Kuffner, R. Zimmer, and T. Lengauer. Pathway analysis in metabolic databases via differential metabolic display. *Bioinformatics*, 16(9):825–836, 2000.

34. K.M. Kyoda, M. Muraki, and H. Kitano. Construction of a generalized simulator for multi-cellular organisms and its application to SMAD signal transduction. In R. B. Altman, A. K. Dunker, L. Hunter, and T. E. Klein, editors, *Pacific Symposium on Biocomputing*, volume 5, pages 317–328, World Scientific Press, Singapore, 2000.

35. H. Matsuno, A. Doi, M. Nagasaki, and S. Miyano. Hybrid Petri net representation of gene regulatory network. In R.B. Altman, A.K. Dunker, L. Hunter, and T.E. Klein, editors, *Pacific Symposium on Biocomputing*, volume 5, pages 341–352, World Scientific Press, Singapore, 2000.

36. H. Matsuno, R. Murakani, R. Yamane, N. Yamasaki, S. Fujita, H. Yoshimori, and S. Miyano. Boundary formation by notch signaling in drosophila multicellular systems: experimental observations and gene network modeling by genomic object net. In R.B. Altman, A.K. Dunker, L. Hunter, and T.E. Klein, editors, *Pacific Symposium on Biocomputing*, volume 8, World Scientific Press, Singapore, 2003.

37. H.H. McAdams and A. Arkin. Stochastic mechanisms in gene expression. *Proceedings of the National Academy of Sciences USA*, 94(3):814–819, 1997.

38. H. H. McAdams and L. Shapiro. Circuit simulation of genetic networks. *Science*, 269(5224):650–656, 1995.

39. L. Mendoza, D. Thieffry, and E. R. Alvarez-Buylla. Genetic control of flower morphogenesis in Arabidopsis thaliana: a logical analysis. *Bioinformatics*, 15(7–8):593–606, 1999.

40. R. Milner. π-nets: a graphical form of the π-calculus. In D. Sannella, editor, *Proceedings of the Fifth European Symposium on Programming (ESOP '94)*, volume 788 of *Lecture Notes in Computer Science*, pages 26–42, Springer-Verlag, Berlin, 1994.

41. R. Milner. *Communicating and Mobile Systems: The π-Calculus*. Cambridge University Press, Cambridge, 1999.

42. P. Nielsen, D. Bullivant, C. Lloyd, D. Nickerson, S. Lett, K. Jim, and P. Noble. The CellML modeling language. http://www.cellml.org/public/specification/, 2000.

43. H. Ogata, S. Goto, W. Fujibuchi, and M. Kanehisa. Computation with the Kegg pathway database. *Biosystems*, 47(1–2):119–128, 1998.

44. F. Orava and J. Parrow. An algebraic verification of a mobile network. *Formal Aspects of Computing*, 4(6):497–543, 1992.

45. A. Pagnoni and A. Visconti. Detection and analysis of unexpected state components in biological systems. In *Proceedings of the first international workshop on Computational Methods in Systems Biology*, Lecture Notes in Computer Science. Springer-Verlag, 2003.

46. T. Pawson and P. Nash. Protein-protein interactions define specificity in signal transduction. *Genes and Development*, 14:1027–1047, 2000.

47. C. Priami, A. Regev, E. Shapiro, and W. Silverman. Application of a stochastic name-passing calculus to representation and simulation of molecular processes. *Information Processing Letters*, 80:25–31, 2001.

48. T. Quatrani. *Visual Modeling with Rational Rose 2000 and UML*. Addison-Wesley, 2000.

49. A. Regev, K. Panina, W. Silverman, L. Cardelli, and E. Shapiro. Bioambients - a calculus for biological compartments. Manuscript in preparation.

50. A. Regev and E. Shapiro. Cellular abstractions. *Nature*, 419:343, 2002.

51. A. Regev, W. Silverman, and E. Shapiro. Representation and simulation of bio-chemical processes using the pi-calculus process algebra. In R.B. Altman, A.K. Dunker, L. Hunter, and T.E. Klein, editors, *Pacific Symposium on Biocomputing*, volume 6, pages 459–470, World Scientific Press, Singapore, 2001.

52. W. Reisig and G. Rozenberg. Informal introduction to Petri nets. *Lecture Notes in Computer Science: Lectures on Petri Nets I: Basic Models*, vol. 1491, 1998.

53. M.G. Samsonova and V.N. Serov. Network: an interactive interface to the tools for analysis of genetic network structure and dynamics. In R.B. Altman, A.K. Dunker, L. Hunter, and T.E. Klein, editors, *Pacific Symposium on Biocomputing*, volume 4, pages 102–111, World Scientific Press, Singapore, 1999.

54. L. Sanchez, J. van Helden, and D. Thieffry. Establishment of the dorso-ventral pattern during embryonic development of Drosophila melanogasater: a logical analysis. *Journal of Theoretical Biology*, 189(4):377–389, 1997.

55. H.M. Sauro. Scamp: a general-purpose simulator and metabolic control analysis program. *Computer Applications in Bioscience*, 9(4):441–450, 1993.

56. S. Schuster, D.A. Fell, and T. Dandekar. A general definition of metabolic pathways useful for systematic organization and analysis of complex metabolic networks. *Nature Biotechnology*, 18(3):326–332, 2000.

57. E. Selkov, Y. Grechkin, N. Mikhailova, and E. Selkov. MPW: the metabolic pathways database. *Nucleic Acids Research*, 26(1):43–45, 1998.

58. E. Shapiro. Concurrent Prolog: a progress report. In E. Shapiro, editor, *Concurrent Prolog (vol. I)*, pages 157–187. MIT Press, Cambridge, MA, 1987.

59. W. Silverman, M. Hirsch, A. Houri, and E. Shapiro. The Logix system user manual, version 1.21. In E. Shapiro, editor, *Concurrent Prolog (vol. II)*, pages 46–78. MIT Press, Cambridge, MA, 1987.

60. T.F. Smith. Functional genomics - bioinformatics is ready for the challenge. *Trends in Genetics*, 14(7):291–293, 1998.

61. L. Stryer. *Biochemistry*. W.H. Freeman, 1995.

62. Z. Szallasi. Genetic network analysis in light of massively parallel biological data. In R.B. Altman, A.K. Dunker, L. Hunter, and T.E. Klein, editors, *Pacific Symposium on Biocomputing*, volume 4, pages 5–16, World Scientific Press, Singapore, 1999.

63. D. Thieffry and D. Romero. The modularity of biological regulatory networks. *Biosystems*, 50(1):49–59, 1999.

64. D. Thieffry and R. Thomas. Qualitative analysis of gene networks. In R.B. Altman, A.K. Dunker, L. Hunter, and T.E. Klein, editors, *Pacific Symposium on Biocomputing*, volume 3, pages 77–88, World Scientific Press, Singapore, 1998.

65. J. van Helden, A. Naim, R. Mancuso, M. Eldridge, L. Wernisch, D. Gilbert D, and S. J. Wodak. Representing and analysing molecular and cellular function using the computer. *Biological Chemistry*, 381(9–10):921–935, 2000.

66. B. Victor and F. Moller. The Mobility Workbench — a tool for the π-calculus. In D. Dill, editor, *CAV'94: Computer Aided Verification*, volume 818 of *Lecture Notes in Computer Science*, pages 428–440. Springer-Verlag, 1994.

67. D. Voet and J. Voet. *Biochemistry. 2nd edition*. John Wiley & Sons, 1995.

68. E. Wingender, X. Chen, E. Fricke, R. Geffers, R. Hehl, I. Liebich, M. Krull, V. Matys, H. Michael, R. Ohnhauser, M. Pruss, F. Schacherer, S. Thiele, and S. Urbach. The transfac system on gene expression regulation. *Nucleic Acids Research*, 29(1):281–283, 2001.

69. I. Xenarios, E. Fernandez, L. Salwinski, X.J. Duan, M.J. Thompson, E.M. Marcotte, and D. Eisenberg. DIP: the database of interacting proteins: 2001 update. *Nucleic Acids Research*, 29(1):239–241, 2001.

70. T.-M. Yi, Y. Huang, M.I. Simon, and J. Doyle. Robust perfect adaptation in bacterial chemotaxis through integral feedback control. *Proceedings of the National Academy of Sciences USA*, 97(9):4649–4653, 2000.

The Topology of Evolutionary Biology

Bärbel M.R. Stadler[1] and Peter F. Stadler[2,3,4]

[1] Max Planck Institute for Mathematics in the Sciences. Leipzig, Germany
   stadler@mis.mpg.de, http://www.tbi.univie.ac.at/~baer/
[2] Lehrstuhl für Bioinformatik, Institut für Informatik, Universität Leipzig
   peter.stadler@bioinf.uni-leipzig.de,
   http://www.bioinf.uni-leipzig.de/~studla/
[3] Institut für Theoretische Chemie und Molekulare Strukturbiologie,
   Universität Wien, Währingerstrasse 17, Wien, Austria
[4] Santa Fe Institute, Santa Fe, New Mexico, USA

Summary. Central notions in evolutionary biology are intrinsically topological. This claim is perhaps most obvious for the discontinuities associated with punctuated equilibrium. Recently, a mathematical framework has been developed that derives the concepts of phenotypic characters and homology from the topological structure of the *phenotype space*. This structure in turn is determined by the genetic operators and their interplay with the properties of the genotype–phenotype map.

1 Introduction

Evolutionary change is the result of the spontaneous generation of genetic variation and the subsequent fixation of variants in the population through natural selection and/or genetic drift.

This is the basic assumption of the neo-Darwinian model. Population genetics appears therefore as a natural framework for studying the evolution of phenotypic adaptation, the evolution of gene sequences, and the process of speciation, see, e.g., [1,2]. Patterns of phenotypic evolution [3], however, such as the punctuated mode (the partially discontinuous nature) of evolutionary change [4], developmental constraints or constraints to variation [5,6], innovation [7], directionality in evolution, and phenotypic stability or homology [8,9] are not adequately described by population genetics models.

The reason for this apparent discrepancy is that selection can determine the fate of a new phenotype only *after* it has been produced or "accessed" by means of variational mechanisms [10,11]. Phenotypes are not varied directly in a heritable fashion, but through genetic mutation and its consequences on development. The accessibility of a phenotype is therefore determined by the *genotype–phenotype map* (GP-map) which determines how phenotypes vary with genotypes [12–14].

The motivation for emphasizing the central role of the GP-map originates from studies in which RNA folding from sequences to secondary structures is used as a biophysically realistic, yet extremely simplified, toy-model of a GP-map. Simulated populations of replicating and mutating sequences under selection exhibit many phenomena known from organismal evolution: neutral drift, punctuated change, plasticity, environmental and genetic canalization, and the emergence of modularity, see, e.g., [14–19]. Laboratory experiments have also generated phenomena consistent with these patterns [20–22].

In this contribution we discuss in some detail the surprisingly far-reaching consequences of formalizing evolutionary theory starting from accessibility. This approach is to a large extent the consequence of the realization that there is a significant discrepancy between the mathematical setup of population genetics (the current implementation of the neo-Darwinian model of evolution) and computational case studies of simple evolution processes mentioned above. Population genetics theory typically assumes that the set of possible phenotypes is organized into a highly symmetric and regular space equipped with a notion of distance; most conveniently, a Euclidean vector space [1]. Computational studies using an explicit genotype–phenotype model based on the RNA folding, however, suggest a quite different picture [14, 18, 23]: if phenotypes are organized according to genetic accessibility, the resulting space lacks a metric and is conveniently formalized by an unfamiliar structure that generalizes topological spaces [24–26].

2 Genotype Space

The structure of *genotype space* is uniquely determined by the genetic operators at work: mutation, recombination, genome rearrangements, etc. In the case of point mutations and constant length genomes the situation is straightforward. Naturally, sequences that differ by a single mutation are *neighbours* in "sequence space" [27, 28]. The sequence space can thus be represented as a graph, Fig. 1(a). The same is true for rearrangements, Fig. 1(b).

In the RNA example of Fig. 2, a genotype is a sequence of nucleotides encoded by A, C, G, and U of a given length n. The genotype space consists of all 4^n sequences. Two sequences x and y are neighbours if they are different in just one position, i.e., if their Hamming distance is $d_H(x, y) = 1$. The resulting graph, a generalized hypercube (Fig. 1), is highly symmetric.

The situation becomes more complicated, however, when recombination (crossover) is considered [29]. The analog of the adjacency relation of the graph is the recombination set $\mathcal{R}(x, y)$, which is defined as the set of all (possible) recombinants of two parents x and y. Recombination sets satisfy (at least) the following two axioms:

(X1) $\{x, y\} \in \mathcal{R}(x, y)$,
(X2) $\mathcal{R}(x, y) = \mathcal{R}(y, x)$.

Condition *(X1)* states that replication may occur without recombination, and *(X2)* means that the role of the two parents is interchangeable. Often a third condition

(X3) $\mathcal{R}(x, x) = \{x\}$

is assumed, which is, however, not satisfied by models of unequal crossover [25, 30]. Functions $\mathcal{R} : X \times X \to \mathcal{P}(X)$ satisfying *(X1)*, *(X2)*, and *(X3)* were considered recently as *transit functions* [31] and as *P-structures* [32, 33].

On the other hand, classical models of population genetics and quantitative genetics (tacitly) assume a Euclidean vector space as the natural framework for studying the evolution of phenotypic adaptation, the evolution of gene sequences, and the process of speciation, see, e.g., [1, 2]. This begs the question whether there is a mathematical framework that contains graphs, recombination sets, and Euclidean vector spaces as special cases. After all, accessibility in terms of the genetic operators is the common mechanism that "creates" the structure of genotype space.

We base our discussion on the notion of *accessibility* [14, 18, 24]. Let us write $x \curvearrowright_U y$ to mean that x is accessible from y "at level U". Whether x can be obtained from y in practice depends on how long we are willing to wait. Alternatively, we might be interested whether we obtain x from y within a fixed time with a certain probability. The symbol $\curvearrowright_U$ emphasizes that we consider accessibility w.r.t. a user-defined criterion, which at this point is deliberately kept vague.

The relation $x \curvearrowright_U y$ is represented in an equivalent way by the set $U = \{(x, y) | x \curvearrowright_U y\}$. Then $U[x] = \{y | x \curvearrowright_U y\}$ can be interpreted as the U-neighbourhood of x: it is the set of all y that have access to x at level U. The composition $\curvearrowright_{U'} \circ \curvearrowright_{U''}$ of two accessibility relations $\curvearrowright_{U'}$ and $\curvearrowright_{U''}$ is defined by $x(\curvearrowright_{U'} \circ \curvearrowright_{U''})y$ if there is a $z \in X$ such that $x \curvearrowright_{U'} z$ and $z \curvearrowright_{U'} y$. It is again an accessibility relation, albeit at a weaker level. Note that we do not require that two relations $\curvearrowright_{U'}$ and $\curvearrowright_{U''}$ are comparable. There is a natural partial order, however, which is determined by the inclusion relation of the associated sets U' and U'' respectively: if $U' \subset U''$ then $\curvearrowright_{U'}$ is a more stringently defined accessibility relation than $\curvearrowright_{U''}$

From the mathematical point of view it is natural to consider the collection $\mathcal{U}$ of all accessibility relations on a given genotype space. This construction follows the spirit of [24] with small technical differences. What are the natural properties $\mathcal{U}$? We propose:

(U0) $X \times X \in \mathcal{U}$ (ergodicity).
(U1') $U', U'' \in \mathcal{U}$ implies $U' \circ U'' \in \mathcal{U}$.
(U1) $U \in \mathcal{U}$ and $U \subseteq U'$ implies $U' \in \mathcal{U}$.
(U2) $x \curvearrowright_U x$ for all $x \in X$ and all $U \in \mathcal{U}$.

The ergodicity hypothesis says that at **some** level everything is accessible from everywhere, if we just wait long enough or if we are content with sufficiently small probabilities. Axiom *(U1')* simply allows us to "combine" accessibility relations to multi-step processes that define weaker accessibility relations, and *(U2)* states the trivial observation that every point is accessible from itself.

Axiom *(U1)* allows us to construct a (weaker) accessibility relation from a given one by adding arbitrary pairs – after all, by axiom *(U0)* arbitrary pairs are accessible at some level! If we assume that $\mathcal{U}$ is non-empty and axioms *(U1)* and *(U2)* are satisfied, then *(U0)* and *(U1')* hold as well. The set-system $\mathcal{U}$ on $X \times X$ is a generalized version of a *uniformity*. In the theory of uniform space [34] one usually assumes additional axioms, which, however, do not seem to be satisfied naturally in the case of genetic accessibility relations. These are:

>   *(U3)* $U, U' \in \mathcal{U}$ implies $U \cap U' \in \mathcal{U}$.
>   *(U4)*   For each $U \in \mathcal{U}$ there is a $V \in \mathcal{U}$ with $V \circ V \subseteq U$.
>   *(U5)* $U \in \mathcal{U}$ implies $U^{-1} = \{(x,y)|(y,x) \in U\} \in \mathcal{U}$.

One speaks of *pre-uniformities* if *(U0–U3)* are satisfied, if *(U4)* also holds we have a *quasi-uniformity*, while for a uniformity *(U0–U5)* must hold. The (generalized) uniform structure $\mathcal{U}$ is associated with a generalized topology on X. In the following section we will briefly review this connection and the underlying mathematical framework.

3 Generalized Topological and Uniform Spaces

Textbooks on topology such as [35,36] usually start by defining a *topology* on a set X by means of a collection $\mathcal{O} \subseteq \mathcal{P}(X)$ of "open sets" (the power-set $\mathcal{P}(X)$ is the set of all subsets of X). For our purposes it is more convenient to use the (equivalent) collection $\mathcal{C} = \{A|(X \setminus A) \in \mathcal{O}\}$ of closed sets as the primitive concept. A pair $(X, \mathcal{C})$ is a *topological space* if $\mathcal{C}$ satisfies the following four axioms [5]:

>   *(I0)* $\emptyset \in \mathcal{C}$.
>   *(I1)* $X \in \mathcal{C}$.
>   *(I2)*   If $A_i \in \mathcal{C}$ for all $i \in I$, then $\bigcap\{A_i|i \in I\} \in \mathcal{C}$.
>   *(I3)*   If $A, B \in \mathcal{C}$ then $A \cup B \in \mathcal{C}$.

Here, I is an arbitrary, possibly infinite, index set. In lattice theory more general so-called *intersection structures* are considered that fulfill only *(I2)*, see, e.g., [37]. The *closure function* cl $: \mathcal{P}(X) \to \mathcal{P}(X)$ defined by

$$\mathsf{cl}(A) = \bigcap\{B \in \mathcal{C}|A \subseteq B\} \tag{1}$$

associates with each set $A \subseteq X$ its "closure" $\mathsf{cl}(A)$. The closure function cl has two important properties: (i) it is *isotone*, i.e., $A' \subseteq A$ implies $\mathsf{cl}(A') \subseteq \mathsf{cl}(A)$, and (ii) it is *idempotent*, i.e., $\mathsf{cl}(\mathsf{cl}(A)) = \mathsf{cl}(A)$. Given an isotone and idempotent closure function cl, one recovers the associated intersection structure by setting $\mathcal{C} = \{C = \mathsf{cl}(A)|A \in \mathcal{P}(X)\}$. One can therefore just as well use the closure function cl to define a topology by means of Kuratowski's axioms [38]:

---

[5] The corresponding axioms for open sets are obtained by exchanging unions and intersections.

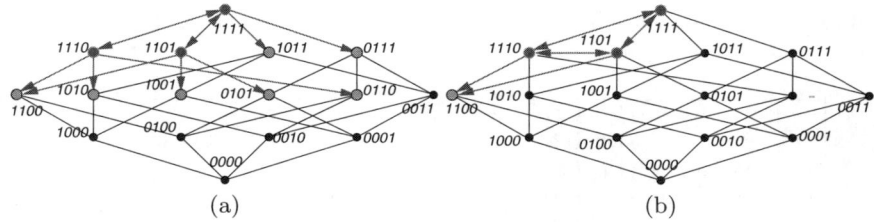

Fig. 1. Closure in Sequence Space: A population $A = \{1111, 1110, 1101\}$ and its closure $\mathsf{cl}(A)$ is shown for (a) point mutations, which give rise to the usual Hamming graph structure, and (b) 1-point recombination. In the latter case we show the embedding of A and $\mathsf{cl}(A)$ in the Hamming graph since the resulting neighbourhood space itself does not have a convenient representation. The homeomorphism of recombination spaces and Hamming graphs is discussed in detail in [29].

(K0) $\mathsf{cl}(\emptyset) = \emptyset$.
(K1) $A \subseteq B$ implies $\mathsf{cl}(A) \subseteq \mathsf{cl}(B)$.
(K2) $A \subseteq \mathsf{cl}(A)$.
(K3) $\mathsf{cl}(A \cup B) \subseteq \mathsf{cl}(A) \cup \mathsf{cl}(B)$ holds for all $A, B \subseteq X$.
(K4) $\mathsf{cl}(\mathsf{cl}(A)) = \mathsf{cl}(A)$ holds for all $A \subseteq X$.

The dual of the closure function is the *interior function* int defined by $\mathsf{int}(A) = X \setminus \mathsf{cl}(X \setminus A)$ and hence $\mathsf{cl}(A) = X \setminus \mathsf{int}(X \setminus A)$. A set N is a *neighbourhood* of a point $x \in X$ if and only if $x \in \mathsf{int}(N)$. We write $\mathcal{N}(x)$ for the collection of all neighbourhoods of x.

Kuratowski's closure axioms can be translated into the language of neighbourhoods where *(Ni)* is equivalent to *(Ki)* above:

(N0) $X \in \mathcal{N}(x)$ for all $x \in X$.
(N1) $N \in \mathcal{N}(x)$ and $N \subseteq N'$ implies $N \in \mathcal{N}(x)$.
(N2) $N \in \mathcal{N}(x)$ implies $x \in N$.
(N3) $N', N'' \in \mathcal{N}(x)$ implies $N' \cap N'' \in \mathcal{N}(x)$.
(N4) $N \in \mathcal{N}(x)$ if and only if $\mathsf{int}(N) \in \mathcal{N}(x)$.

Generalized topological spaces have been used in various applied domains of computer science, such as digital image processing, information representation, the semantics of modal logic, handwriting recognition, and artificial chemistry, see, e.g., [39–46] for a few examples.

A generalized neighbourhood function $\mathcal{N}$ can be constructed from a generalized uniform structure $\mathcal{U}$ in a natural way. For each $x \in X$ and each $U \in \mathcal{U}$ we define the sets

$$U[x] = \{y \in X \,|\, (x, y) \in U\} \qquad \text{for each } U \in \mathcal{U}, \tag{2}$$

and consider the collection

$$\mathcal{U}[x] = \{N \,|\, \exists U \in \mathcal{U} : U[x] \subseteq N\}. \tag{3}$$

Table 1. Correspondence between generalized uniformities and generalized topologies.

	(U0–U1)	↔	(N0–N1) isotone space
	(U0–U2)	↔	(N0–N2) neighbourhood space
pre-uniformity	(U0–U3)	↔	(N0–N3) pretopology
quasi-uniformity	(U0–U4)	↔	(N0–N4) topology
semi-uniformity	(U0–U3,U5)	↔	symmetric pretopology [47]
uniformity	(U0–U5)	↔	completely regular topology

It is not hard to verify that $\mathcal{N}_\mathcal{U} : X \to \mathcal{P}(X)$, $x \mapsto \mathcal{U}[x]$ is a neighbourhood function on X. Conversely, given a neighbourhood function $\mathcal{N}$ we may construct a corresponding generalized uniformity as the collection $\mathcal{U}_\mathcal{N}$ of all sets U that contain a set of the form

$$U = \big\{(x,y)|x \in X \text{ and } y \in N_x \text{ for a fixed } N_x \in \mathcal{N}(x)\big\}. \tag{4}$$

$\mathcal{U}_\mathcal{N}$ is the *generalized uniformization* of the neighbourhood system $\mathcal{N}$. The axioms for the generalized uniformities translate into properties of the resulting closure spaces as listed in Table 1. The correspondence of pre-uniformities and pretopologies is described in [24]; for a proof that every topology is quasi-uniformizable and for the characterization of uniformizable spaces see [34].

4 Phenotype Space

Accessibility at the phenotypic level is the crucial determinant for evolution since it is the phenotype that is subject to selection. We have seen in Sect. 2 that the structure of genotype space is determined by physical processes, namely mutation and recombination, acting on genes. The phenotype, on the other hand, is therefore not modified directly, but indirectly through the modification of the genome from which it arises (We simplify here by disregarding, for example, epigenetic inheritance [48].) The accessibility relation at the phenotypic level therefore has to be understood as a consequence of the interplay of genotypic accessibility and the GP-map.

The motivation for emphasizing the central role of the GP-map arose from a series of computer simulations in which fitness was modelled as a function of RNA secondary structures [14–19].

The RNA model, see Fig. 2, is a simplified, yet biophysically realistic, model for a GP-map that has the advantage that its predictions are amenable to experimental tests, see, e.g., [49]. In nature, RNA molecules can act both as genotype (e.g., as the genome of certain viruses) and as phenotype (the 3D structure can bind specifically to other molecules or even perform catalytic

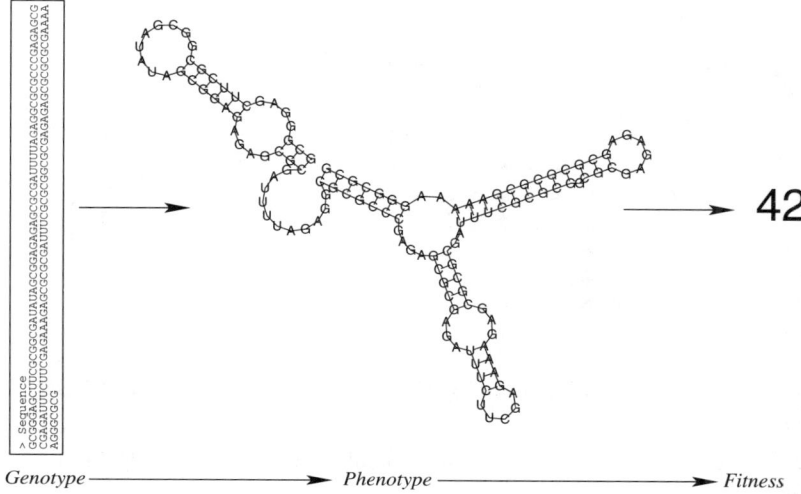

Genotype ⟶ *Phenotype* ⟶ *Fitness*

Fig. 2. An RNA molecule folds by first establishing the secondary structure, i.e., an outerplanar graph in which every nucleotide (letter) is connected to its sequence neighbours and to at most one other nucleotide with which it forms hydrogen bonds. The three-dimensional structure of the molecule is formed only in a second step [54]. The main part of the energy of structure formation can be explained in terms of the base pair stacking at the secondary structure level [55]. The secondary structure of an RNA molecule is therefore a useful model for a biophysically realistic GP-map which serves as the basis for fitness landscapes that are obtained by assigning a fitness value to each secondary structure graph.

functions). These functions can then be "evaluated" by the environment, e.g., in a replication experiment [50] or in a SELEX process [51]. RNA secondary structures are a convenient computational model because they can also be computed efficiently given only the sequence information by means of a dynamic programming algorithm [52, 53].

A generic feature of the GP-map φ at least of biopolymers is *redundancy*, i.e., there are many more sequences than structures (at a resolution of practical interest). Computational studies for both RNA [16] and protein [56] show quite convincingly that the set $\varphi^{-1}(\alpha)$ of sequences that fold into a common phenotype α forms a connected network in sequence space [16], Fig. 3.

In practice, any useful notion of nearness will have to correlate with the likelihood of a transition from one phenotype β to a "neighbouring" one α. For concreteness, let us consider a finite pretopological space, i.e., a directed graph, that might arise from a particular mutation operator. The possible mutants of a phenotype α then form the graph-theoretical boundary $\mathrm{bd}\varphi^{-1}(\alpha)$ of the neutral network of α. The set $\varphi^{-1}(\beta) \cap \mathrm{bd}\varphi^{-1}(\alpha)$ is the subset of all these mutants that have phenotype β. Since the neutral networks are — at least in

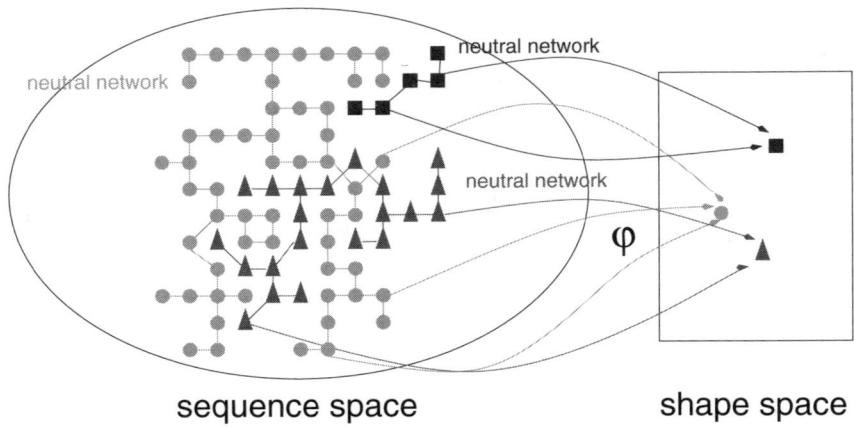

Fig. 3. Schematic representation of the neutral networks arising in RNA and protein folding.

the RNA and protein case — approximately homogeneous we can estimate the probability of reaching β from α as

$$\chi(\beta \leftarrow \alpha) = \frac{|\varphi^{-1}(\beta) \cap \mathrm{bd}\varphi^{-1}(\alpha)|}{|\mathrm{bd}\varphi^{-1}(\alpha)|} \qquad (5)$$

where $|A|$ denotes the cardinality of the set A. Equation (5) in turn defines a collection of accessibility relations $\curvearrowright_p$ by setting

$$\beta \curvearrowright_p \alpha \iff \chi(\beta \leftarrow \alpha) \geq p. \qquad (6)$$

An approach to formulation accessibility in terms of so-called probabilistic closure spaces or, alternatively, fuzzy closure space is briefly discussed in [57].

Depending on the size of the underlying neutral nets, nearness between two phenotypes does not have to be symmetric. Consider a shape α with a very large, and a shape β with a very small, neutral network. It might be possible that almost all sequences in $\mathrm{bd}\varphi^{-1}(\beta)$ fold into shape α, i.e., $\chi(\alpha \leftarrow \beta) \approx 1$, and $\alpha \curvearrowright_p \beta$ for a large fixed value of p. On the other hand, we have $\chi(\beta \leftarrow \alpha) \approx 0$ because $|\varphi^{-1}(\beta)| \ll |\varphi^{-1}(\mathrm{bd}\alpha)|$, i.e., $\beta \not\curvearrowright_p \alpha$. This effect can indeed be observed in the case of RNA secondary structures [14, 18]. As a consequence, such asymmetries may introduce a directionality in evolution that is purely structural and hence independent of fitness.

5 Continuity

The notion of continuity lies at the heart of topological theory. Its importance is emphasized by a large number of equivalent definitions, see, e.g., [58, 59].

Let (X, cl) and (Y, cl) be two isotone spaces. Then $f : X \to Y$ is continuous if one (and hence all) of the following equivalent conditions holds:

(i) $\mathsf{cl}(f^{-1}(B)) \subseteq f^{-1}(\mathsf{cl}(B))$ for all $B \in \mathcal{P}(Y)$.
(ii) $f^{-1}(\mathsf{int}(B)) \subseteq \mathsf{int}(f^{-1}(B))$ for all $B \in \mathcal{P}(Y)$.
(iii) $B \in \mathcal{N}(f(x))$ implies $f^{-1}(B) \in \mathcal{N}(x)$ for all $x \in X$.
(iv) $f(\mathsf{cl}(A)) \subseteq \mathsf{cl}(f(A))$ for all $A \in \mathcal{P}(X)$.

We say that $f : X \to Y$ is *continuous in* x if $B \in \mathcal{N}(f(x))$ implies $f^{-1}(B) \in \mathcal{N}(x)$. Obviously, $f : X \to Y$ is continuous if it is continuous in each $x \in X$. Continuity is preserved under function composition. If $f : X \to Y$ and $g : Y \to Z$ are continuous (in x and $y = f(x)$), then $g \circ f : X \to Z$ is also continuous (in x).

The least stringent form of accessibility in phenotype space is defined by setting $\alpha \frown \beta$ iff there is a sequence $y \in \varphi^{-1}(\beta)$ that has a potential offspring with phenotype α, i.e.,

$$\mathsf{cl}(\alpha) = \varphi(\mathsf{cl}(\varphi^{-1})(\alpha)). \tag{7}$$

In fact, this closure function on phenotype space is the most restrictive (finest) one with the property that φ is continuous. We have argued in the previous section that in most cases a finer closure structure will reflect the practical accessibilities in phenotype space. It follows that the GP-map will in general not be continuous everywhere. We shall return to this topic in Sect. 7.

6 Fitness and Fitness Landscapes

Fitness landscapes were introduced in the 1930s by Sewall Wright [60,61] as a means of visualizing evolutionary adaptation. In this picture a population moves uphill on a kind of "potential function" due to the combined effects of mutation and selection, Fig. 4. We have seen in the previous sections, however, that the horizontal — phenotypic — axis, in many cases, is not a real line but a neighbourhood space. Alternatively, we may want to regard fitness landscapes as the graphs of functions from the genotype space with its neighbourhood structure into the real numbers $\mathbb{R}$. This latter case is usually studied in simulations of molecular evolution.

Fitness and energy landscapes have become a unifying theme in fields as diverse as drug design, spin glass physics, molecular structure, protein folding, combinatorial optimization, and evolutionary theory, see, e.g., [62] for a recent review. In each case, there is a function f, e.g., a molecular index, a Hamiltonian, a cost-function, or a fitness, that evaluates each member $x \in X$ of a (usually very large) configuration set X. These configurations can be (organic) molecules, spin configurations, conformations of a polypeptide chain, tours of a Travelling Salesman Problem, or genotypes of living organisms.

Characteristic properties of fitness landscapes such as local minimum and saddle points are inherently topological notions as well [63,64]. For instance,

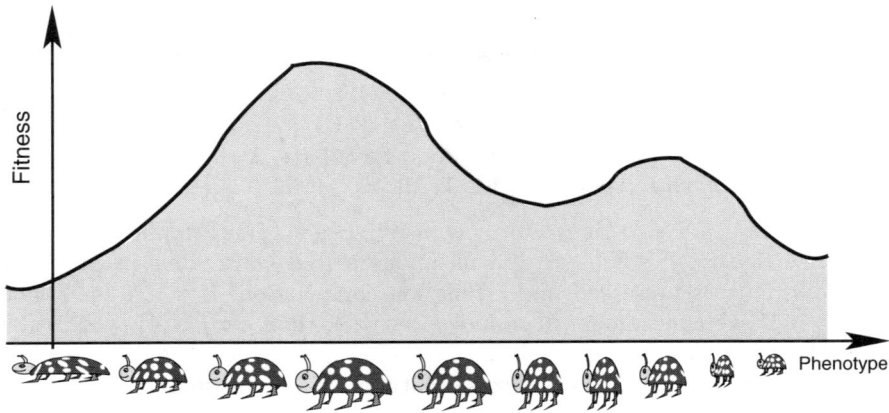

Fig. 4. Fitness landscapes assign fitness values to each (in this case) phenotypic state. The action of selection drives a population to local maximum of the fitness landscape.

a point $x \in X$ is a local minimum of $f : X \to \mathbb{R}$ if there is a neighbourhood $N \in \mathcal{N}(x)$ such that $f(y) \geq f(x)$ for all $y \in N$. It is instructive to consider also the following alternative definition. For each neighbourhood $U \in \mathcal{N}(x)$ there is a neighbourhood $N' \subseteq U$ such that $f(x) \leq f(y)$ for all $y \in N'$. Of course, in pretopological spaces both versions are equivalent: if $f(y) \geq f(x)$ for all y in a neighbourhood N then $N' = U \cap N$ is another, smaller, neighbourhood of x which obviously satisfies $f(x) \leq f(y)$ for all $x \in N'$. In general neighbourhood spaces, however, the second definition is strictly stronger because the intersection $U \cap N$ of two neighbourhoods is in general no longer a neighbourhood of x. This example highlights the pitfalls that appear because of the generality of the framework: "obvious" equivalences of properties that are familiar from topology textbooks might no longer be valid.

Related approaches to defining saddle points are discussed in [26]. The topological approach to fitness landscapes is largely unexplored. It appears promising, however, because it provides a common framework for the discrete landscapes of combinatorial optimization and for the continuum models that are used in evolutionary computation [65] and the analysis of chemical potential energy surfaces [66,67].

7 Evolutionary Trajectories

An evolutionary trajectory can be regarded as a function f from the time axis into phenotype space, where $f(t)$ represents the phenotype (e.g., the dominating phenotype in a population) at time t. (Alternatively, one might want to consider "populations", i.e., multisets $P \subseteq' Y$ of phenotypes.) Computer sim-

Fig. 5. Evolutionary trajectory. Here we show the time dependence of fitness for an adaptive walk with neutrality on an RNA landscape with a fitness function of the form $F(\alpha) = F_0 - d(\alpha, \alpha_0)$, where α_0 is a randomly chosen target structure and $d(\,.\,,\,.\,)$ denotes the number of base pairs by which two structures differ. Each attempted move in genotype space corresponds to one step on the time axis. Novel phenotypes (secondary structures) are indicated below in the string notation used by the **Vienna RNA Package** [53]. Discontinuous transitions w.r.t. the Fontana-Schuster topology are marked by a + sign.

ulations [16, 17] using the RNA model described in Fig. 2 reveal a pattern of periods of stasis with intermittent bursts of adaptive evolution that is reminiscent of the structure of the fossil record and consistent with *in vitro* evolution experiments [20, 21], see also [68]. The analysis of the patterns of changes along evolutionary trajectories obtained from RNA simulations lead to a notion of continuous versus discontinuous phenotypic transitions [14,18,69]. The topological language outlined here is the result of attempting to recast these findings in a more traditional mathematical framework [23, 24].

The time axis is usually described as well-known topological space, namely the real line $\mathbb{R}$ endowed with its standard topology. In the case of computer simulations and samples from the fossil record, which intrinsically represent time in the form of discrete steps, it is more natural to use the pretopology corresponding to the directed infinite path graph

$$\cdots \rightarrow \bullet \rightarrow \bullet \rightarrow \bullet \rightarrow \bullet \rightarrow \bullet \rightarrow \cdots$$

Let us write $\mathbb{T}$ for the generalized topological space that represents the time axis, and let $\xi : \mathbb{T} \rightarrow (X, \mathsf{cl})$ be a continuous function from the time axis into genotype space. Continuity means here simply that we assume that subsequently sampled genotypes are (easily) accessible from their predecessors. The composition $\varphi \circ \xi : \mathbb{T} \rightarrow (Y, \mathsf{cl})$ describes the sequence of phenotypes as a function of time, i.e., $\varphi(\xi(t))$ is the phenotype at time t. Note that $\varphi \circ \xi$ cannot be guaranteed to be continuous everywhere, since the GP-map φ will in general not be continuous.

The simplest approximation of an evolutionary process is the adaptive walk [70]. A population is represented as a single genotype which could be interpreted as the consensus genotype. In each time step a mutant is explored and accepted if the fitness does not decrease. Thus the fitness $f = F \circ \varphi \circ \xi$ is a monotonically increasing function of time. A typical trajectory is shown in Fig. 5. It does not differ qualitatively from the trajectories obtained in a more complex, population-based setting. In the RNA case, at least, there is a rather clear-cut distinction between "continuous transitions" (where $\varphi \circ \xi$ is locally continuous at a time t_0 where the phenotype changes) and discontinuous transitions. In particular, there is a close correspondence between continuity of a transition and the kind of structural changes that the RNA structure undergoes at the transition point. For more details we refer to [14, 18].

8 Characters as Factors of Phenotype Space

The correspondence of structural changes and transition type suggests that the topological language might be suitable to deal with a much more general question: Can one use the information about the evolutionary process that is represented by the topological structure of phenotype space to define a biologically meaningful character concept? In [24, 71] we proposed that the most promising avenue is to start with Lewontin's notion of "quasi-independence" [72] which was introduced to clarify the mechanistic assumptions underlying the adaptationist research program.

Explaining a character state as an adaptation caused by natural selection requires the assumption that the character state can be produced by mutation without significantly affecting the functionality and/or structure of the rest of the body. This notion does **not** assume that genetic and mutational variation among characters is stochastically independent (it may in fact be strongly correlated). All that is assumed is that genetic variation can be produced at not too low rate so that natural selection can adjust one character without permanently altering other attributes of the phenotype. We can therefore interpret the notion of quasi-independence as a statement about the accessibility relations in the phenotypic configuration spaces, namely that the phenotype space can be represented as a product of generalized topological space, Fig. 6.

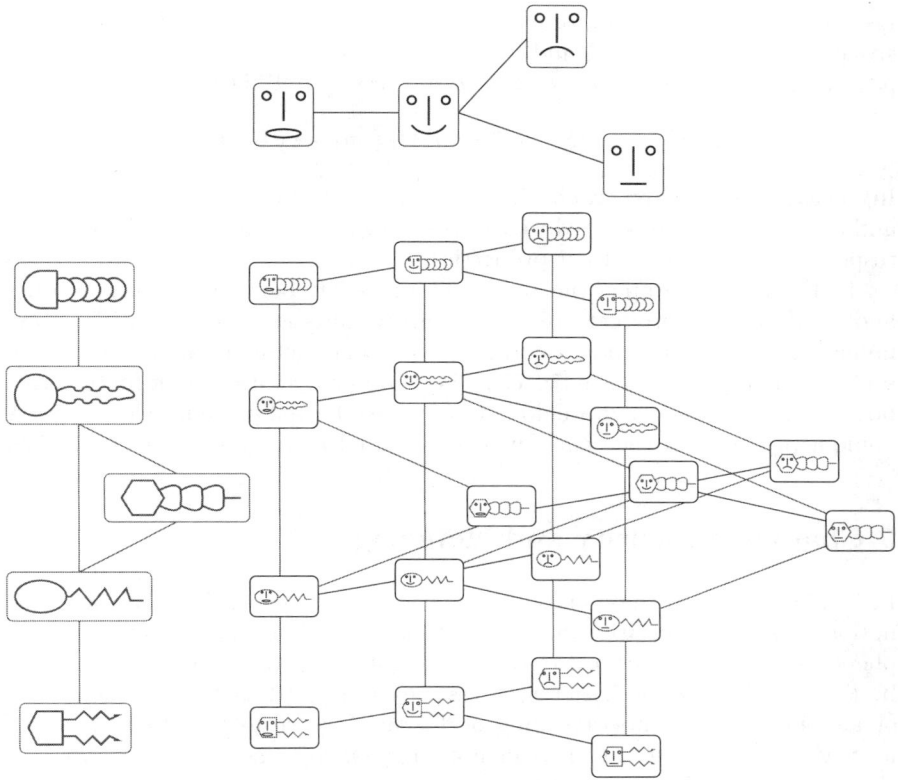

Fig. 6. Characters as factors of a graph. In this case we have two factors w.r.t. the Cartesian graph product, the "faces" and the "bodies" of the "animals". These are identified as the unique prime factors of their phenotype space graph. Note that while the representations of the "bodies" suggest "heads" and "tails" as finer subdivisions, these are not characters in our setting, because states of "heads" and "tails" do not vary independently here.

In the following few paragraphs we outline very briefly how this approach works at the formal level. First we recall the definition the usual topological product and one of its interesting variants.

Let $(X, \mathcal{N})$ and $(Y, \mathcal{M})$ be two isotone spaces defined by their neighbourhood functions. Then the *(canonical) product space* $(V, \mathcal{V}) = (X, \mathcal{N}) \times (Y, \mathcal{M})$ is defined on the set $V = X \times Y$ and has the following neighbourhood function $\mathcal{V}$: $U \in \mathcal{V}$ iff there is $N \in \mathcal{N}$ and $M \in \mathcal{M}$ such that $N \times M \subseteq U$. The space $(W, \mathcal{W}) = (X, \mathcal{N}) \square (Y, \mathcal{M})$ is the *inductive product* where $W = X \times Y$ and $U \in \mathcal{W}$ iff there is $N \in \mathcal{N}$ and $M \in \mathcal{M}$ such that $(N \times \{y\}) \cup (\{x\} \times M) \subseteq U$. When the neighbourhood function associated with each space is clear we write simply $V = X \times Y$ and $W = X \square Y$, respectively. The closure function of the inductive product can be characterized as follows. Consider two isotone spaces (X, c_X) and (Y, c_Y) and a subset $S \subseteq X \times Y$. Then $(p, q) \in \mathsf{cl}(S)$ iff $p \in c_X(\{x | (x, q) \in S\})$ or $q \in c_Y(\{y | (p, y) \in S\})$ [47, 17.D.7], i.e.,

$$\mathsf{cl}(S) = (c_X(S^{-1}[q]) \times \{q\}) \cup (\{p\} \times c_Y(S[p])) . \tag{8}$$

In the case of connected graphs, i.e., finite pretopological spaces, the canonical and the inductive product reduce to the strong and Cartesian graph products, respectively; see [73] for their properties.

In the case of graphs the factorizations w.r.t. both the strong product and the Cartesian product are unique (up to automorphisms) [74, 75]. Polynomial time algorithms for computing the prime factor decomposition of an undirected graph are known [76, 77]. This is probably true also for finite neighbourhood spaces although we do not yet have a formal proof for this claim. A "unique prime factor theorem" for closely related structures is given in [78, 79].

9 Character Identity and Homology

In [24, 71] we argue that quasi-independence is a local rather than a global notion. Local factorization means that the variational neighbourhood of a phenotype can be described only in a neighbourhood of a given phenotype by the combination of character states, i.e., the coordinates of "dimensions" or factors. The biological meaning of "locally factorizable" is that there are no variational limitations on realizing all possible combinations of *adjacent* character states. The range of phenotypes that can be described as a combination of states of a given set of characters, on the other hand, may of course be limited. For instance, it may be possible to describe all fish species by a combination of character states of the set of "fish characters", but there is no such set of characters which would describe the phenotypic disparity of all metazoans.

In mathematical terms the problem becomes to identify a factorizable subspace Q of (X, cl). In terms of graphs this means that we have to find induced subgraphs that are factorizable. Ideally, given a graph G one would like to find maximal induced subgraphs that have non-trivial prime factor decompositions. Unfortunately, an efficient algorithm for this *Induced Subgraph Factorization Problem* is not known at present.

The original definition of homology by Owen identified two characters as homologous if they are "*the same*" in some unspecified way. The meaning of "sameness" was implicitly defined through the morphological criteria used to distinguish between superficial and essential similarity, i.e., between analogy and homology. This notion was reinterpreted by Darwin with reference to a common ancestor. It can be called the "historical homology concept" since it is defined solely on the basis of historical, genealogical relationships. It does not, however, clarify what "character identity" means [80]. Alternatively, one can define homologues as clusters of observable attributes that remain stable during adaptive evolution by natural selection. This "biological homology concept" [8,9] is, in its definition, independent of relatedness by common descent and thus has an unclear relationship to the historical homology concept.

Both homology concepts and their relationship can be accommodated in the topological theory of character identity [71]. In the previous section we have identified characters with local factors in phenotype space. The question of character identity hence can be rephrased as follows: Suppose we are given two points x and y and factorizations of their neighbourhoods: can we identify factors of the neighbourhoods that correspond to each other? The first step towards answering this question is to clarify the relationship between different subspaces. More formally, consider a neighbourhood space (X, cl) and let (U, c_U) be a factorizable subspace, i.e., there are spaces (U_1, c_1) and (U_2, c_2) such that $(U, c_U) = (U_1, c_1) \times (U_2, c_2)$. Then every set of the form $A = A_1 \times A_2$ with $A_1 \subseteq U_1$ and $A_2 \subseteq U_2$ is also a factorizable subspace [71]. In particular, every point $x \in \mathsf{int}(U)$ is locally factorizable, i.e., every neighbourhood of x contains a factorizable neighbourhood. This property can then be used to establish the identity of characters (i.e., factors of local factorizations) at different points x and y, provided there is a connected factorizable set H such that $x, y \in \mathsf{int}(H)$, see Fig. 7a. The assumption of connectivity is crucial for the uniqueness of the factorization [73]. Overlapping factorizable regions may then be used to extend character identity to pairs of points that are not contained in a common factorizable region. The necessary condition is that there is a point $z \in \mathsf{int}(A) \cap \mathsf{int}(B)$ such that the local restrictions of the factors of A and B to sufficiently small neighbourhoods of z coincide. For the technical details we refer to [71].

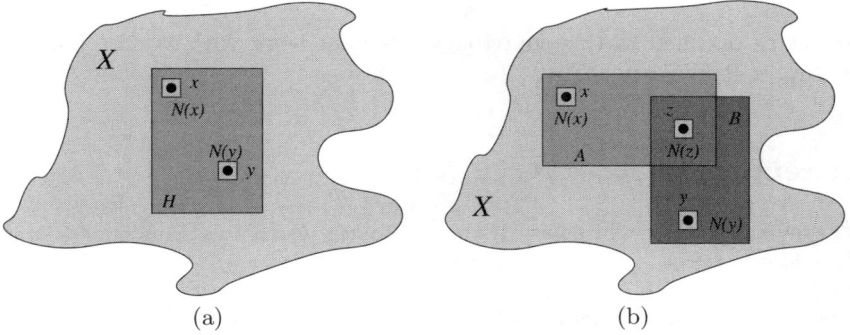

(a)   (b)

Fig. 7. A factorizable region H establishes the identity of characters between two points $x, y \in \mathsf{int}(H)$. In a second step overlapping factorizable regions A and B can mediate character identity via points in their common interior $z \in \mathsf{int}(A) \cap \mathsf{int}(B)$.

The identity of variational characters is therefore well defined (via overlapping regional factorizations) and determines a class of (in most cases) variationally connected phenotypes sharing a certain factor. Phenotypes which share a certain factor/character can therefore evolve into each other without going through states where the character is not defined. The notion of character identity based on quasi-independence is thus fully consistent with the historical homology concept: continuity of descent is *sufficient* to establish character identity. It is not a necessary condition, however, because nothing in the general theory prevents two lineages from evolving phenotypes which have the same variational character – however unlikely this scenario might be.

10 Concluding Remarks

In this contribution we have discussed a mathematical framework that is capable of describing important aspects of macroevolution within the Darwinian framework. The language is general enough to be applicable in the same way to both the discrete setting of sequences and the continuum domain of population genetics; nevertheless it is powerful enough, for example, to construct a theory for continuity in evolution and to yield a meaningful notion of character and homology. Another interesting consequence to developmental evolution is discussed in [81].

We have focused on the formal aspects of this language rather than on real or potential applications. The reason is that the kind of variational data that would be necessary for a non-trivial example are not readily accessible and that a crucial technical step, namely the factorizable induced subgraph problem, remains still to be solved.

Acknowledgments

The theory outlined in this contribution is joint work with Walter Fontana and Günter P. Wagner.

References

1. Futuyma, D.J.: Evolutionary Biology. Sinauer Associates, Sunderalnd, Massachusetts (1998)
2. Graur, D., Li, W.H.: Fundamentals of Molecular Evolution. Sinauer Associates, Sunderland, Massachusetts (2000)
3. Schlichting, C.D., Pigliucci, M.: Phenotypic Evolution: A Reaction Norm Perspective. Sinauer Associates, Sunderland, Massachusetts (1998)
4. Eldredge, N., Gould, S.J.: no title. In Schopf, T.J.M., ed.: Models in Paleobiology. Freeman, San Francisco (1972) 82–115
5. Maynard-Smith, J., Burian, R., Kauffman, S.A., Alberch, P., Campbell, J., Goodwin, B., Lande, R., Raup, D., Wolpert, L.: Developmental constraints and evolution. Quart. Rev. Biol. **60** (1985) 265–287

6. Schwenk, K.: A utilitarian approach to evolutionary constraint. Zoology **98** (1995) 251–262

7. Müller, G.B., Wagner, G.P.: Novelty in evolution: Restructuring the concept. Annu. Rev. Ecol. Syst. **22** (1991) 229–256

8. Wagner, G.P.: The biological homology concept. Annu. Rev. Ecol. Syst. **20** (1989) 51–69

9. Wagner, G.P.: The origin of morphological characters and the biological basis of homology. Evolution **43** (1989) 1157–1171

10. Fontana, W., Buss, L.W.: "The arrival of the fittest": Towards a theory of biological organization. Bull. Math. Biol. **56** (1994) 1–64

11. Schuster, P.: Artificial life and molecular evolutionary biology. In Moran, F., Moreno, A., Merelo, J.J., Chacon, P., eds.: Advances in artificial life. Proceedings of Third European Conference on Artificial Life, Canada 1995. Volume 929 of Lecture Notes in Artificial Intelligence. Springer-Verlag, Berlin (1995) 3–19

12. Lewontin, R.C.: The Genetic Basis of Evolutionary Change. Columbia University Press, New York (1974)

13. Wagner, G.P., Altenberg, L.: Complex adaptations and the evolution of evolvability. Evolution **50** (1996) 967–976

14. Fontana, W., Schuster, P.: Continuity in evolution: On the nature of transitions. Science **280** (1998) 1451–1455

15. Fontana, W., Schnabl, W., Schuster, P.: Physical aspects of evolutionary optimization and adaption. Phys. Rev. A **40** (1989) 3301–3321

16. Schuster, P., Fontana, W., Stadler, P.F., Hofacker, I.L.: From sequences to shapes and back: A case study in RNA secondary structures. Proc. R.Soc. B **255** (1994) 279–284

17. Huynen, M.A., Stadler, P.F., Fontana, W.: Smoothness within ruggedness: The role of neutrality in adaptation. Proc. Natl. Acad. Sci. USA **93** (1996) 397–401

18. Fontana, W., Schuster, P.: Shaping space: The possible and the attainable in RNA genotype-phenotype mapping. J. Theor. Biol. **194** (1998) 491–515

19. Ancel, L., Fontana, W.: Plasticity, evolvability and modularity in RNA. J. Exp. Zool. (Mol. Dev. Evol.) **288** (2000) 242–283

20. Spiegelman, S.: An approach to experimental analysis of precellular evolution. Q. Rev. Biophys. **4** (1971) 213–253

21. Lenski, R.E., Travisano, M.: Dynamics of adaptation and diversification: A 10,000-generation experiment with bacterial populations. Proc. Natl. Acad. Sci. USA **91** (1994) 6808–6814

22. Szostak, J.W., Ellington, A.D.: *In Vitro* selection of functional RNA sequences. In Gesteland, R.F., Atkins, J.F., eds.: The RNA World. Cold Spring Harbor Laboratory Press, Plainview, New York (1993) 511–533

23. Cupal, J., Kopp, S., Stadler, P.F.: RNA shape space topology. Artif. Life **6** (2000) 3–23

24. Stadler, B.M.R., Stadler, P.F., Wagner, G.P., Fontana, W.: The topology of the possible: Formal spaces underlying patterns of evolutionary change. J. Theor. Biol. **213** (2001) 241–274

25. Stadler, B.M.R., Stadler, P.F., Shpak, M., Wagner, G.P.: Recombination spaces, metrics, and pretopologies. Z. Phys. Chem. **216** (2002) 217–234

26. Stadler, B.M.R., Stadler, P.F.: Generalized topological spaces in evolutionary theory and combinatorial chemistry. J. Chem. Inf. Comput. Sci. **42** (2002) 577–585

27. Maynard-Smith, J.: Natural selection and the concept of a protein space. Nature **225** (1970) 563–564
28. Eigen, M., Schuster, P.: The Hypercycle. Springer-Verlag, New York, Berlin (1979)
29. Gitchoff, P., Wagner, G.P.: Recombination induced hypergraphs: a new approach to mutation-recombination isomorphism. Complexity **2** (1996) 37–43
30. Shpak, M., Wagner, G.P.: Asymmetry of configuration space induced by unequal crossover: Implications for a mathematical theory of evolutionary innovation. Artif. Life **6** (2000) 25–43
31. Changat, M., Klavžar, S., Mulder, H.M.: The all-path transit function of a graph. Czech. Math. J. **51** (2001) 439–448
32. Stadler, P.F., Wagner, G.P.: The algebraic theory of recombination spaces. Evol. Comput. **5** (1998) 241–275
33. Stadler, P.F., Seitz, R., Wagner, G.P.: Evolvability of complex characters: Population dependent Fourier decomposition of fitness landscapes over recombination spaces. Bull. Math. Biol. **62** (2000) 399–428
34. Page, W.: Topological Uniform Structures. Dover Publications, Mineola, New York (1994)
35. Gaal, S.A.: Point Set Topology. Academic Press, New York (1964)
36. Steen, L.A., Seebach, Jr., J.A.: Counterexamples in Topology. Holt, Rinehart & Winston, New York (1970)
37. Davey, B.A., Priestley, H.A.: Introduction to Lattice and Order. Cambridge University Press, Cambridge (1990)
38. Kuratowski, C.: Sur la notion de limite topologique d'ensembles. Ann. Soc. Polon. Math. **21** (1949) 219–225
39. Eckhardt, U., Latecki, L.: Digital topology. Technical Report 89, Hamburger Beitr. z. Angew. Math. A (1994)
40. Smyth, M.B.: Semi-metric, closure spaces and digital topology. Theor. Comput. Sci. **151** (1995) 257–276
41. Pfaltz, J.: Closure lattices. Discrete Math. **154** (1996) 217–236
42. Galton, A.: Continuous motion in discrete space. In Cohn, A.G., Giunchiglia, F., Selman, B., eds.: Principles of Knowledge Representation and Reasoning: Proceedings of the Seventh International Conference (KR2000), Morgan Kaufmann, San Francisco (2000) 26–37
43. Marchand-Maillet, S., Sharaiha, Y.M.: Discrete convexity, straightness, and the 16-neighborhood. Comput. Vision Image Understanding **66** (1997) 316–329
44. Largeron, C., Bonnevay, S.: A pretopological approach for structural analysis. Inf. Sci. **144** (2002) 169–185
45. F. LeBourgeois, M. Bouayad, H.E.: Structure relation between classes for supervised learning using pretopology. In Fifth International Conference on Document Analysis and Recognition (1999) 33–36
46. Speroni di Fenizio, P., Banzhaf, W., Ziegler, J.: Towards a theory of organizations. In Proceedings of the Fourth German Workshop on Artificial Life GWAL'00. (2002)
47. Čech, E.: Topological Spaces. Wiley, London (1966)
48. Jablonka, E., Lamb, R.M.: Epigenetic Inheritance and Evolution. Oxford University Press, Oxford (1995)
49. Schultes, E., Bartel, D.: One sequence, two ribozymes: Implications for the emergence of new ribozyme folds. Science **289** (2000) 448–452

50. Mills, D., Peterson, R., Spiegelman, S.: An extracellular Darwinian experiment with a self-duplicating nucleic acid molecule. Proc. Natl. Acad. Sci. USA **58** (1967) 217
51. Klug, S.J., Famulok, M.: All you wanted to know about SELEX. Mol. Biol. Rep. **20** (1994) 97–107
52. Zuker, M., Sankoff, D.: RNA secondary structures and their prediction. Bull. Math. Biol. **46** (1984) 591–621
53. Hofacker, I.L., Fontana, W., Stadler, P.F., Bonhoeffer, L.S., Tacker, M., Schuster, P.: Fast folding and comparison of RNA secondary structures. Monatsh. Chem. **125** (1994) 167–188
54. Shelton, V.M., Sosnick, T.R., Pan, T.: Applicability of urea in the thermodynamic analysis of secondary and tertiary RNA folding. Biochemistry **38** (1999) 16831–16839
55. Mathews, D., Sabina, J., Zucker, M., Turner, H.: Expanded sequence dependence of thermodynamic parameters provides robust prediction of RNA secondary structure. J. Mol. Biol. **288** (1999) 911–940
56. Babajide, A., Hofacker, I.L., Sippl, M.J., Stadler, P.F.: Neutral networks in protein space: A computational study based on knowledge-based potentials of mean force. Folding Des. **2** (1997) 261–269
57. Stadler, P.F.: The genotype phenotype map. Konrad Lorenz Institute Workshop on Biological Information. (2002)
58. Hammer, P.C.: Extended topology: Continuity I. Port. Math. **25** (1964) 77–93
59. Gnilka, S.: On continuity in extended topologies. Ann. Soc. Math. Pol., Ser. I, Commentat. Math. **37** (1997) 99–108
60. Wright, S.: The roles of mutation, inbreeding, crossbreeeding and selection in evolution. In Jones, D.F., ed.: Proceedings of the Sixth International Congress on Genetics. Volume 1. (1932) 356–366
61. Wright, S.: "Surfaces" of selective value. Proc. Natl. Acad. Sci. USA **58** (1967) 165–172
62. Reidys, C.M., Stadler, P.F.: Combinatorial landscapes. SIAM Rev. **44** (2002) 3–54
63. Flamm, C., Hofacker, I.L., Stadler, P.F., Wolfinger, M.T.: Barrier trees of degenerate landscapes. Z. Phys. Chem. **216** (2002) 155–173
64. Stadler, P.F.: Fitness landscapes. In Lässig, M., Valleriani, A., eds.: Biological Evolution and Statistical Physics. Springer-Verlag, Berlin (2002) 187–207
65. Rechenberg, I.: Evolutionstrategie. Frommann-Holzboog, Stuttgart (1973)
66. Heidrich, D., Kliesch, W., Quapp, W.: Properties of Chemically Interesting Potential Energy Surfaces. Volume 56 of Lecture Notes in Chemistry. Springer-Verlag, Berlin (1991)
67. Mezey, P.G.: Potential Energy Hypersurfaces. Elsevier, Amsterdam (1987)
68. Schuster, P.: Evolution in silico and in vitro: The RNA model. Biol. Chem. **382** (2001) 1301–1314
69. Fontana, W.: Modelling "evo-devo" with RNA. BioEssays **24** (2002) 1164–1177
70. Kauffman, S.A.: The Origin of Order. Oxford University Press, New York, Oxford (1993)
71. Wagner, G., Stadler, P.F.: Quasi-independence, homology and the unity of type: A topological theory of characters. J. Theor. Biol. **220** (2003) 505–527
72. Lewontin, R.C.: Adaptation. Sci. Am. **239** (1978) 156–169
73. Imrich, W., Klavžar, S.: Product Graphs: Structure and Recognition. Wiley, New York (2000)

74. Dörfler, W., Imrich, W.: Über das starke Produkt von endlichen Graphen. Österreich. Akad. Wiss. Math.-Natur. Kl. S.-B. II **178** (1970) 247–262

75. McKenzie, R.: Cardinal multiplication of structures with a reflexive multiplication. Fundam. Math. **70** (1971) 59–101

76. Feigenbaum, J., Schäffer, A.A.: Finding the prime factors of strong direct products of graphs in polynomial time. Discrete Math. **109** (1992) 77–102

77. Imrich, W.: Factoring cardinal product graphs in polynomial time. Discrete Math. **192** (1998) 119–144

78. Lovász, L.: Operations with structures. Acta Math. Acad. Sci. Hung. **18** (1967) 321–328

79. Lovász, L.: Unique factorization in certain classes of structures. In: Mini-Conference on Universal Algebra, Szeged 1971, Bolyai Janos Math. Soc. (1971) 24–25

80. Wagner, G.P.: Homology and the mechanisms of development. In Hall, B.K., ed.: Homology: The Hierarchical Basis of Comparative Biology. Academic Press, San Diego, California (1994) 273–299

81. Wagner, G.P.: What is the promise of developmental evolution? Part II: a causal explanation of evolutionary innovations may be impossible. J. Exp. Zool. (Mol. Dev. Evol.) **291** (2001) 305–309

Models of Genome Evolution

Yi Zhou[1] and Bud Mishra[2,3]

[1] Biology Department, New York University
joey@cs.nyu.edu
[2] Courant Institute of Mathematical Sciences, New York University
mishra@cs.nyu.edu
[3] Watson School of Biological Sciences, Cold Spring Harbor Laboratory

Summary. The evolutionary theory, "evolution by duplication", originally proposed by Susumu Ohno in 1970, can now be verified with the available genome sequences. Recently, several mathematical models have been proposed to explain the topology of protein interaction networks that have also implemented the idea of "evolution by duplication". The power law distribution with its "hubby" topology (e.g., P53 was shown to interact with an unusually large number of other proteins) can be explained if one makes the following assumption: new proteins, which are duplicates of older proteins, have a propensity to interact only with the same proteins as their evolutionary predecessors. Since protein interaction networks, as well as other higher-level cellular processes, are encoded in genomic sequences, the evolutionary structure, topology, and statistics of many biological objects (pathways, phylogeny, symbiotic relations, etc.) are rooted in the evolution dynamics of the genome sequences. Susumu Ohno's hypothesis can be tested *"in silico"* using Polya's urn model. In our model, each basic DNA sequence change is modelled using several probability distribution functions. The functions can decide the insertion/deletion positions of the DNA fragments, the copy numbers of the inserted fragments, and the sequences of the inserted/deleted pieces. Moreover, those functions can be interdependent. A mathematically tractable model can be created with a directed graph representation. Such graphs are Eulerian and each possible Eulerian path encodes a genome. Every "genome duplication" event evolves these Eulerian graphs, and the probability distributions and their dynamics themselves give rise to many intriguing and elegant mathematical problems. In this chapter, we explore and survey these connections between biology, mathematics and computer science in order to reveal simple, and yet deep, models of life itself.

1 Introduction

The genome of an organism is a collection of its genes, encoded by four chemical *bases* in its DNA (DeoxyriboNucleic Acid), and forms the genetic core of a cell. The genes ultimately encode for the proteins (chains of amino acids) and, in turn, the genes themselves are regulated by transcription factors and other operons, many of which are proteins. The sequences of amino acids, specified by the DNA through transcription and translation processes, determine the three-dimensional structure and biochemical properties of the proteins as well as the nature of their interactions. Furthermore, mRNA stability, protein degradation, post-translational modifications, and many other biochemical processes tightly regulate the time-constants involved in the resulting biochemical machinery. Proteins also associate in complexes to form *dimers* (pairs of proteins), *trimers* (triplets), and *multimers*. An *isoform* of a protein is a slightly different protein with a closely related sequence, and often shares similar functional properties, e.g., enzymatic reactions, but is regulated differently. However, we know that this complex machinery of life has evolved over several billion years through random mutations and rearrangements of the underlying genomes while being shaped by the selection processes. We ask the following questions: Are the processes altering, experimenting on, and correcting the genomes completely untamed and haphazardly? If not, what signatures have they left on the genomes, proteomes, pathways, organs, and organisms? What structures have they imposed on these biological elements and their interactions? We posit that a better understanding of biology is hinged on a deep information-theoretic study of evolving genomes and their roles in governing metabolic and regulatory pathways.

We begin with the following account. Genomes are not static collections of DNA materials. Various biochemical and cellular processes – including point mutation, recombination, gene conversion, replication slippage, DNA repair, translocation, imprinting, and horizontal transfer – constantly act on genomes and drive the genomes to evolve dynamically. These alterations in the genomic sequences can further lead to the corresponding changes in the higher-level cellular information (transcriptomes, proteomes, and interactomes), and are crucial in explaining the myriad of biological phenomena in the higher-level cellular processes. However, until recently, the lack of sufficient historical data and the complexity of biological processes involved have hampered the development of a rigorous, faithful, and yet simple abstract model for genome evolution.

Present genomes can be viewed as a snapshot of an ongoing genome evolution process. Although it would be ideal, it is usually impossible to base genome evolution studies on ancient genome samples. Fortunately, various historical evolutionary events leave their "signatures" in the present sequences, which can be deciphered by statistical analyses on a family of genomes that are currently available. With the development of high-throughput experimental technology, the flow of information at different levels of biology (genome, proteome, transcriptome, and interactome) is increasing dramatically. By analysing and comparing this data, we are now able to look for the structure of cellular processes and the dynamics of the evolution process driving it.

A survey of the literature reveals many interesting statistical analyses of various kinds on genomic and proteomic data. Among the large collection of results, it is worthwhile to note that many interesting statistical characteristics are shared by data from all organisms, from different cellular processes, as well as at various scales. For example, research during the last decade reveals long-range correlation (LRC) between single nucleotides in the genomic sequences of various species from different kingdoms [1, 20], and in different regions of the genomic sequences. Furthermore, the LRC is persistent in all the genomic sequences examined. This indicates that on the single-nucleotide level, genomes have evolved independently to share a common scale-free global structure. On a slightly larger scale than single nucleotides, the short words in DNA sequences (mers, oligonucleotides) and protein sequences (short peptides) also seem to display similar generic statistical properties. The frequency distributions of the short words in the genomic and proteomic sequences from various organisms are found to follow a power law [14, 18], a feature often found in linguistic studies. Those general properties are further reflected in the higher-level cellular processes. As the large-scale metabolism networks and protein interaction networks in some model organisms become available, e.g., metabolic networks in *E. coli* [8] or protein interaction networks in *S. cerevisiae* [3] and *H. pylori* [11], the topology of those networks is analysed [9,19] and is found to be characteristic of a group of graphs known as scale-free networks [4]. Scale-free networks are characterized by their "hubby" structures associated with a power-law distribution of their connectivities, and can be created by an evolution process following a "rich gets richer" rule. All those statistical features (the positive correlation between single nucleotides, the over-representation of high-frequency words in genomes and proteomes, the "hubbiness" due to highly-connected nodes in the protein and genetic networks) can be viewed as the different aspects of an underlying generic structure. (see Sect. 3 for more examples). Different organisms preserve a common structure in their cellular processes despite drastically different evolutionary environments. This common structure may reflect the most fundamental processes in biology.

The positive feedback mechanism suggested by the highly correlated structures found in various data is reminiscent of the "evolution by duplication" theory originally proposed by Ohno in the 1970s [15]. Based on this theory, we develop a mathematical model to explain the observations in our mer-frequency distribution analysis. The model is an extension of Polya's urn model [13], and considers genome evolution as a stochastic process with three main events: *substitution*, *deletion*, and *duplication*. We study a simpler version of the model in numerical simulation as well as a more realistic, thus more complicated, version of the model in a large-scale *in silico* evolution simulation. The simple model fits the real-world data for mer-frequency distributions. These results suggest that despite the highly diversified evolutionary environment for different organisms, the essential composition of the evolutionary dynamics is commonly shared. A simple stochastic process (*substitution*, *deletion*, and *duplication*) can describe the recurrent pattern in the statistical signatures of different organisms. The model is extremely intriguing as it suggests that all the complexities found in life can be the result of a simple stochastic evolutionary process.

2 Evolution by Duplication

Susumu Ohno proposed an evolutionary theory called "evolution by duplication" [15]. Although not explicitly stated, his theory suggested a "rich gets richer" rule in genome evolution. The theory argues that the evolutionary advantage of evolution by duplication lies in the promise that with an extra copy, the selection pressure on the gene is somewhat relaxed. Since the original function can be maintained efficiently by either copy of the duplicated gene, the other copy can undergo various modifications, increasing the chance of the organism obtaining a new advantageous gene. Gene duplication can speed up the search for higher fitness in various ways [23]: it can adjust gene dosage; attain a permanent heterozygous advantage by incorporating two former alleles into the genome; allow more specialized functions by differential regulation of the duplicated genes; or create a new gene with a diverged function.

The duplication process mentioned in the theory can be well explained by molecular biology. There are various molecular mechanisms that can cause DNA duplication of different size ranges. For example, during DNA replication, replication slippage [6] can introduce small insertions or deletions locally, when the newly replicated DNA fragments misalign to the template. The misalignment, usually triggered by tandem repeats or secondary structure in the template DNA strand, causes the DNA polymerase to pause, dissociate, and continue an erroneous strand extension after re-association. During meiotic cell division, recombinations between two DNA molecules occur through cross-overs between corresponding homologous regions. Unequal cross-overs between two DNA molecules bearing successive repeated fragments will result in the duplication or deletion of the repeated units in the daughter cells [23]. Another process that can introduce duplications or deletions of relatively large sizes and globally in the genome involves mobile DNA elements (insertion elements, transposons, and retrotransposons) [23]. The mobile elements can be either excised or copied from their original positions, and subsequently inserted elsewhere in the genome where target sequences can be found. The frequencies and sizes of the deletions or insertions vary with specific elements. Since the target sequences are widely distributed in the whole genome, the mobile elements can essentially affect sequence changes on the whole-genome range. In summary, duplications and deletions in genomes can be fully justified by the well-known molecular mechanisms.

However, the duplication dynamics do not proceed completely unopposed – the cell also possesses DNA repair machineries to counteract the changes made in the sequences and prevent genomes from changing too rapidly [7]. For example, the mismatch repair (MMR) mechanism is mainly responsible for correcting most of the deletions and insertions of various sizes. Therefore, duplications, deletions, and other changes in genomes are the results of the interactions between the molecular mechanisms leading to genomic sequence changes and the surveillance system of the cell.

If we assume that the target gene of every duplication is randomly chosen from the genes that are already in the genome, then we have a realization of Polya's urn model [13]. Therefore, under the "evolution by duplication" theory, genome evolution can be viewed as a stochastic duplication process that can lead to a highly correlated structure with an over-abundance of some elements.

Several mathematical models recently proposed to explain the topology of protein interaction networks have also implemented the idea of "evolution by duplication" [2,10,17]. The power-law distribution with its "hubby" topology (e.g., P53 was shown to interact with an unusually large number of other proteins [12]) can be explained if one makes the assumption that new proteins which are duplicates of older proteins have a propensity to interact only with the same proteins as their evolutionary predecessors. Since the protein interaction networks, as well as other higher-level cellular processes, are encoded in genomic sequences, the evolution of their topology is rooted in the genomic sequence changes. Therefore, we believe that a more general model of "evolution by duplication" at the genomic level should explain the common pattern observed at various scales and different cellular information levels, and may be exploited prudently in the design of better bioinformatics algorithms.

3 Genomic Data Analysis

Large-scale genomic data analysis is the essential starting point in the search for the main players during genome evolution. Previous researches have provided statistical evidence favoring a general genome evolution dynamic, "evolution by duplication". Here, we report further discoveries that support such a hypothesis. We have examined the statistical properties of the distribution of short words in various whole genomes and proteomes. Our results confirm and extend the previous conclusion that there is an over-representation of high-frequency words in all the sequences studied. Furthermore, our analysis of the distribution of the end-points of putative large segmental duplications in the human genome provides convincing evidence that duplications tend to occur more often around the prior duplication sites. In other words, the already duplicated segments are more likely to be duplicated again, thus the already over-represented segments tend to be more over-represented, while the other segments are more likely to be suppressed. The duplication dynamics implied by the end-point distribution analysis may explain our observations on protein domain family sizes, which follow a power-law distribution and are characterized by an over-representation of larger families.

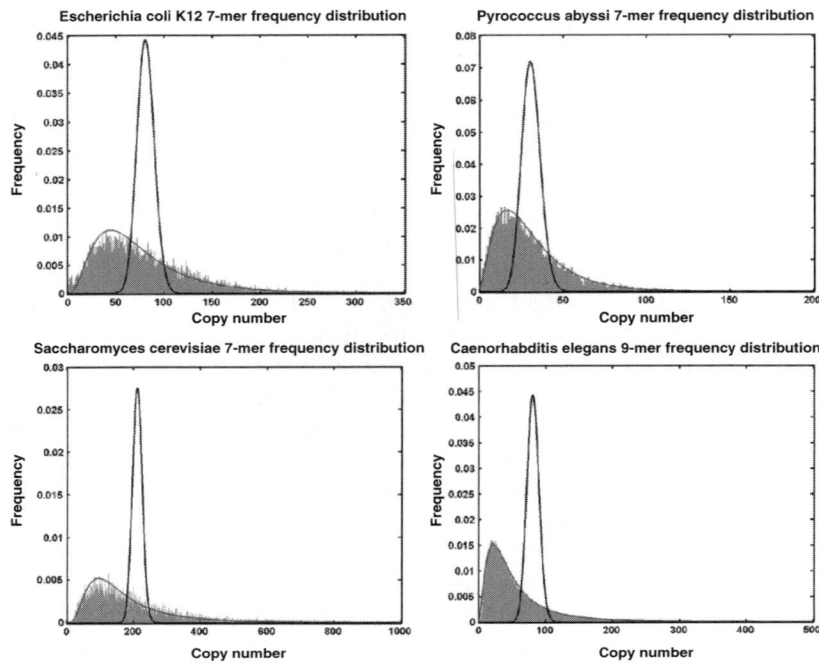

Fig. 1. Mer-frequency distribution examined in some genomes. The plots show the non-overlapping mer frequency distribution (shaded bars) in the genomes of a eubacteria (*E. coli* K12), an archea (*P. abyssi*), and two eukaryotes (*S. cerevisiae* and *C. elegans*). When compared with the expected distribution from a random sequence of the same length (black line), the distributions from real sequences consistently show an over-representation of high-copy mers. Our simulation results (gray line) from the simple graph model closely fit the "real" mer-frequency distribution. Given that we have only two free parameters (q = single point mutation probability and p_1/p_0 = ratio of probabilities of duplication over deletion, see below) in the model, the data-fitting is extremely convincing.

3.1 Mer-frequency Distribution Analysis

To study the statistics of short words in different whole genomic sequences, we performed a large-scale non-overlapping mer-frequency distribution analysis. The experiment was conducted on all the reasonable mer-sizes, covering almost all the presently available whole genomic sequences and including various organisms from all the kingdoms. To avoid the complication of inversions, we treated two inversely complementary mers as one species. (For example, 5′-ATCG-3′ and 5′-CGAT-3′ are counted as one mer species, i.e., their frequencies are combined.) Therefore, for mer-size l, there are $\frac{4^l}{2}$ species of l-mers. From our results, it is clear that the mer-frequency distributions from all the genomic data examined deviate from the random distribution (see Fig. 1 for some of the results). Furthermore, they are all characterized by the same type of deviation – over-representation of high-frequency mers.

We have also looked at the mer-frequency distribution in just the coding sequences, and the distribution of amino-acid word frequency in the corresponding proteome sequences. (For a length of n, there are 20^n species of different amino-acid words.) Both results share the same type of deviation from the random distribution that is observed in the whole genomic sequences (data not shown).

A simple simulation of "evolution by duplication" was also performed, where a short random sequence (1000bp) was allowed to evolve to a final length of 500Kb by duplicating fragments randomly chosen from itself. The deviation in the mer-frequency distribution of the final sequence from a random sequence closely resembles the pattern seen in real genomes (see Figure 2). Therefore, the particular statistics of mers in genomes and short amino-acid words in proteomes can be simply due to the duplication processes during genome evolution.

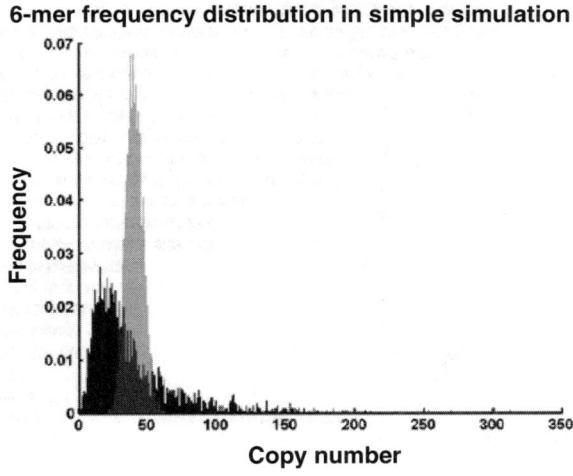

6-mer frequency distribution in simple simulation

Fig. 2. The 6-mer frequency distribution of the resulting sequence of a simple "evolution by duplication" simulation. The initial condition is a random sequence of length 1000bp. The sequence is evolved through multiple iterations until it reaches a length of 500Kb. In each iteration, a fragment of length uniformly randomly distributed from 1 to 100bp is randomly chosen from the sequence, duplicated, and reinserted randomly into the sequence. The dark bars in the plot show the 6-mer frequency distribution of the final sequence from the simulation. The gray bars show the 6-mer frequency distribution of a random sequence of the same length.

3.2 Analysis of the End-Points of Potential Large Segmental Duplication Fragments in Human Genome

As mentioned above, the results of various statistical analyses on genomic data have suggested that there is a generic evolution dynamic dominated by duplication. To justify this hypothesis, it is important to study the dynamic of duplication processes. Although the exact molecular mechanisms that cause duplications are not fully understood, we can approach the problem indirectly by looking at the distribution of the most recent segmental duplications in genomes.

Recently, intensive large segmental duplications (both intra- and inter-chromosomal) have been reported in the assembled human genome [21], and the potential large duplicated regions (>500bp, >95% identity) have been

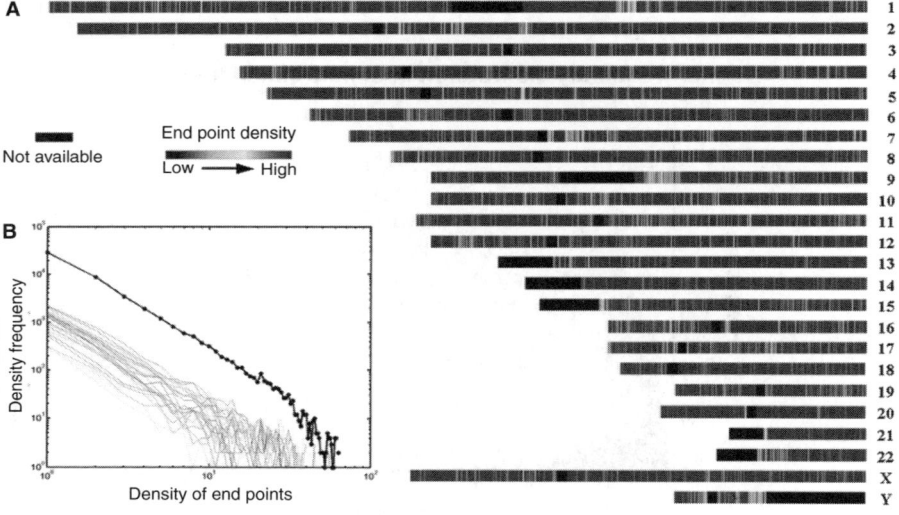

Fig. 3. The distribution of the potential duplication "hot-spots" on the human genome. A. The distribution of the duplicated segment end-points on the chromosomes (over windows of size 1Kb). The "hot-spot" density is coded (see the gray-scale bar). The dark areas represent chromosomal regions where no reference sequences are available. There is a tendency for areas with high densities to cluster together on the chromosomes. B. The distribution of the end-point densities on a log-log scale. The X axis shows the number of end-points in non-overlapping windows of size 1Kb (density of end-points), starting from 1. The Y axis indicates the number of non-overlapping 1Kb windows containing a given number of end-points (density frequency). It is clear that the logarithmic plots form a linear relationship, both in the whole-genome range (thick black points) and in the individual chromosome range (multiple thin gray lines).

mapped out in pairs under the standard sequence homology criteria [22]. Although the exact molecular mechanism is unclear, one could hypothesize that some of the processes involve single- or double-strand DNA breakage at the time of duplication, like homologous or heterologous recombination, and transposition. The end-points of the mapped-out duplicated segments are good candidate sites where such processes initiate or terminate at the time of duplication.

Since the end-points can be viewed as the signatures left by the duplication process over the genome evolutionary history, their distribution along the genome could reveal some dynamic features of the duplication process. To verify this hypothesis about the "density" distribution of the end-points, we first fragmented the genome into non-overlapping windows of a fixed size. The number of end-points covered by each window was treated as its local "density" over the corresponding genomic region. When we plotted the histogram of the end-point density over a chromosome or over the whole genome, we discovered another power-law distribution (Fig. 3). This implies that duplications tend to happen more often at the previous duplication sites, which is driven by a positive feedback dynamic. To further verify this interpretation, we performed a correlation test (detrended fluctuation analysis) [5] on the series of densities along the genome in relation to the distances between them. The result of the test indicated a positive correlation between neighbouring densities of end-points. Such a positive correlation suggests that over the evolutionary history, consecutive segmental duplications occur favorably near or on some previously duplicated segments, and are absent elsewhere.

3.3 Protein Domain Family Size Distribution

In our mer-frequency distribution analysis, the chosen mer sizes range up to 12 nucleotides. In comparison to other functional elements in the genome, the mers are of the smallest scale. To check whether the generic structure observed on such a small scale also persists on a bigger scale, we studied the protein domain family size distribution.

The protein domain families in different organisms are extracted from the protein family database InterPro [16]. The sizes of the domains are mostly above 50 amino acids (150 nucleotides) – a bigger scale than that of the mer analysis. The analysis of domain families is based on sequence signature and homology. A family of protein domains found by this method can be viewed as a cluster of amino acid sequences that share enough similarity with each other and have maintained their critical sequences. When the histograms of the sizes of those protein domain families from various organisms were plotted on a log-log scale, a linear relationship was observed in all cases, including in *E. coli* K12, *P. abyssi*, *S. cerevisiae*, and *H. sapiens* (Fig. 4). Therefore, the domain family size distribution, or, more generally, the cluster size of homologous amino acid sequences, seems to follow a power-law distribution. Here, on this larger scale, we observed the same deviation pattern from the

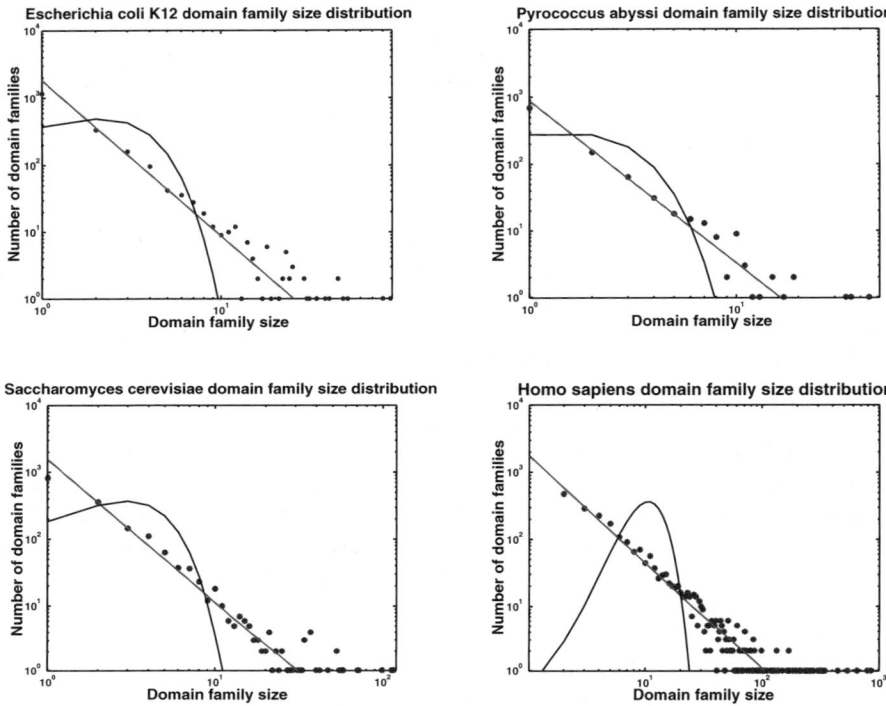

Fig. 4. Protein domain family size distribution examined in some genomes. The plots show the protein domain family size distribution in the corresponding proteomes (small circles). The protein domain family data is extracted from the Inter-Pro database. The plots are on a log-log scale. The almost linear shape (straight line) on such a plot indicates a power-law relationship between a domain family size and the number of domain families of that size. Therefore, the protein domain family size distributions are also characterized by an over-representation of large size families when compared with uniformly random distributions (black curves).

random distribution as in the small-scale mer distribution analysis, which further confirms the effects of duplication processes during genome evolution.

4 Evolutionary Models

The results from genomic data analyses at different scales, at different levels, and in different organisms repeatedly show the same pattern (over-representation of high-frequency elements). This consistently suggests a generic evolutionary dynamic involving positive feedback as postulated by Ohno's theory of "evolution by duplication." If we assume that the target fragment of every duplication is uniformly randomly chosen from the genome, then the fragments that are already over-represented in the genome will have a higher probability of being duplicated again. Similarly, the fragments that are initially under-represented in the genome will be further suppressed. To further verify the theory quantitatively, we develop a mathematical model based on the theory.

4.1 Graph Model

We develop a Eulerian graph model to explain the frequency distribution of non-overlapping mers (all the different species of oligonucleotides of a particular size) in various genomic sequences. The model aims to capture the parsimonious processes needed to recover the dynamics involved in genome evolution. Yet, it preserves enough fidelity to validate biological reality. The processes included in the model are: *duplication*, *deletion*, and *substitution*. The parsimony of the model can be inferred from the fact that the omission of any of the three processes renders the model unsuitable to fit real genomic data. In the model, a genome is represented by a Eulerian graph. Each mer species of a particular length is represented by a node. Whenever two non-overlapping mers are immediately adjacent to each other in the genome, they are connected by an additional directed edge. Without loss of generality, the edges are always directed from the 5′ end to the 3′ end. Therefore, the number of directed edges from node i to node j ($k_{i,j}$) indicates how many times the ith mer is immediately adjacent to the 5′ end of the jth mer. We use k_i to represent both the out-degree ($k_i{}^{out}$) and the in-degree ($k_i{}^{in}$) of the node i, since due to the Eulerian property of the graph, each node has identical in- and out-degrees, each being equal to the copy number of the corresponding mer in the genome. For mers of size l, and a genome of length L, the graph will have a total of $N = \frac{4^l}{2}$ nodes and $E = \frac{L}{l} = \sum_{i=1}^{N} k_i$ edges. Such graphs are Eulerian and each possible Eulerian path in the non-trivial (non-singleton) connected component encodes a genome. The genomes represented by the same graph share the same mer-frequency distributions but not necessarily the same arrangement of mers.

The evolution of a genome is modelled as a stochastic evolution process of the graph that goes through multiple iterations. The model assumes that all the presently existing genomes originated from a proto-genome, which is very small and has randomly distributed mers. Thus, the initial graph is a random graph with a small average degree. In each iteration, one of the three possible processes occurs: *duplication* of a chosen mer (with probability p_1), *deletion* of a chosen mer (with probability p_0), or *substitution* of a chosen mer by another mer (with probability q) (Fig. 5). Therefore,

$$p_1 + p_0 + q = 1. \tag{1}$$

To avoid extinction, we let $p_1 > p_0$. Biological processes that can cause duplications or deletions include homologous or heterologous recombination

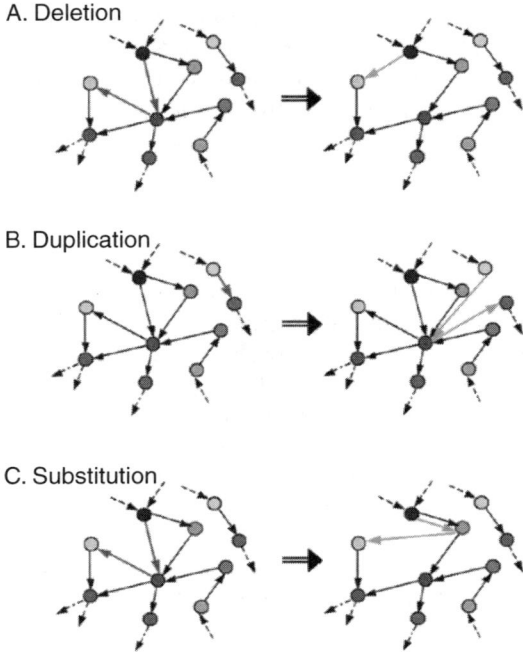

Fig. 5. The three processes during graph evolution: *deletion, duplication,* and *substitution.* In each process, the target node (indicated by the "central" circle) is chosen with preference for nodes with larger degrees. If the ith node has degree k_i, the probability of it being chosen is proportional to $k_i/(\sum_{i=1}^{N} k_i)$. For deletion (**A**), one incoming edge and one outgoing edge of the target node are randomly chosen, and deleted from the graph. A new forward edge is added from the root node of the deleted incoming edge to the head node of the deleted outgoing edge. For duplication (**B**), an edge is randomly selected from the graph, and deleted. Two forward edges are added from the root node of the deleted edge to the target node, and from the target node to the head node of the deleted edge, respectively. For substitution (**C**), besides the target node as the substituted node, another node is randomly chosen from the graph uniformly as the substituting node. One incoming and one outgoing edge of the target node are randomly selected, and rewired to the selected substituting node. Note that all the processes during graph evolution keep the out-degree and in-degree of a node identical.

and DNA polymerase slippage. Substitutions can be caused by random point mutation. During graph evolution, let $k_i{}^t$ and E^t indicate the copy number of the ith mer and the total number of mers in the evolving genome at the tth iteration. If we assume that the target mer for any process is chosen uniformly randomly from the genome, then the probability of the ith mer species being chosen for a process in the next iteration is proportional to its frequency in the genome in the current iteration ($\propto \frac{k_i{}^t}{E^t}$). With these assumptions, we implemented the "rich gets richer" rule in dynamics. If, for simplicity, a mer chosen for substitution is assumed to change into any other mer with equal probability during substitution, the conditional probabilities describing how the copy number of the ith mer changes in each iteration can be expressed as follows[4]:

$$P(k_i^{\,t} = n \mid k_i^{\,t-1} = n - 1) = p_1 \frac{n-1}{E^{t-1}} + (1 - \frac{n-1}{E^{t-1}})q\frac{1}{N-1}, \tag{2}$$

$$P(k_i^{\,t} = n \mid k_i^{\,t-1} = n) = 1 - p_1 \frac{n}{E^{t-1}} - p_0 \frac{n}{E^{t-1}}$$
$$-q\frac{n}{E^{t-1}} - (1 - \frac{n}{E^{t-1}})\frac{q}{N-1}, \tag{3}$$

$$P(k_i^{\,t} = n \mid k_i^{\,t-1} = n + 1) = p_0 \frac{n+1}{E^{t-1}} + q\frac{n+1}{E^{t-1}}. \tag{4}$$

We are now able to write down the difference equation describing the expected probability distribution for the copy number of the ith mer:

$$
\begin{aligned}
P(k_i^{\,t} = n) &= P(k_i^{\,t-1} = n - 1)P(k_i^{\,t} = n \mid k_i^{\,t-1} = n - 1) \\
&\quad + P(k_i^{\,t-1} = n)P(k_i^{\,t} = n \mid k_i^{\,t-1} = n) \\
&\quad + P(k_i^{\,t-1} = n + 1)P(k_i^{\,t} = n \mid k_i^{\,t-1} = n + 1) \\
&= P(k_i^{\,t-1} = n - 1)\left(p_1 \frac{n-1}{E^{t-1}} + (1 - \frac{n-1}{E^{t-1}})\frac{q}{N-1} \right) \\
&\quad + P(k_i^{\,t-1} = n)\left(1 - \frac{n}{E^{t-1}} - (1 - \frac{n}{E^{t-1}})\frac{q}{N-1} \right) \\
&\quad + P(k_i^{\,t-1} = n + 1)\left(p_0 \frac{n+1}{E^{t-1}} + q\frac{n+1}{E^{t-1}} \right) \\
&= P(k_i^{\,t-1} = n - 1)\left((p_1 - \frac{q}{N-1})\frac{n-1}{E^{t-1}} + \frac{q}{N-1} \right) \\
&\quad + P(k_i^{\,t-1} = n)\left(1 - (1 - \frac{q}{N-1})\frac{n}{E^{t-1}} - \frac{q}{N-1} \right) \\
&\quad + P(k_i^{\,t-1} = n + 1)\left(p_0 \frac{n+1}{E^{t-1}} + q\frac{n+1}{E^{t-1}} \right).
\end{aligned}
\tag{5}
$$

Since the total number of mers in a genome is usually very large, and each mer species only accounts for a very small fraction of the genome, we assume that the copy number of each mer species evolves independently. Therefore, the above equation can be viewed as an expression of the copy number distribution of all possible mers in a genome. This assumption is validated by Monte Carlo simulations.

---

[4] We approximate the probability of a specific mer being chosen to substitute another mer during substitution as $\frac{1}{N-1}$ (instead of $\frac{1}{3l}$). This approximation holds when mer size l is small.

4.2 Model Fitting

The simple graph model is applied to fit the mer-frequency distributions from various genomes. Since the process is non-stationary, we use numerical simulation in the model fitting. The initial condition is set as a proto-genome: a random sequence of size 1Kb containing uniformly distributed mers. The iteration proceeds until the final genome size reaches the real genome size under study. Some of the fitted mer-frequency distribution results can be seen in Fig. 1. The model has two degrees of freedom in its parameter space. Here we choose to estimate q and $\frac{p_1}{p_0}$. The parameters for the optimal fit to the data from genome analysis are estimated so that the sum of the absolute differences between the real data and the data produced by the model are minimized. Some of the fitted parameter values are shown in Table 1.

Table 1. Graph model parameters (q, p_1/p_0) fitted to the mer-frequency distribution data (6- to 9-mer) from the whole-genome analysis.

Mer size	6-mer		7-mer		8-mer		9-mer	
	q	p_1/p_0	q	p_1/p_0	q	p_1/p_0	q	p_1/p_0
M. genitalium	0.0176	1.1	0.0436	1.1	0.1587	1.5	0.3222	1.5
M. pneumoniae	0.0319	1.1	0.1151	1.5	0.2309	1.5	0.4363	1.5
P. abyssi	0.0269	1.1	0.0672	1.2	0.1778	1.4	0.3897	1.5
P. horikoshii	0.0234	1.1	0.0443	1.1	0.139	1.3	0.3456	1.5
P. furiosus	0.0213	1.1	0.0384	1.1	0.1119	1.2	0.3114	1.5
H. pylori	0.018	1.1	0.032	1.1	0.0925	1.3	0.2262	1.5
H. influenzae	0.0202	1.1	0.0366	1.1	0.1364	1.5	0.2802	1.5
S. tokodaii	0.018	1.1	0.032	1.1	0.0925	1.3	0.2262	1.5
S. subtilis	0.0187	1.1	0.0326	1.1	0.1139	1.4	0.2585	1.5
E. coli K12	0.0207	1.1	0.0334	1.1	0.0698	1.1	0.2389	1.5
S. cerevisiae	0.0113	1.1	0.0176	1.1	0.0459	1.2	0.1311	1.4
C. elegans	–	–	0.0076	1.1	0.0115	1.1	0.0275	1.2

The fitted parameters in the table show some interesting properties. The optimal relative substitution probabilities, q values, of a particular genome increase monotonically with the mer-size (l). This may reflect the scaling effect in this simple model introduced by fixing the size of duplication or deletion as the size of one mer. In the related biological processes, while one substitution changes one mer to another, one duplication or deletion may change the copy numbers of more than one mer. In a particular duplication or deletion event, when the mer-size increases, the corresponding number of mers being affected by the process decreases. Therefore, the relative probabilities of substitution of larger mers tend to be larger than those of the smaller mers. It is also noticeable that the values of the parameter $\frac{p_1}{p_0}$ increase along with the mer-sizes in each genome. This suggests that duplication probability p_1 decays more slowly than deletion probability p_0 when the mer size increases. Such a behaviour indicates that duplications of large regions occur more often than deletions of large regions. Furthermore, the ratios $\frac{p_1}{p_0}$ are consistently larger than 1, which validates our assumption of $p_1 > p_0$ in the model.

The organisms listed in Table 1 are ordered by their genome sizes (from 580Kb for *M. genitalium* to 97Mb for *C. elegans*). There is a slight tendency for q values to be smaller when the genome sizes become larger. This may reflect the fact that the sizes of duplication or deletion units increase with the corresponding genome size. However, there are exceptions: for example, *M. genitalium* has a genome of size 580Kb, and its q values are smaller than many listed organisms with larger genome sizes, such as *M. pneumoniae* (816Kb), *P. abyssi* (1.8Mb), etc. This observation may be explained by a higher substitution rate in *M. genitalium*.

Under biologically reasonable assumptions, the parameters of the model can be used to estimate the size distribution of the duplication and deletion events in real genomes. Previous research [24, 25] has shown that the size distribution of the insertion and deletion regions in the genomes examined follows a power law. The exponents of the power law are the key characterizing factors of the size distributions, and are of the greatest interest as they reveal the link between the genome dynamics and genome statistics. Unfortunately, direct estimation of the exponents (e.g., from sequence comparison) not only requires complex and expensive computation, but also imposes strong constraints on data sources so as to minimize ambiguity. However, for a specific genome, our model and the mer distribution data are sufficient to determine the exponents reasonably well, as described below. We start with the following assumptions. During genome evolution, the averaged rate of point mutation in each time interval is μ per nucleotide; the probability of duplicating a fragment of size x in each time interval is $f_1 x^{-b_1}$; the probability of deleting a fragment of size x in each time interval is $f_0 x^{-b_0}$. (Here, f_1 and f_0 are normalization constants; b_1 and b_0 are the exponents for the power-law distributions of duplication and deletion sizes, respectively.) The relationships between the model parameters and the size of the mer (l) they are fitted for in a specific genome are shown in Eqs. (6), (7), and (8).

When the model and its parameters are fitted to a specific genome for a sufficiently large number of mer-sizes, the exponents (b_1 and b_0) can be estimated by a linear regression using the relationship between the parameter ratios and mer sizes (l) from Eqs. (6), (7), and (8). This approach allows us to infer the size distribution of duplication and deletion events over the evolutionary history of that genome.

$$\frac{q}{p_1} = \frac{P(\text{an } l \text{ mer gets substituted})}{P(\text{an } l \text{ mer gets duplicated})} = \frac{\mu l}{\int_{x \geqslant l}^{\infty} f_1 x^{-b_1} (x/l) dx}$$

$$= \frac{\mu l^2}{\int_{l}^{\infty} f_1 x^{1-b_1} dx}$$

$$= \frac{\mu (b_1 - 2)}{f_1} l^{b_1} \propto l^{b_1} \tag{6}$$

$$\frac{q}{p_0} = \frac{P(\text{an } l \text{ mer gets substituted})}{P(\text{an } l \text{ mer gets deleted})} = \frac{\mu l}{\int_{x \geqslant l}^{\infty} f_0 x^{-b_0} (x/l) dx}$$

$$= \frac{\mu (b_0 - 2)}{f_0} l^{b_0} \propto l^{b_0} \tag{7}$$

$$\frac{p_1}{p_0} = \frac{P(\text{an } l \text{ mer gets duplicated})}{P(\text{an } l \text{ mer gets deleted})} = \frac{\int_{x \geqslant l}^{\infty} f_1 x^{-b_1} (x/l) dx}{\int_{x \geqslant l}^{\infty} f_0 x^{-b_0} (x/l) dx}$$

$$= \frac{f_1 (b_0 - 2)}{f_0 (b_1 - 2)} l^{b_0 - b_1} \propto l^{b_0 - b_1}. \tag{8}$$

4.3 Polya's Model

Although the parsimonious model described above captures the most important elements during genome evolution, it omits most of the details. To get a more comprehensive and specific understanding of genome evolution, we develop a more realistic model. The model will mainly include the parsimonious rules but apply them in a more interactive way. The model is an extension of Polya's urn model on a string. In this model, the same three main events in evolution are considered: *duplication*, *deletion*, and *substitution* (Fig. 6). Similar to the simple model, genome evolution is modelled as a stochastic process that goes through multiple iterations. Within each iteration, one of the three events happens with a certain probability. However, unlike in the simple model, the details of the events can also be manipulated (Fig. 6). In every iteration, a set of probability distributions is applied to decide the changes in the *in silico* evolution. All the probability distribution functions can be interdependent, as well as independent.

As the model approaches its resemblance to reality, it becomes increasingly complex, thereby making the explicit mathematical approach infeasible. Therefore, large-scale *in silico* experiments are needed. Finally, genome evolution is also a population process. To understand the genome evolution more completely, we also plan to simulate it in a population model which integrates natural selection and polymorphism effects.

5 Conclusion

Among the few fundamental "dogmas" at the core of biological sciences, a central and most elegant one is likely to be "evolution by duplication". For

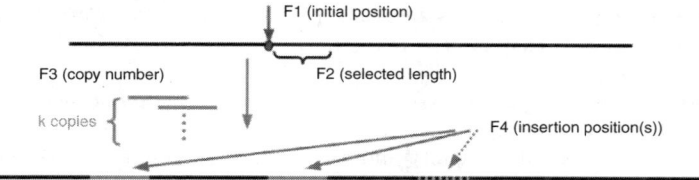

Fig. 6. The more realistic model in our *in silico* simulation. During each iteration in the simulation, just like in the simple graph model, one of the three events (deletion, duplication, or substitution) happens with probabilities p_0, p_1, and q. A probability distribution (F_1) decides the initial position in the genome where a chosen event will happen. Another probability distribution function (F_2) controls the size of the fragment chosen from the existing genome if either a duplication or a deletion event happens. A third probability distribution function (F_3) decides the copy number for the duplication. (When the copy number is one, a translocation happens.) And a fourth probability distribution function (F_4) decides the insertion site(s) for the duplicated fragment(s) during a duplication event. The distribution functions can be interdependent. The model is a realization of Polya's urn on a string.

many years, this theory is likely to remain intriguing and mysterious in its pervasive power in explaining many seemingly unrelated biological phenomena. Our efforts to understand it better will continue to raise many beautiful mathematical and computational questions requiring many novel techniques.

For several years, we have focused on these problems, and have developed many computational techniques, not discussed in this chapter but briefly mentioned here: *Valis*, a computational environment and language allowing us to rapidly prototype genome-analysis algorithms and visualization tools; *Genome Grammar*, a highly memory and run-time efficient tool for large-scale *in silico* evolution simulations; *Simpathica*, a tool for understanding biological processes involved in genome evolution and their effects on pathways.

References

1. Peng, C.K. et al: Long-range correlations in nucleotide sequences. Nature **356**, 168–170 (1992)
2. Gomez, S.M., Rzhetsky, A.: Birth of scale-free molecular networks and the number of distinct DNA and protein domains per genome. Bioinformatics **17**, 988–996 (2001)
3. Fields, S., Schwikowski, B., Uetz, P.: A network of protein-protein interactions in yeast. Nature Biotechnology **18**, 1257-1261 (2000)
4. Albert, R., Barabasi, A.-L.: Statistical mechanics of complex networks. Reviews of Modern Physics **74**, 48–97 (2002)
5. Havlin, S. et al: Mosaic organization of DNA nucleotides. Physical Review E **49**, 1685–1689 (1994)

6. Ehrlich, S.D., Viguera, E., Canceill, D.: Replication slippage involves DNA polymerase pausing and dissociation. EMBO Journal **20**, 2587–2596 (2001)
7. Lilley, D.M.J., Eckstein, F.: *DNA Repair* (Springer, Berlin Heidelberg New York 1998)
8. Albert, R. et al: The large-scale organization of metabolic networks. Nature **407**, 651–654 (2000)
9. Barabasi, A.L. et al: Lethality and centrality in protein networks. Nature **411**, 41–42 (2001)
10. Gerstein, M., Qian, J., Luscombe, N.M.: Protein family and fold occurrence in genomes: power-law behavior and evolutionary model. Journal of Molecular Biology **313**, 673–681 (2001)
11. Rain, J.C. et al: The protein-protein interaction map of *Helicobacter pylori*. Nature **409**, 211–215 (2001)
12. Vogelstein, B., Lane, D., Levine, A.J.: Surfing the P53 network. Nature **408**, 307–310 (2000)
13. Johnson, N.L.: *Urn models and their application* (Wiley 1977)
14. Ganapathiraju, M. et al: Comparative n-gram analysis of whole-genome protein sequences. In: *HLT'02: Human Language Technologies Conference,* San Diego, California, USA, 2002.
15. Ohno, S.: *Evolution by Gene Duplication* (Springer, Berlin Heidelberg New York 1970)
16. Apweiler, R. et al: The InterPro database, an integrated documentation resource for protein families, domains and functional sites. Nucleic Acids Research **29**, 37–40 (2000)
17. Sole, R.V., Pastor-Satorra, R., Smight, E.: Evolving protein interaction networks through gene duplication. Santa Fe Institute Working Paper 02-02-008 (2002)
18. Mantegna, R.N. et al: Linguistic features of noncoding DNA sequences. Physical Review Letters **73**, 3169–3172 (1994)
19. Sneppen, K., Maslov, S.: Specificity and stability in topology of protein networks. Science **296**, 910–913 (2002)
20. Buldyrev, S.V. et al: Fractal landscapes and molecular evolution: modeling the myosin heavy chain gene family. Biophysical Journal **65**, 2673–2679 (1993)
21. Eichler, E.E.: Recent duplication, domain accretion and the dynamic mutation of the Human genome. Trends in Genetics **17**, 661–669 (2001)
22. Bailey, J.A. et al: Recent segmental duplications in the human genome. Science **297**, 1003–1007 (2002)
23. Graur, D., Li, W-H.: *Fundamentals of Molecular Evolution* (Sinauer 2000)
24. Gu, X., Li, W.-H.: The size distribution of insertions and deletions in human and rodent pseudogenes suggests the logarithmic gap penalty for sequence alignment. Journal of Molecular Evolution **40**, 464–473 (1995)
25. Ophir, R., Graur, D.: Patterns and rates of indel evolution in processed pseudogenes from humans and murids. Gene **205**, 191–202 (1997)

Natural Computing Series

W.M. Spears: **Evolutionary Algorithms. The Role of Mutation and Recombination.**
XIV, 222 pages, 55 figs., 23 tables. 2000

H.-G. Beyer: **The Theory of Evolution Strategies.** XIX, 380 pages, 52 figs., 9 tables. 2001

L. Kallel, B. Naudts, A. Rogers (Eds.): **Theoretical Aspects of Evolutionary Computing.**
X, 497 pages. 2001

G. P un: **Membrane Computing. An Introduction.** XI, 429 pages, 37 figs., 5 tables. 2002

A.A. Freitas: **Data Mining and Knowledge Discovery with Evolutionary Algorithms.**
XIV, 264 pages, 74 figs., 10 tables. 2002

H.-P. Schwefel, I. Wegener, K. Weinert (Eds.): **Advances in Computational Intelligence.
Theory and Practice.** VIII, 325 pages. 2003

A. Ghosh, S. Tsutsui (Eds.): **Advances in Evolutionary Computing. Theory and
Applications.** XVI, 1006 pages. 2003

L.F. Landweber, E. Winfree (Eds.): **Evolution as Computation.** DIMACS Workshop,
Princeton, January 1999. XV, 332 pages. 2002

M. Hirvensalo: **Quantum Computing.** 2nd ed., XIII, 214 pages. 2004 (first edition
published in the series)

A.E. Eiben, J.E. Smith: **Introduction to Evolutionary Computing.** XV, 299 pages. 2003

A. Ehrenfeucht, T. Harju, I. Petre, D.M. Prescott, G. Rozenberg: **Computation in Living
Cells. Gene Assembly in Ciliates.** XIV, 202 pages. 2004

R. Paton, H. Bolouri, M. Holcombe, J. H. Parish, R. Tateson (Eds.): **Computation in Cells
and Tissues. Perspectives and Tools of Thought.** Approx. 350 pages. 2004

L. Sekanina: **Evolvable Components. From Theory to Hardware Implementations.**
XVI, 194 pages. 2004

R. W. Morrison: **Designing Evolutionary Algorithms for Dynamic Environments.**
XII, 148 pages, 78 figs. 2004

G. Ciobanu (Ed.): **Modelling in Molecular Biology.** X, 310 pages. 2004

M. Amos: **Theoretical and Experimental DNA Computation.** Approx. 200 pages. 2004

Printing: Strauss GmbH, Mörlenbach
Binding: Schäffer, Grünstadt